S0-BYD-364

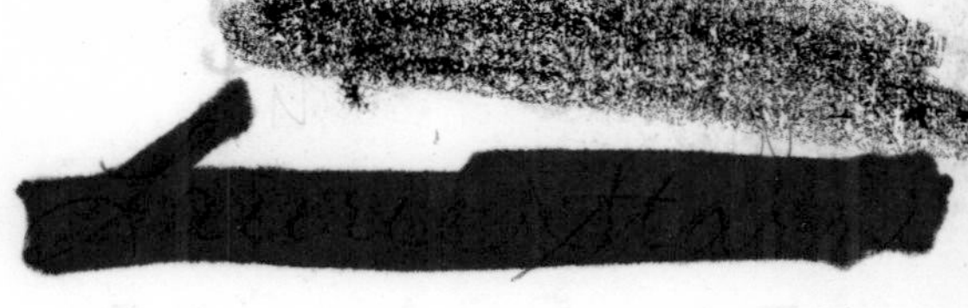

SYSTEMS SIMULATION

SYSTEMS SIMULATION
the art and science

ROBERT E. SHANNON

University of Alabama in Huntsville
Huntsville, Alabama

Prentice-Hall, Inc.

Englewood Cliffs, New Jersey

Library of Congress Cataloging in Publication Data

Shannon, Robert E.
Systems simulation: the art and science.

Includes bibliographies and index.
1. Simulation methods. I. Title.
T57.62.S47 601'.84 75-1174
ISBN 0-13-881839-8

10 9 8

Printed in the United States of America.

Prentice-Hall International, Inc., *London*
Prentice-Hall of Australia, Pty. Ltd., *Sydney*
Prentice-Hall of Canada, Ltd., *Toronto*
Prentice-Hall of India Private Limited, *New Delhi*
Prentice-Hall of Japan, Inc., *Tokyo*
Prentice-Hall of Southeast Asia (Pte.) Ltd., *Singapore*

CONTENTS

appendix A

CASE STUDIES 275

appendix B

RANDOM VARIATE GENERATION 347

appendix C

STATISTICAL TABLES 369

INDEX 381

PREFACE

Simulation is one of the most powerful analysis tools available to those responsible for the design and operation of complex processes or systems. The concept of simulation is both simple and intuitively appealing. It allows the user to experiment with systems (real and proposed) where it would be impossible or impractical otherwise. Every engineer and manager should be familiar with this technique if his education is to be complete. Therefore, this book is intended for two groups of readers—those who will design simulation models and those who will use the results of studies based upon them.

Simulation modeling is heavily based upon computer science, mathematics, probability, and statistics. Yet simulation modeling and experimentation remain very much intuitive processes. The basic sciences which underlie simulation are well presented in other books; therefore, this is not a book on computer programming or statistics. We leave the detailed treatment of those aspects to others, attempting to instead present the process of designing a model, running experiments with it, and analyzing the results. Since this poorly understood process is as much art as science, we offer few firm rules or fixed outlines. We can only make the reader aware of the critical issues, give some guidelines for resolving them, and warn of possible pitfalls. This will be an unsatisfactory approach to those readers who believe and want to be told that there is a precise and correct way to do everything. But anyone who has ever used simulation will recognize that the issues addressed here are vital ones which spell the difference between success and failure.

Like all powerful tools which depend heavily upon art in their application, simulation is capable of giving either very good or very bad results, depending upon how it is used. It can either enlighten or mislead. It is therefore important that the manager or decision-maker who will use the results be aware of the implications of certain assumptions made, the strengths and weaknesses, the benefits and pitfalls. We believe that the only way a manager can intelligently take advantage of any quantitative technique without disappointing or disasterous results is to understand the nuances of the underlying assumptions and philosophy. He need not be able to design, program, or even run the model. But he must understand what is going on if he is not going to treat the model as a magic box or be completely at the mercy of the analyst.

This book is designed for use in a three-semester-hour course in engineering, management science, computer science, or business administration programs. It is assumed that the reader is familiar with statistics and elementary computer programming. Problem sets and exercises which can be used at the discretion of the instructor appear at the ends of chapters. For students with a good background in statistics and computer programming, a real world study problem or term project is recommended as the major focal point. True facility in simulation techniques can only be gained through experience.

I believe that this book differs significantly from others in the field in several ways:

1 It can be used by students with different degrees of sophistication in mathematics, probability, statistics, and computer programming. The exercises are such that the instructor can choose homework assignments to suit the student's skills. Thus, the book should appeal to a much wider range of readers.
2 It does not spend a lot of time and space on material that already appears in numerous other places. The literature is liberally cited so the reader can get all of the details he desires from primary sources.
3 The major thrust of the book is upon the art of modeling rather than examining in detail already existing models. Case studies of existing models are presented in an appendix.
4 The problems of managing the research effort and obtaining user acceptance are discussed. This aspect of simulation is totally ignored by other simulation books.

Chapter 1 is devoted to a discussion of general modeling philosophy and an overall view of simulation. Chapter 2 discusses some of the techniques that have been found useful in analyzing the system to be studied and the design of the model. Methods of determining the value of parameters and establishing the form of the interactions between components are presented. Chapter 3 is devoted to a discussion of computer programming languages and how to choose the most appropriate one to fit the particular problem. Chapter 4 attempts to familiarize the reader with the terminology and concepts of experimental design. This is done in order to allow the reader to explore the detailed literature in this field for help in specific cases. Chapter 5 presents some of the problems and pitfalls encountered in running and analyzing the experiments. In Chapter 6, we discuss the significant problem of validating the model and the inferences drawn from the experiment; this is done both from a philosophical and practical viewpoint. Finally, Chapter 7 confronts the problems of managing the research activity and of gaining user acceptance.

Unlike most books on simulation modeling, this one does not make extensive use of example models in the main discussion sections. It is not at all clear that teaching models is the same as teaching modeling. Although some things can be learned by examining models developed by others, I personally believe that this is of limited teaching value. Example models have therefore been grouped together in Appendix A, where six existing models are described.

The methods of generating pseudo-random numbers and stochastic variates on a computer have appeared in numerous books and technical articles. In addition, most computer facilities have such subroutines as a part of their library. We have therefore not repeated their derivation but merely present them as computational aides in Appendix B in case they are not already available.

Finally, I believe that one of the most significant difficulties in gaining acceptance of quantitative methods by managers is the problem of semantics. Management scientists and operations researchers have developed a vocabulary all their own. While this is helpful for discussions among themselves, it proves to be a tremendous barrier to communication when dealing with managers. We have therefore tried to present and define as many of the alternate terms for each concept as possible.

I would like to express my appreciation to the many individuals and organizations who have made it possible to complete this project. In particular I would like to acknowledge the help, assistance, and encouragement extended by Dr. Jafar Hoomani, Dean of the School of Science and Engineering, Dr. Robert A. Brown, Chairman of the Department of Industrial and Systems Engineering, and my colleague Dr. Richard M. Wyskida, all of The University of Alabama in Huntsville. Special thanks is owed to Dr. J. Tinsley Oden of the University of Texas at Austin, whose example and encouragement did much to keep me moving when discouragement set in. The comments and suggestions of Dr. Joe H. Mize of Oklahoma State University, Dr. Martin K. Starr of Columbia University, and Dr. Roy Czernikowski of the Rochester Institute of Technology were also very helpful. The contractual support through the years provided by the Research Development and Engineering Laboratory of the U.S. Army Missile Command under the direction of Dr. John L. McDaniel provided the experience and background which made this book possible. My gratitude must be extended to Miss Liney Monroe, who typed the manuscript over and over, and to Mrs. Judy DuVall who made the corrections. Finally, I must acknowledge in a special way several of my students, including Major William B. Whipple and Pat R. Odom, who made major contributions to Chapters 3 and 4, respectively, as well as James H. Donnelly who edited the first draft and compiled the solutions manual.

I would like to dedicate this book to my beautiful wife, Marion, whose faith in me and encouragement made this project possible, and to Kelly and Teddy, the joys of my life.

Robert E. Shannon

Huntsville, Alabama

chapter one

FUNDAMENTALS OF MODELING

"When I use a word," Humpty Dumpty said, in a rather scornful tone, "it means just what I choose it to mean—neither more nor less."

"The question is," said Alice, "whether you can make words mean so many different things."

"The question is," said Humpty Dumpty, "which is to be master—that's all."

Lewis Carroll

1.1 Simulation

Management today is becoming increasingly difficult as the man-organized systems of our society grow more complex. This complexity derives from the interrelations among the various elements of our organizations and the physical systems with which they interact. Although this complexity has existed for a long time, we are just beginning to appreciate its importance. We now recognize that changing one aspect of a system may very well produce changes or create the need for changes in other parts of the system; hence, the science of *systems analysis** has evolved to help managers and engineers study and understand the ramifications of such changes. Particularly since the advent of electronic computers, one of the most important and useful tools for analyzing the design and operation of complex processes or systems is simulation. To *simulate*, according to *Webster's Collegiate Dictionary*, is "to feign, to attain the essence of, without the reality."

In substance, every model or representation of a thing is a form of simulation. Simulation is a very wide-open and somewhat

* "Operations research" and "management science" are terms closely related to "systems analysis" and are often used interchangeably.

ill-defined subject of great importance to those responsible for the design of systems as well as those responsible for their operation. With some trepidation, and at the risk of seeming to be egotistical, we shall forego exploring various definitions offered by other authors and provide our own. *Simulation is the process of designing a model of a real system and conducting experiments with this model for the purpose either of understanding the behavior of the system or of evaluating various strategies (within the limits imposed by a criterion or set of criteria) for the operation of the system.* Thus we understand the process of simulation to include both the construction of the model and the analytical use of the model for studying a problem. By a *model* of a real system we mean a representation of a group of objects or ideas in some form other than that of the entity itself, and here the term "real" is used in the sense of "in existence or capable of being brought into existence." Thus systems in the preliminary or planning stage can be modeled as well as those already in existence.

Many authors use the term "simulation" in a much narrower sense than the one preceding. Under our definition, simulation may or may not involve a stochastic model and Monte Carlo experimentation. In other words, the inputs to the model and/or the functional relationships among its various components may or may not involve a random element subject to the rules of probability. Furthermore, we do not restrict our definition of simulation to experiments conducted on electronic computer models. Many useful simulations can be and are run with only paper and pen or with the aid of a desk calculator. Simulation modeling is, therefore, an experimental and applied methodology which seeks to:

1 describe the behavior of systems;

2 construct theories or hypotheses that account for the observed behavior;

3 use these theories to predict future behavior, that is, the effects that will be produced by changes in the system or in its method of operation.

Unlike most technologies, which can be classified according to the discipline (e.g., physics or chemistry) in which they have their roots, simulation is applicable in all disciplines. Simulation as we know it today received its original impetus from the aerospace programs, but even a casual survey of the literature indicates the broad field of present applications. For example, books have been written on the use of simulation in business [12, 22, 25], economics [24, 26], marketing [2, 21], education [3], politics

[7], social science [8, 13], behavioral science [17, 30], international relations [14, 15], transportation [18], manpower, law enforcement, urban studies, global systems, and many, many others. In addition, countless technical articles, papers, reports, and masters and doctoral theses in almost every field of social, economic, technical, and humanistic endeavor attest to the widespread impact and growth of the use of simulation on almost every aspect of our lives.

1.2 Simulation example

Before we go any further, perhaps we should look at a very simple example to demonstrate the concept of simulation. Consider a single-channel queueing (waiting line) system such as a checkout stand at a small gift shop. Assume that the time between arrivals of the customers is uniformly distributed from 1 thru 10 minutes (for simplicity we round off all times to the nearest whole minute.) Let us assume further that the amount of time required to service each customer is also uniformly distributed from 1 thru 6 minutes. We are interested in the average time a customer spends in the system (both waiting and being serviced) and the percentage of time that the check-out clerk is not occupied with work.

To simulate the system, we need to generate an artificial experience that would be characteristic of the situation. To do so, we must devise a method of generating the artificial arrival of a string of customers and the service time required for each. In one method that we might use, we could start by borrowing ten poker chips and a single die from one of our gambling friends. We could then number the poker chips from one to ten, place them in a hat, and mix them up by shaking the hat. By drawing a poker chip out of the hat and reading off the number, we could then represent the time between the arrival of the present customer and the last one. Rolling our die and reading off the number of spots on the upper face would then represent how long it takes to service him. Thus, by repeating this sequence (replacing the poker chip and shaking the hat between draws), we could generate a series of arrivals and corresponding service times. Our problem then becomes one of bookkeeping. Table 1.1 shows what a sample of 20 customers might look like.

Obviously, to obtain statistical significance we would have to use a much larger sample size, and we have ignored several important things, such as starting conditions, which will be discussed later. The important point is that we have used two devices for generating random numbers (numbered poker chips and a die) to produce artificial (simulated) experience with a system in order to examine certain of its behavior characteristics.

Table 1.1 *Simulation of check-out stand.*

Customer	Time Since Last Arrival (min)	Service Time (min)	Arrival Clock Time	Service Begins	Service Ends	Customer Wait (min)	Clerk Idle (min)
1	—	1	0:00	0:00	0:01	1	0
2	3	4	0:03	0:03	0:07	4	2
3	7	4	0:10	0:10	0:14	4	3
4	3	2	0:13	0:14	0:16	3	0
5	9	1	0:22	0:22	0:23	1	6
6	10	5	0:32	0:32	0:37	5	9
7	6	4	0:38	0:38	0:42	4	1
8	8	6	0:46	0:46	0:52	6	4
9	8	1	0:54	0:54	0:55	1	2
10	8	3	1:02	1:02	1:05	3	7
11	7	5	1:09	1:09	1:14	5	4
12	3	5	1:12	1:14	1:19	7	0
13	8	3	1:20	1:20	1:23	3	1
14	4	6	1:24	1:24	1:30	6	1
15	4	1	1:28	1:30	1:31	3	0
16	7	1	1:35	1:35	1:36	1	4
17	1	6	1:36	1:36	1:42	6	0
18	6	1	1:42	1:42	1:43	1	0
19	7	2	1:49	1:49	1:51	2	6
20	6	2	1:55	1:55	1:57	2	4
					TOTALS	68	55

Average customer wait time $= \frac{68}{20} = 3.40$ min.

Percent of time clerk idle $= \frac{55}{117}(100) = 47\%$.

1.3 "Model" defined

A model is a representation of an object, system, or idea in some form other than that of the entity itself. Its purpose is usually to aid us in explaining, understanding, or improving a system. A model of an object may be an exact replica of the object (although executed in a different material and to a different scale), or it may be an abstraction of the object's salient properties. Because simulation is only one type of modeling, albeit very important, we set the stage for later discussion of simulation modeling by considering modeling in general terms first.

The functions of a model are usually considered to be those of prediction and comparison—to provide a logical way to forecast the outcomes that follow alternative actions; and, hopefully, to indicate a preference among them. Although this usage of models is important, it is by no means their only purpose. As we shall see in the following sections, model building provides a systematic, explicit, and efficient way for various experts and decision makers to focus their judgment and intuition. By introducing a precise framework, a model can also serve as an effective means of communication and as an aid to thought.

Many people think the use of models is recent. Modeling is not new, however; the conceptualization and development of models has played a vital part in mankind's intellectual activity since he began to try to understand and manipulate his environment. He has always used the idea of models to attempt to represent and express ideas and objects. Modeling has ranged from such acts of communicating as wall paintings and building idols to writing complex systems of mathematical equations for the flight of a rocket through outer space. As a matter of fact, the progress and history of science and engineering are most accurately reflected in the progress of man's ability to develop models of natural phenomena, ideas, and objects.

Almost without exception, science writers state that one of the major elements required in attacking a problem is the construction and use of a model. Such a model may take many forms, but one of the most useful and certainly the most used is mathematical, expressing through a set of equations the essential features of the system or phenomena under study. Unfortunately, it is not always possible to create a mathematical model in the narrow and restricted sense. In studying most industrial and military systems, we can define goals, specify restrictions, and discern that the design follows the laws of engineering and/or economics. The essential relationships can be discovered and represented mathematically in one form or another. In contrast, an attack on problems of air pollution, crime prevention, urban renewal, or medical care involves working with goals that are obscure as well as conflicting, and with choices determined by political and social factors. Thus, our definition must include and allow qualitative as well as quantitative models.

1.4 Function of models

The concept of representing some object, system, or idea with a model, is so general that it is difficult to classify all the functions models fulfill. Elmaghraby [9] recognizes at least five legitimate and common uses:

1 an aid to thought
2 an aid to communication
3 purposes of training and instruction
4 a tool of prediction
5 an aid to experimentation

A model's usefulness as an aid to thought is evident. Models can help us organize and sort out hazy concepts and inconsistencies. For example, the construction of a network representation or PERT (program evaluation and review technique) model of a complex systems design job forces us to think through what steps are necessary and in what sequence. Such a model helps show us interrelationships, needed accomplishments, timing, required resources, etc. The very act of trying to represent our verbalizations and thoughts in some other form often points out inconsistencies and ambiguities. Properly done, model building forces us to organize, evaluate, and examine the validity of thoughts.

As an aid to communication, well thought out models have no peer. "One picture is worth a thousand words" testifies to this function. All verbal languages tend to be ambiguous and indistinct when it comes to presenting complex ideas or descriptions. Properly conceived models can help to remove this ambiguity by providing a more efficient and effective mode of communication. A model's advantage over verbal descriptions is its concise representation of a situation. It helps make the over-all structure more comprehensible and reveals important cause and effect relationships.

Models have been and continue to be widely used as training and instruction aids. Psychologists have long recognized the importance of a person's learning certain skills under conditions in which he is not over-motivated. He needs to practice without pressure. A crisis situation is a poor time and place to try to learn new skills; therefore, models are often ideal for training a person to cope with various eventualities before they occur. Most readers are already familiar with such applications as full-scale mockups or models of space vehicles used for training astronauts, models for automobile driver training, and business games for training executives.

Perhaps one of the most important uses of models, practically and historically, is predicting the behavior characteristics of the entity modeled. It is not economically feasible to build a supersonic jet to determine its flight characteristics, but its performance can be predicted by simulation. Simulation was used to check out Apollo 13's emergency measures before they were implemented; they allowed the astronauts to return safely after the oxygen-tank explosion. Most of the models with which we are concerned in this book are tools for prediction.

Finally, the use of models makes possible controlled experimentation in situations where direct experiments would be impractical or prohibitive in cost. Direct experimentation upon a system usually consists of varying certain parameters of the system while holding all others constant and observing the results. For most of the systems with which the systems analyst is concerned, this is impractical, too costly, or both. When it is too costly and/or impossible to experiment with the real world system itself, a model of the system may often be constructed and experiments run on the model with relative ease and low cost. By experimenting on a model of a complex system, we often can learn more about its internal interactions than we could through manipulation of the real world system itself, owing to our control of and the measurability of the model organizational structure, its policies, and the ease of parameter variation.

We can summarize the preceding discussion by recognizing that all these uses form a dichotomy. A model may serve one of two major purposes: either *descriptive*, for explaining and/or understanding; or *prescriptive*, by predicting and/or duplicating behavior characteristics. The latter usually implies the former but not vice versa, i.e., a prescriptive model useful in design is almost always descriptive of the entity being modeled, but a descriptive model is not necessarily useful for design purposes. Perhaps this is one reason why economic models (which have tended to be descriptive) have had little impact upon manipulating economic systems and little use as tools to aid top management, whereas operations research models have had an acknowledged significant impact in these areas.

In his book *Industrial Dynamics* [11], J. W. Forrester gives an excellent discussion on differences in models in the physical sciences, engineering, and the social sciences. He points out that the physical scientists have been trying, with great success, to model natural phenomena, whereas the engineers and social scientists have been trying to model systems that man determines. He believes that the difference in usefulness between engineering and social science models is largely the way the tools of model building have been used and the different emphasis on end objectives. In engineering, models serve as aids in designing new or improved systems, whereas in the social sciences and economics, they explain existing systems. A model that is useful in design must also explain, but apparently models undertaken *solely* to explain often fail even in their primary purpose.

1.5 Classification of simulation models

We can classify models in general and simulation models in particular in a number of ways. Unfortunately, none is completely satisfactory, although each serves a particular purpose. Some of these classification schemes are as follows:

1 static (cross-section) vs. dynamic (time-series)
2 deterministic vs. stochastic
3 discrete vs. continuous
4 iconic vs. analog vs. symbolic

Following the suggestion of A. J. Rowe [27], we like to think of simulation models as a continuous spectrum, starting with exact models or mockups of reality and proceeding to completely abstract mathematical models (see Fig. 1.1). The models at the beginning of the spectrum are often called physical or iconic models because they resemble the system being studied. For many years engineers have used full-scale mockups and reduced or scaled models for testing. The armed forces, National Aeronautics and Space Administration, and commercial airlines have also used full-scale mockups for training purposes. Static physical models, such as architectural models or plant layout models, help us visualize space relationships. An example of a dynamic physical model would be either a pilot plant (reduced scale) to study a new chemical process prior to going into full-scale production, or a scale model airplane tested in a wind tunnel to examine dynamic stability. The distinguishing characteristic of a physical (also sometimes called *iconic*) model is that it in some sense "looks like" the entity being modeled. Physical models may be full-scale mockups (such as trainers), scaled down (such as a model of the solar system), or scaled up (such as a model of an atom). They may also be two or three dimensional. They may be used for demonstration (e.g., a world globe) or for indirect experimentation. Scaled templates used in plant layout studies is an example of a two-dimensional, scaled down, physical model for experimentation.

Analog models are those in which a property of the real object is represented by a substituted property that often behaves in a similar manner. The problem is sometimes solved in the substituted state and the answer translated to the original properties. An electronic analog computer in which the voltage through a network might represent the flow of goods through a system is an excellent example of an analog simulation model. Another is the slide rule, in which the measured property is represented by logarithmic lengths along a scale.

A graph is yet another analog model; in which distance represents properties such as time, age, number, etc. A graph can also show the relationship between different quantities, and can

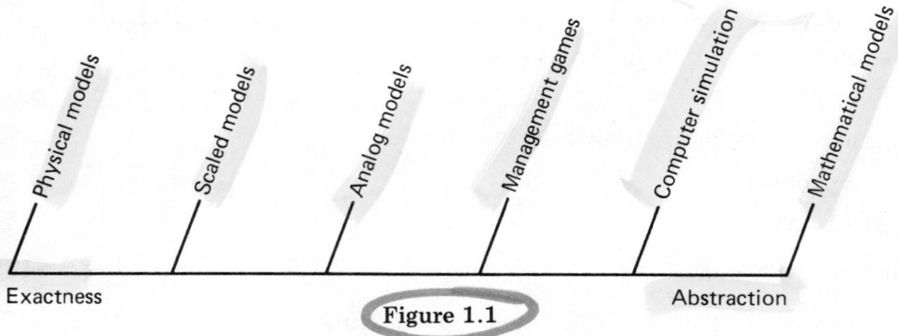

Figure 1.1

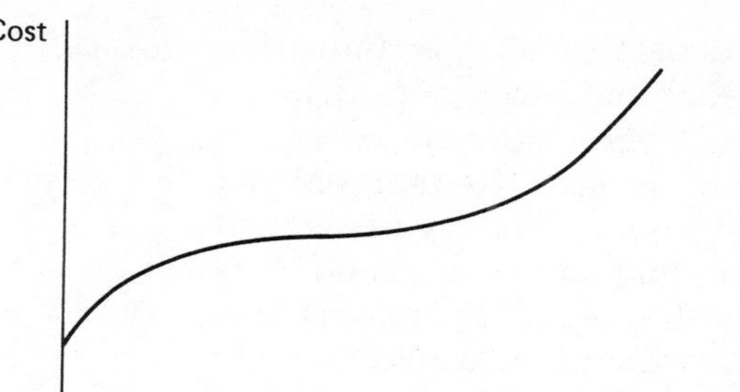

Figure 1.2 Production cost curve.

predict changes in some quantities when other quantities are changed. For example, Fig. 1.2 shows how the cost of producing a certain item may be related to volume of production. We thus see how cost is related to volume, and we can predict what will happen to cost if we increase or decrease volume. For certain relatively simple problems, a graph can actually solve the problem in question. From Fig. 1.2, we can derive the marginal cost curve.

If the problem is to determine the optimal production volume for a given price (i.e., the volume that maximizes net profit), we do so by inserting unit price on the same graph. Optimal volume corresponds to the point where price and marginal cost intersect. Graphical solutions are also possible for certain linear programming and game problems. Sometimes graphs are employed together with mathematical models, where one provides input to the other.

Schematic models other than graphs are also useful analog models; the organization chart is a familiar example. Using boxes and lines, it represents the formal chains of authority and communication existing among members of an organization at the time the chart is drawn. As we shall see in Chapter 2, flow process charts in which various occurrences—such as operations, delays, inspections, storage, etc.—are represented by flow lines and symbols are also widely used in systems studies.

As we proceed across the spectrum, we reach those models in which a human component and the computer interact. Such simulations are often called games (management, war, planning). Since the decision processes of management personnel or military officers are difficult to model, we often find it expedient not even to try. In so-called management games, the human interacts with the computer output (which simulates all other aspects of the system) and makes decisions on information received. His decisions are then fed back into the computer as inputs to the system. Extending this further, we reach the completely computerized simulations that people generally visualize when the term simulation is used. The computer may be a part of all the simulation models up to this portion of the spectrum, but not necessarily.

Symbolic or mathematical models are those in which a symbol, rather than a physical device, is used to represent an entity. Thus, in a mathematical model we might use symbols such as x and y to represent production volume and cost instead of a measured scale or some physical entity. Representing systems by a set of differential equations is a familiar example. Because they are the most abstract and hence the most general models, mathematical models are widely used in systems studies. Very real dangers and potential pitfalls, however, are inherent in the use of mathematical models. Symbolic models are always an abstract idealization of the problem in which simplifying assumptions are required if the model is to be solvable. Therefore, special care must be taken to insure that the model is a valid representation of the problem. More will be said about this later.

In trying to model a complex system, the researcher will usually resort to a combination of more than one of the pure types just discussed. Any system or subsystem can be represented in several ways, which will vary widely in complexity and detail. Most systems studies will result in several different models of the same system. Usually, simple models lead to more complex models, as the researcher analyzes and better understands the problem.

1.6 Advantages and disadvantages of simulation

All simulation models are so-called input-output models. That is, they yield the output of the system given the input to its interacting subsystems. Simulation models are therefore "run" rather than "solved" in order to obtain the desired information or results. They are incapable of generating a solution on their own in the sense of analytical models; they can only serve as a tool for the analysis of the behavior of a system under conditions specified by the experimenter. Thus, simulation is not a theory but a methodology of problem solving. Furthermore, simulation is only one of several valuable problem solving approaches available to the systems analyst. Since it is necessary and desirable to fit the tool or technique to the problem rather than vice versa, this naturally brings up the question, when is simulation useful?

We have defined simulation as experimentation with a model of a real system. An experimental problem becomes apparent when a need develops for specific information about a system that is not available from existing known sources. Barish [4] points out that direct experimentation upon the real life system eliminates many of the difficulties in obtaining a good match between the model and actual conditions; however, the disadvantages of direct experimentation are sometimes great.

1 It could disrupt company operations.

2 If people are an integral part of the system, the so-called "Hawthorne effect" may affect results—i.e., the fact that people are being observed may modify their behavior.

3 It may be very difficult to maintain the same operating conditions for each replication or run of the experiment.

4 It may be more time consuming and more costly to obtain the same sample size (and therefore statistical significance).

5 It may not be possible to explore many types of alternatives in real life experimentation.

Therefore, the analyst should consider the use of simulation when one or more of the following conditions exist:

1 A complete mathematical formulation of the problem does not exist or analytical methods of solving the mathematical model have not yet been developed. Many waiting line (queueing) models are in this category.

2 Analytical methods are available, but the mathematical procedures are so complex and arduous that simulation provides a simpler method of solution.

3 Analytical solutions exist and are possible but are beyond the mathematical ability of available personnel. The cost of designing, testing, and running a simulation should then be evaluated against the cost of obtaining outside help.

4 It is desired to observe a simulated history of the process over a period of time in addition to estimating certain parameters.

5 Simulation may be the only possibility because of the difficulty in conducting experiments and observing phenomena in their actual environment—e.g., studies of space vehicles in interplanetary flight.

6 Time compression may be required for systems or processes with long time frames. Simulation affords complete control over time, since a phenomenon may be speeded up or slowed down at will. Analysis of urban decay problems is in this category.

An additional advantage of simulation is its powerful educational and training application. The development and use of a simulation model allows the experimenter to see and play with the system. This, in turn, should greatly assist him in understanding and gaining a feel for the problem, thus aiding the process of innovation. In one of the rare articles dealing with this aspect of simulation, McKenney [20] has the following to say:

> When the manager had achieved a viable understanding and began to manipulate the model, he continuously gained new insights into his operation. He desired the model to test a variety of alternatives so he could evaluate these new insights. In essence, he was using the model to amplify his manipulative skill by explicitly identifying all important ramifications of a given change. Because of the complexity of the system it may have been possible for him to do this on the real system, but very tedious, and he probably would have made errors. Thus, he turned to the model as an evaluator of his new insights. It is conjectured the model design will never be stabilized, but continue to develop in response to the manager's new understanding.

The idea of simulation is intuitively appealing to both managers and systems analysts because of its simplicity. Therefore, one tends to adopt the simulation approach to almost every problem one encounters. Even though simulation is considered a brute force approach or court of last resort by those with extensive mathematical training, the fact remains that it is one of the most widely used tools of the management scientist or operations researcher. Table 1.2 shows the results of a survey taken among a sample of nonacademic, full members of the Operations Research Society of America by Shannon and Biles [29]. Only probability and cost effectiveness rank as having greater value or utility to practicing researchers. Others reinforce these results, such as Weston [33], who surveyed the 1000 largest U.S. firms (as listed by *Fortune* magazine) as to the relevance of

Table 1.2 *Utility of O.R. techniques to practitioners [29].*

Topic	Value
Probability theory (and statistical inference)	0.182
Economic analysis (cost effectiveness)	0.150
Simulation	0.143
Linear programming	0.120
Inventory	0.097
Waiting line (queueing)	0.085
Network analysis (sequencing)	0.072
Replacement analysis	0.042
Gaming theory	0.040
Dynamic programming	0.031
Search techniques	0.020
Nonlinear programming	0.018
	1.000

Table 1.3 *Quantitative tools most frequently employed in corporate planning [33].*

Topic	Frequency	%
Simulation studies	60	29
Linear programming	43	21
Network analysis (including PERT and CPM)	28	14
Inventory theory	24	12
Nonlinear programming	16	8
Dynamic programming	8	4
Integer programming	7	3
Queueing theory	7	3
Other	12	6
	205	100

certain techniques for corporate planning. His results appear in Table 1.3. The wide use of simulation as an approach to problem solving is further evident in the U.S. Army Missile Command's expenditure of 74 million dollars in fiscal year 1968 for simulation studies.

Therefore, despite its lack of mathematical sophistication and elegance, simulation is one of the most widely used quantitative techniques employed in management problem solving. Most managers and analysts are primarily interested in getting an answer to their immediate problems, and the end justifies the means! But it is precisely this concern for the ends that causes us to ask whether they can actually be achieved in the most efficient and effective manner by simulation. The answer is oftentimes no, for the following reasons:

1 Development of a good simulation model is often expensive and time consuming, requiring a high caliber of talent that may not be readily available. Forrester [11] indicates that 3 to 10 years may be required to develop a good corporate planning model.

2 A simulation can appear to reflect accurately a real world situation when, in truth, it does not. Several intrinsic problems in simulation can yield wrong results if not resolved correctly.

3 Simulation is imprecise, and we cannot measure the degree of this imprecision. Analysis of the sensitivity of the model to changing parameter values can only partially overcome this difficulty.

4 Simulation results are usually numerical, and given to any number of decimal points the experimenter chooses. Thus, there arises the danger of "deification of the

numbers"—i.e., attributing to the numbers a greater degree of validity than is justified.

The preceding discussion indicates that although simulation is an extremely valuable and useful approach to problem solving, it is certainly not a panacea for all of management's problems. The development and use of simulation models are still to a very large extent arts rather than sciences. Thus, as with other arts, it is not so much the technique that determines success or failure, but rather how the technique is used. Although simulation is an art, it can be mastered readily by those possessing the necessary talents of ingenuity, insight, and resourcefulness.

1.7 Structure of simulation models

Before one begins the process of developing a model, one should understand the structural building blocks from which models are constructed. Although a model may be very complicated mathematically or physically, its underlying structure is very simple. We can represent this underlying structure mathematically as

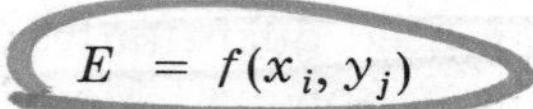

$$E = f(x_i, y_j)$$

where

- E is the effect of the system's performance
- x_i are the variables and parameters that we can control
- y_j are the variables and parameters that we cannot control
- f is the relationship between x_i and y_j, which gives rise to E

This obvious oversimplification is not too much help, except that it shows the performance of the system to be affected by variables under our control as well as by those which are not. In a slightly expanded view, we find that almost every model consists of some combination of the following ingredients:

1. components
2. variables
3. parameters
4. functional relationships
5. constraints
6. criterion functions

By *components*, we mean the constituent parts that when taken together make up the system. We also sometimes refer to the components as the *elements* or *subsystems.* For example, in a model of a missile or space vehicle, the components might be such things as the propulsion system, the guidance system, the control system, the structural system, etc. In a model of a city, the components might be the educational system, health system, transportation system, etc. In an economic model, the components might be individual firms, individual consumers, etc. *A system is defined as a group or set of objects united by some form of regular interaction or interdependence to perform a specified function.* Components are the objects constituting the system of interest.

Parameters are a quantity to which the operator of the model may assign arbitrary values, as distinguished from a *variable*, which can assume only those values that the form of the function makes possible. Another way of looking at this is that parameters, once set, are constant and do not vary. For example, in an equation such as $y = 3x$, the number 3 is a parameter and x and y are variables. We could just as easily say $y = 16x$ or $y = 30x$. Statistical analysis often involves attempts to determine these unknown but fixed parameters for a set of data. If we are considering a given set of data or a given population, then measures of central tendency, such as the mean, median, or mode, are population parameters, as are the measures of variability, such as the range, variance, and standard deviation. Thus, for a poisson distribution, where the probability of x is given by $P(x) = e^{-\lambda} (\lambda^x/x!)$ where λ is a parameter of the distribution whereas x is the variable and e is a constant.

We can recognize two types of variables in a model of a system, *exogenous* and *endogenous.* Exogenous variables are also called input variables, i.e., they are variables originating or produced outside of the system or resulting from external causes. Endogenous variables are those produced within the system or resulting from internal causes. We also refer to endogenous variables as either *status* variables (to indicate their state or condition within the system) or *output* variables (when they leave the system). Statisticians sometimes call exogenous variables independent, and endogenous variables dependent.

Functional relationships describe variables and parameters in such a way as to show their behavior within a component or between components of a system. These relationships or operating characteristics are either deterministic or stochastic in nature. Deterministic relationships are identities or definitions that relate certain variables or parameters where a process output is uniquely determined by a given input. Stochastic relationships, on the other hand, are those in which the process characteristically has an uncertain output for a given input. Both types of relationship usually take the form of a mathematical equation relating the endogenous and status variables to the exogenous variables. Usually these relationships can only be hypothesized or inferred from statistical and mathematical analysis.

Constraints are limitations imposed on the values of the variables or on the way in which resources can be allocated or expended. These constraints can be either self-imposed by the designer or system-imposed by the nature of the system. Examples of self-imposed constraints would be the setting of maximum and minimum employment levels for consideration or setting an upper limit on the amount of funds available for capital expenditure. In a physical system such as a missile, a self-imposed constraint might be the setting of a minimum range or maximum weight. Most specifications for systems constitute self-imposed constraints. System-imposed constraints are imposed by the very nature of the system. For example, one cannot sell more items than the system can produce, nor could one design a system that violated natural law. Thus, one type of constraint is owing to the unalterable laws of nature whereas the other is man-made and subject to change. This concept is important for the analyst to remember, because he should review the man-made constraints constantly throughout the study, to relax or tighten them if necessary.

The *criterion function* is an explicit statement of the objectives or goals of the system and how they are to be evaluated. Ackoff and Sasieni [1] point out two types of objectives—*retentive* and *acquisitive*. Retentive objectives are those which deal with keeping or preserving either resources (time, energy, skills, etc.) or states (comfort, safety, employment levels, etc.). Acquisitive goals concern acquiring resources (profits, personnel, customers, etc.) or attaining states (share of market, deterrent position, etc.) that the organization or manager seeks. The statement of the criterion should be an unambiguous specification of the goals and objectives against which decisions are to be measured. *Webster's New Collegiate Dictionary* defines criterion as "a standard of judging; a rule or test by which anything is tried in forming a correct judgment respecting it." This explicit and unambiguous statement of the criterion is critically important for two reasons. First, it has a paramount influence on the design and manipulation of the model. Second, an erroneous statement of the criterion will usually lead to wrong conclusions. The criterion function is usually an integral part of the model, and the manipulation of the model is guided by attempts to optimize or satisfy the criteria established.

Even the most restricted portions of the real world are too complex to be comprehended fully and described by human effort. Almost all problem situations are enormously complex, containing an almost infinite number of elements, variables, parameters, relationships, constraints, etc. When attempting to build a model, we could include an infinite number of facts and spend an endless amount of time gathering detailed facts about any situation and defining the relationships among them. Consider the simple act of taking a piece of paper and writing a letter. One could study the detailed chemical composition of the paper, lead, and eraser; the effect of atmospheric conditions on the moisture content of the paper, and its effect upon the friction of the pencil lead as it moves across the paper; the statistical distribution of the letters in the sentences of the text; etc. However, if the only aspect of the situation that is of interest is whether a letter is sent, none of these details is pertinent. Consequently, we must ignore most of the actual features of an event under study and abstract from the real situation only those aspects that make up an idealized version of the real event. All models are simplifications and abstractions of the real world. If done correctly, this idealization provides a useful approximation of the real situation, or at least certain parts of the real situation.

The similarity of a model to the thing it represents is called the degree of isomorphism. To be *isomorphic* is to be identical or of like form, and requires two conditions. First, there must be a one-to-one correspondence between the elements of the model and the item being represented. Second, the exact relationships or interactions between the elements must be preserved. The degree of isomorphism of a model is relative, and most models are homomorphic rather than isomorphic. By *homomorphic*, we mean like in form but different in fundamental structure, and only superficially resembling the different groups of elements. Homomorphic models are the result of both simplification and abstraction.

To develop an idealized homomorphic model, we usually break up the system into a number of smaller parts for separate treatment, i.e., to analyze the problem. Doing so depends on the existence of parts or elements that are approximately independent of one another or that mutually interact in relatively simple ways. Thus we can first analyze the performance of an automobile by examining the motor, transmission, power drive, suspension, etc., even though these are not completely independent.

Closely related to the preceding analysis is the process of simplification. Simplification is a concept that is readily acceptable to most people and entails the stripping away of unimportant details or the assumption of simpler relationships. For instance, we often assume a linear relationship between two variables even though we may suspect or even know that the true relationship is curvilinear. We assume that at least over a limited range of variation such an approximation will be satisfactory. Thus, an electrical engineer works with models of circuits in which the values of resistors, capacitors, etc., are assumed to have constant values—a simplification because we know that the electrical characteristics of many of these components vary as a function of temperature, humidity, age, etc. The mechanical engineer works with models in which gases are perfect, pressures are adiabatic, and conduction is uniform. For the majority of practical cases, such approximations or simplifications are good enough and yield usable results.

The management scientist also simplifies to get usable models. He assumes his variables to be either deterministic (a highly simplified version of reality) or random variables of known probability distribution functions, such as normal, poisson, exponential, etc. He also often assumes linearity of relationships between variables when he knows this is not exactly correct. This is often necessary and justifiable to construct mathematically tractable models.

The other aspect of analysis is abstraction, a concept that is neither as easy to explain nor as easy to understand as simplification. An abstraction comprises or concentrates in itself the essential qualities or behavior of a thing but not necessarily in the same form or detail as in the original. Most models are abstractions in that they try to represent the qualities and behavior of the entity being modeled in a different form or manner. Thus an organization chart attempts to abstract or show the working relationships among the different groups or individual members. That it can depict only a cursory view of the true relationships does not negate its utility for certain purposes.

When we have analyzed and modeled the parts or elements of a problem we then try to begin combining them. In other words, we may construct an approximation of a complex real situation by synthesizing from relatively simple parts. Two warnings are important. First, the parts used must be the correct ones, and second, their interactions must be correctly hypothesized (discussed in detail later when we consider validation and verification). If properly done, this process of analysis, abstraction, simplification, and synthesis will result in a model that approximates the behavior of the real system under study. Remember, however, that the model is only an approximation, and therefore will not behave exactly as the real world. We optimize a model, not the real system. Whether there is any relationship between the performance of our model and that of the real world depends upon how valid and reasonable have been our analysis, abstraction, simplification, and synthesis.

1.9 The art of modeling

The process by which a systems engineer or management scientist derives a model of a system he is studying can best be described as an intuitive art. Any set of rules for developing models has limited usefulness at best and can only serve as a suggested framework or approach. In attempting to make the process of modeling explicit, we are trying to systematize the intuition and experience of those who have gone before. Unfortunately, all scientific research is reported and communicated in the form of a logical reconstruction of events, which aims at justifying the implications produced. This logical reconstruction has little if any relationship to the manner in which the research was actually conducted. No scientific report shows the false starts, the erroneous assumptions pursued and later abandoned, the frustrations of failures, and the flashes of sudden insight. They merely report the sequence of events and justifications of how the researcher would have gone about it, now that he has the benefit of hindsight and knows the final outcome. One of the few exceptions is J. D. Watson's book entitled *The Double Helix* [32]. Watson received a great deal of abuse from other scientists for showing the process of science and research as it really is.

Of course, the danger for the inexperienced modeler is that, finding nothing in the literature except logically constructed hindsight descriptions, he assumes that this is a description of the process of discovery. Then, when things don't work out according to the book, he becomes easily discouraged and disillusioned. The experienced modeler knows that the mental process of developing a model is very different from that reported in textbooks and the literature. But how is this to be conveyed to the beginner? As W. T. Morris [23] points out, "It is not at all clear that the teaching of models by exposing the inexperienced to the ad hoc, contributes much to the development of creative model building ability. The teaching of modeling is not the same as the teaching of models. We seldom encounter a model which is already available in fully satisfactory form for a given management situation."

The approach to the successful building of models appears to proceed on the basis of elaboration and enrichment. One begins with a very simple model and attempts to move in an evolutionary fashion toward a more elaborate model that reflects the complex situation more clearly. Analogy or association with previously well-developed structures appears to play an important role in determining the starting point for this process of elaboration and enrichment. The process of elaboration and enrichment involves a constant interaction and feedback process between the real world situation and the model. There is a constant interplay between the modification of the model and a confrontation with the data generated. As each version of the model is tested and attempts to validate it are made, a new version is produced that leads to a retesting and revalidation. As long as the model is computationally tractable, the analyst may seek further enrichment or complication of the assumptions. When it becomes intractable or cannot be solved, he resorts to simplification and further abstraction.

Thus, the art of modeling consists in an ability to analyze a problem, abstract from it its essential features, select and modify basic assumptions that characterize the system, and then enrich and elaborate the model until a useful approximation results. Morris [23] suggests seven guidelines:

1 Factor the system problem into simpler problems.
2 Establish a clear statement of the objectives.
3 Seek analogies.
4 Consider a specific numerical instance of the problem.
5 Establish some symbols.
6 Write down the obvious.
7 If a tractable model is obtained, enrich it. Otherwise simplify.

Generally speaking, one can simplify by doing one of the following, whereas enrichment implies just the opposite:

1 making variables into constants
2 eliminating or combining variables
3 assuming linearity
4 adding stronger assumptions and restrictions
5 restricting the boundaries of the system

The evolutionary nature of model building is inevitable and desirable; hence, we should not think of the project as the design of one grand single model. As goals are achieved and problems solved, new problems are identified or greater realism desired, which leads to model revisions and better solutions. This process of starting with a simple model and then elaborating and enriching it also has some positive implications for the process of implementation. The speed and direction of the evolution is dependent upon two major factors. The first of these is obviously the inherent flexibility of the model and the second is the relationship of the model builder to the model user. By working closely together through the process of evolution, the model builder and model user can create an environment of mutual trust and understanding which will help to ensure that the end result meets the relevant goals, objectives, and criteria.

The art of modeling can be mastered by those who possess the necessary skills of ingenuity, insight, and resourcefulness, as well as an extensive exposure to the systems and physical phenomena they are trying to model. There is no hard and fast rule about how the problem is originally formulated, i.e., how one looks at it in the first place. There are no magic formulas for deciding what should be included in the model in the form of variables and parameters, descriptive relationships and constraints, and criterion for judgment of effectiveness. Remember that nobody solves the problem; rather, everybody solves the model that he has constructed of the problem. This concept helps to keep the model and the art of modeling in the proper perspective.

1.10 Criteria for a good model

We have defined simulation as the process of designing a model of a real system and conducting experiments with it for the purpose of understanding either the behavior of the system or the evaluation of various strategies being considered for the operation of the system. This definition suggests several important characteristics of the nature and domain of good simulation; (1) it is concerned with the operation of systems; (2) it is concerned with the solution of real world problems; and (3) it is performed as a service for the benefit of those in control of the system or at least those interested in its behavior. Let us examine what this implies in terms of criteria for a good simulation model.

Simulation is concerned with the operation of a system. A *system* is a group or set of objects united by some form of regular interaction or interdependence to perform a specified function. The system may be a military weapons system, an industrial enterprise, an organization, a transportation network, a hospital, an urban development project, or a man and the machine he operates. An *operation* is a set of coordinated acts required for the accomplishment of some specified objective. Thus, the systems of interest are goal or objective directed. This requires us to adhere vigorously to concern for the goals or purpose of the system when we try to model it. We must constantly remember the objectives of the system and of the model so that the appropriate model will result.

Since simulation is concerned with the solution of real world problems, we must be sure that the end result accurately portrays the true situation. Thus, a model that can give us absurd answers should immediately be suspect. Any model should be evaluated at the extreme ranges of the values of the parameters and variables. If a ridiculous answer results, we need to return to the drawing board. The model should also be capable of answering "what if . . ." questions, since these are the type of questions that will be most useful to us in understanding a problem and seeking to evaluate the possible alternative actions available to us.

Finally, always remember the user of the information generated. There is no defensible justification for the development of a simulation model if it is not ultimately used and usable by the decision maker. This can be the person responsible for the design of the system or the person responsible for its operation, but there must ultimately be a user or we have wasted our time and talents. The seeking of knowledge for knowledge's sake is a lofty and commendable goal, but very few managers will long support an operations research, management science, or systems analysis group whose results are not being used.

We can therefore state certain criteria that any good simulation model should be able to meet [19]. A good simulation model should be:

1 simple to understand by the user
2 goal or purpose directed
3 robust, in that it does not give absurd answers
4 easy for the user to control and manipulate, i.e., it should be easy to communicate with
5 complete on important issues
6 adaptive, with an easy procedure for model modification or updating
7 evolutionary, in that it should start simply and become more complex, in conjunction with the user

The need for most of these criteria is immediately evident, but will be discussed more fully in later chapters devoted to implementation and management of the modeling effort. Suffice it to say here that if the model is to be used, great care and planning must be employed throughout its development in considering the needs and psychology of the ultimate user. Simulation modeling should be a learning process for both the modeler and the user. In fact, this may very well be the most beneficial aspect of the use of simulation in problem solving.

1.11 The simulation process

Assuming that a simulation is to be used to investigate the properties of a real system, the following stages may be distinguished:

1 *System Definition*—Determining the boundaries, restrictions and measure of effectiveness to be used in defining the system to be studied.

2 *Model Formulation*—Reduction or abstraction of the real system to a logic flow diagram.

3 *Data Preparation*—Identification of the data needed by the model, and their reduction to an appropriate form.

4 *Model Translation*—Description of the model in a language acceptable to the computer to be used.

5 *Validation*—Increasing to an acceptable level the confidence that an inference drawn from the model about the real system will be correct.

6 *Strategic Planning*—Design of an experiment that will yield the desired information.

7 *Tactical Planning*—Determination of how each of the test runs specified in the experimental design is to be executed.

8 *Experimentation*—Execution of the simulation to generate the desired data and to perform sensitivity analysis.

9 *Interpretation*—Drawing inferences from the data generated by the simulation.

10 *Implementation*—Putting the model and/or results to use.

11 *Documentation*—Recording the project activities and results as well as documenting the model and its use.

The preceding steps assume that the problem can best be solved by simulation. As we have already stated, this may not be the most effective way. It has often been said that a simulation is a last resort or brute force approach to problem solving. It is certainly true that whenever a problem can be reduced to a simple model and solved analytically, there is no need for a simulation. One should investigate all the possible tools available to handle each particular problem and optimize between cost and results. One should convince himself that a simple analytical model is inadequate before he considers the possibility of a simulation.

Since it is necessary and desirable to fit the tool to the problem, the decisions as to which tool or method to use must follow the formulation of the problem. The decision to use simulation should not be considered irrevocable. As more data and

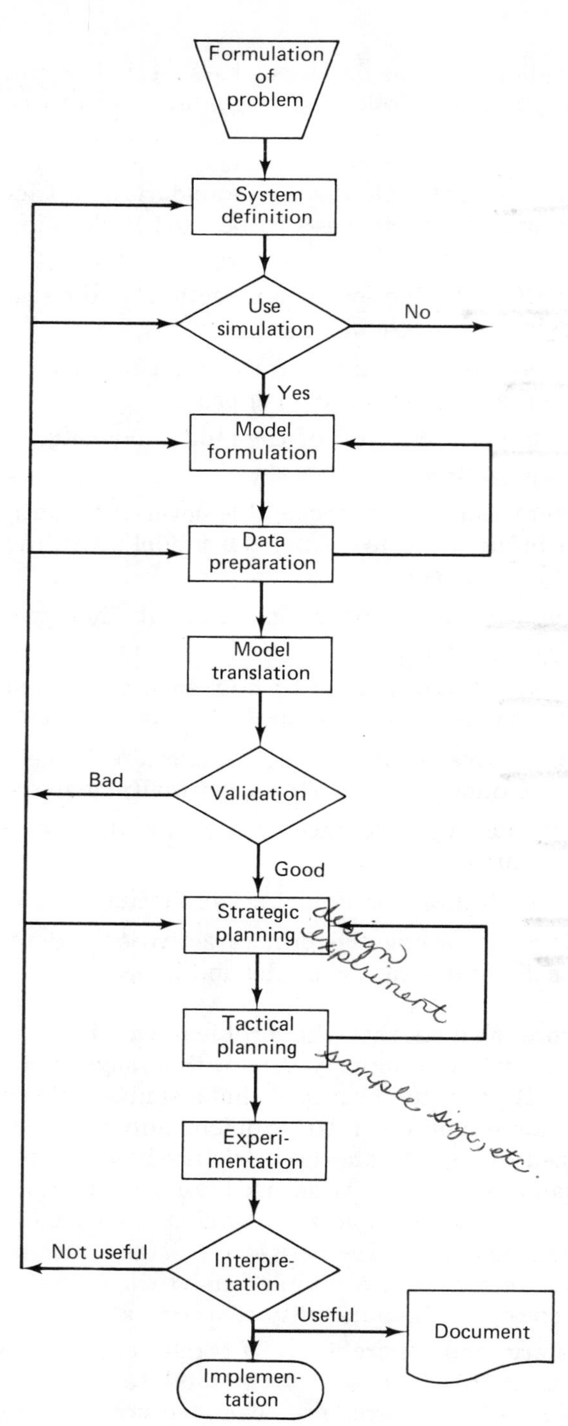
Formulation of problem
System definition
Use simulation
No
Yes
Model formulation
Data preparation
Model translation
Bad
Validation
Good
Strategic planning
Tactical planning
Experimentation
Not useful
Interpretation
Useful
Document
Implementation

Figure 1.3

understanding of the problem are gained, the validity of using simulation should be reevaluated. Because large computers and large sample sizes are so often involved, the cost of a simulation is nearly always high compared with solving a small analytical model. The probable cost and time of the simulation should always be weighed against the value of the information it is likely to produce.

The steps or elements of a simulation and their relationships are shown in the flowchart of Fig. 1.3 The initiation of the project occurs when someone in the organization decides that a problem exists and needs investigation. An individual (normally from the organizational group with the problem) is assigned to undertake some preliminary investigations. At some point it will be recognized that quantitative techniques may be useful in studying the problem, and a systems analyst will be brought into the picture. Thus the definition and formulation of the problem begin.

1.12 Problem formulation and model definition

Albert Einstein once stated that the proper formulation of a problem was even more essential than its solution. To find an acceptable or optimal solution to a problem, one must first know what the problem is. As simple and apparent as that statement may sound, too many management scientists completely ignore the obvious. Millions of dollars are spent each year in coming up with elegant and sophisticated answers to the wrong questions. In discussing the use of quantitative methods in Department of Defense planning, Charles J. Hitch [16] said the following:

> It is my experience that the hardest problems for the systems analyst are not those of analytical technique. In fact the techniques we use in the Office of the Secretary of Defense are usually rather simple and old fashioned. What distinguishes the useful and productive analyst is his ability to formulate (or design) the problem . . .

Unlike examples in this and other textbooks, most practical problems are initially communicated to a management science team in a vague, imprecise way. In many cases management may not have or may not be able to identify its own problem correctly. It knows that it has *a* problem, but may not have identified the *true* problem. Therefore, a systems analysis project normally begins by conducting an orientation study of the system under the decision maker's control. The research team must understand and articulate a set of germane objectives and goals. Experience indicates that the formulation of a problem is a continuing process throughout the study. New information concerning restrictions, objectives, and alternatives are continuously generated by the investigation. Such information should be used periodically to update the formulation and definition of the problem.

An important part of problem formulation is the definition of the system to be studied. All systems are themselves subsystems of other larger systems. Therefore, we must specify the purpose and restrictions under which we create our abstraction or formal model. Bartee [5] says that a problem can be defined as a state of unsatisfied need. A situation becomes a problem when the effect of the system's operation does not provide the desired results. When the desired results are not achieved, a need develops to modify the system or the environment in which it operates. Mathematically, we can define a problem as

$$P_t = |D_t - A_t|$$

where

P_t is the problem at time t

D_t is the desired state at time t

A_t is the actual state at time t

Therefore, the first step in defining the system to be studied is to make an analysis of the need environment. This analysis begins with the specification of goals and establishment of boundary conditions (i.e., what is and is not a part of the system to be studied). Two functional boundaries or interfaces concern us: the boundary that separates the problem from the rest of the universe and the boundary between the system of interest and the environment (i.e., what we consider to be an integral part of the system and what we consider to be the environment in which it operates). We can describe what happens within the boundary of a system in many different ways. Unless we select the elements and relationships to be considered for a specific purpose, an infinite number of connections and combinations would be possible.

Having specified the goals and objectives of the study and defined the boundaries of the system, we next reduce the real system to a logical flow diagram or static model. We wish to construct a model of the real system that neither oversimplifies the system to the point where the model becomes trivial (or worse, misleading) nor carries so much detail that it becomes clumsy and prohibitively expensive. The danger in starting with a flow diagram of the real operation is that the model may tend to be too detailed and include elements which contribute little or nothing to the understanding of the problem.

The tendency is nearly always to simulate too much detail rather than too little. Thus, one should always design the model around the questions to be answered rather than imitate the real system exactly. Pareto's law says that in every group or collection there exists a vital few and a trivial many. Nothing really significant happens unless it happens to the vital few. The tendency among systems analysts has too often been to transfer all the detailed difficulties in the real situation into the model, hoping that the computer would solve their problems. This approach is unsatisfactory not only because of the increased difficulty of programming the model and the additional cost of longer experimental runs, but also because the truly significant aspects and relationships may get lost in all the trivial details. Therefore, the model must include only those aspects of the system relevant to the study objectives.

For many studies, the modeling effort can end at this point. In a surprisingly large number of cases, when a situation is described in a precise, logical manner, the bottlenecks and defects become apparent, thereby eliminating the need to proceed with the simulation study.

1.13 Model formulation

Every study involves data gathering. Data gathering is usually interpreted to mean gathering numbers, but the gathering of numbers is only one aspect of data gathering. The systems analyst must be concerned with data regarding the inputs and outputs of the system he is studying as well as information about the various components of the system and the interconnections or relationships between them. He is therefore interested in gathering both quantitative and qualitative data, and he must decide what data are needed, whether they are pertinent, whether existing data are valid for his purpose, and how to gather this information. Textbooks usually give the student all of the pertinent information and data without reference to how it was gathered and validated. The student then becomes schizophrenic when faced with his first unstructured problem for which he must determine on his own what data are needed and how to gather them.

The design of a stochastic simulation model always involves a choice of whether to use empirical data directly in the model or to use theoretical probability or frequency distributions. This choice is important and fundamental for three reasons. First, using raw empirical data implies that all one is doing is simulating the past. The use of data from one year would replicate only the performance of that year and not necessarily tell us anything about the expected future performance of the system. The only events possible are those which have transpired. It is one thing to assume that the basic form of the distribution will remain unchanged with time and quite another to assume that the idiosyncracies of a particular year will always be repeated. Second, it is generally more efficient of computer time and storage requirements to use a theoretical frequency or probability distribution rather than to use table look-up procedures for generating the necessary random variates for the model's operation. Third, it is highly desirable, if not almost mandatory, that the analyst determine the sensitivity of his model to the precise form of the probability distributions it contains and the values of the parameters. In other words, tests of the sensitivity of results to data input are critically important.

Thus, decisions regarding the data to be used and their validity, form, and goodness of fit to theoretical distributions and past performance, are all critical to the success of the simulation experiment, and far from being academic points.

Eventually one is faced with the problem of describing the model in a language acceptable to the computer to be used. The rapid acceptance of computer simulation has led to the development of a large number of special programming languages designed to facilitate this translation. In practice, most of the languages suggested have been implemented for only a limited number of machines each. Simulation models are usually very complicated logically, having many interactions among the systems elements, and with many of these interactions changing dynamically during the program. This situation has led researchers to develop programming languages that attempt to make the translation problem easier. Thus simulation languages such as GPSS, Simscript, Simula, Dynamo, etc., are higher order languages than the general purpose ones such as FORTRAN, ALGOL, and BASIC. Any general language can express the desired model; nevertheless, depending upon the characteristics of the model, one of the specialized simulation languages may have distinct advantages.

The major differences among the different simulation languages are (1) the organization of time and activities, (2) the naming and structure of entities, (3) the testing of conditions in activities, (4) the type of statistical tests possible on the data, and (5) the ease of changing model structure.

Although several of the special simulation languages have very positive and useful attributes, the sad fact remains that the choice of which language to use will most likely be resolved by the type of machine available and the languages known to the analyst. But when there is a choice, the correct or best choice is probably dependent upon the degree to which the analyst is involved in simulation. For occasional use, a simple language that is easy to understand and learn may be more valuable than one of the more sophisticated languages, whose extra facilities and features make it more complicated to use.

1.14 Validation of the model

Validation is the process of bringing to an acceptable level the user's confidence that any inference about a system derived from the simulation is correct. It is impossible to prove that any simulator is a correct or "true" model of the real system. Fortunately, we are seldom concerned with proving the "truth" of a model. Instead, we are mostly concerned with validating the insights we have gained or will gain from the simulation. Thus, it is the operational utility of the model and not the truth of its structure that usually concerns us.

Validation of the model is extremely important, because simulators look real and both modelers and users find them easy to believe. Unfortunately, simulators often conceal their assumptions from the casual observer and sometimes even from the modeler. Therefore, if validation and evaluation are not carried out carefully and thoroughly, erroneous results may be accepted with disastrous consequences.

There is no such thing as the "test" for validity. Rather, the experimenter must conduct a series of tests throughout the process of developing the model in order to build up his confidence. Three tests may be used to validate a model. First, we must ascertain that the model has face validity. For example, is it possible for the model to give absurd answers if parameters are carried to extreme values? We must also ask if the results of the model appear to be reasonable. The latter test can be conducted for models of existing systems by a method attributed to Turing [31]. It consists in finding people who are directly involved with the actual system and asking them to compare the results of the simulator with actual outputs from the real system. To make the test a little more scientific, we can present several sets of simulated data and several sets of real system data to determine if experienced people can tell the difference.

The second method of validation is the testing of assumptions and the third is the testing of input-output transformations. These latter two may entail the use of statistical tests of means and variances, analysis of variance, regression, factor analysis, spectral analysis, auto-correlation, chi-square, and nonparametric tests. Since each of these statistical tests make assumptions about the underlying process, the use of each raises questions of validity. Some statistical tests require fewer assumptions than others, but in general the power of the test decreases as the assumptions are relaxed.

Fishman and Kiviat [10] divide the evaluation of simulations into the three categories: (1) verification—insuring that the model behaves the way an experimenter intends; (2) validation—testing the agreement between the behavior of the model and that of the real system; and (3) problem analysis—the drawing of statistically significant inferences from the data generated by the computer simulation. Often a large number of actions are necessary to carry out this evaluation, ranging from testing the model stage-by-stage on a desk calculator before assembling the stages into a machine program, to conducting field experiments. However, the testing itself suffers from the standard problems of empirical research: (1) small samples owing to high cost of data; (2) data that are too aggregated; and (3) data of questionable validity.

The question of validation is thus two-faced: determining that the model behaves in the same fashion as the real life system; validating that the inferences drawn from the experiments with the model are valid or correct. In concept, both these points resolve themselves to the standard decision problem of balancing the cost of each action against the value of the increased information and the consequences of erroneous conclusions.

1.15 Strategic and tactical planning

We have defined simulation as being experimentation via a model to gain information about a real world system. It then follows that we must concern ourselves with the *strategic planning* of how to design an experiment that will yield the desired information. The use of experimental designs has found widespread application in the biological and physical sciences, and now in systems simulation. The purpose of using these designs is twofold: (1) they are economical in terms of reducing the number of experimental trials required, and (2) they provide a structure for the investigators' learning process.

The objective of any experimental investigation, including simulation, is to learn more about the system being studied. The experiment is the process of observation and analysis that provides the information leading to solutions. The experimental design selects a particular approach for gathering original information that will provide enough knowledge about the phenomena or system to allow valid inferences to be drawn about its behavior. Two types of experimental objectives are readily recognized: (1) finding the combination of parameter values that will optimize the response variable, and/or (2) explaining the relationship between the response variable and the controllable factors in the system. A wide range of experimental designs have been developed and are available for both these purposes.

In addition, successful learning requires the full use of prior knowledge in proposing possible hypotheses to be tested or strategies to be evaluated. Good experimental design provides strategies for gathering evidence useful for such synthesis and conjecture. Presently available experimental designs and analytical methods fulfill our needs very well. The mathematical formulations associated with the designs provide many possible alternatives. The techniques for divulging the information contained in the data provided by the designs are well documented and usually easy to perform. Experimental design can thus greatly enhance the synthesis of new knowledge and the conjecture of new ideas and at the same time minimize the time, effort, and cost of the experiment.

In general, *tactical planning* involves questions of efficiency and deals with the determination of how each of the test runs specified in the experimental design is to be executed. Primarily, tactical planning is concerned with the resolution of two problem areas: (1) starting conditions, as they affect reaching equilibrium, and (2) the need to reduce the variance of the answer as far as possible while minimizing the required sample sizes.

The first problem (i.e., starting conditions and their effect upon reaching equilibrium) arises from the artificial nature of model operation. Unlike the real world the model represents, a simulation model operates only periodically. That is, the experimenter starts the model, obtains his observations, and shuts it down until the next run. Each time a run is started, it may take a certain period of time for the model to reach equilibrium conditions representative of the real world system operations. Thus, the initial period of operation of the model is distorted owing to the initial start up conditions. The solution is to (1) exclude data for some initial period from consideration, and (2) choose starting conditions that reduce the time required to reach equilibrium. Reasonable starting conditions can reduce but not eliminate the time required for the simulation model to approach equilibrium conditions. Therefore, it is still necessary to determine when measurement should begin.

The second part of the tactical planning problem deals with the necessity to estimate the precision of experimental results and the confidence attributable to the conclusions or inferences drawn. This immediately brings us face-to-face with such areas as variability, sample size, and replication. In any experiment, we try to obtain as much information as possible from a limited amount of experimentation. Several techniques for reducing the variance of response have been proposed (mostly in connection with survey sampling procedures), which can significantly reduce the required sample size and number of replications of the experiments. The use of very large sample sizes can overwhelm virtually all the tactical problems of simulation but usually at a great cost in computer and analysis time. The more complex is the simulation model, the more important is good tactical planning before running the experiments.

1.16 Experimentation and sensitivity analysis

Ultimately, after all development and planning, we run the model to obtain the desired information. At this stage, we begin to find the flaws and oversights in our planning, and to retrace our steps until we achieve our originally established goals.

Sensitivity analysis is one of the most important concepts in simulation modeling. By this we mean determining the sensitivity of our final answers to the values of the parameters used. Sensitivity analysis usually consists in systematically varying the values of the parameters over some range of interest and observing the effect upon the response of the model. In almost any simulation model, many of the set variables are based upon highly questionable data. In many cases, their values may have been determined solely upon the best guess of experienced personnel or very cursory analysis of minimal data. It is therefore extremely important to determine the degree of sensitivity of the results to the values used. If the answer changes greatly with slight variations in the values of some of these parameters, this may provide the motivation and justification for expenditure of more time and money to obtain more accurate estimates. On the other hand, if the results do not change over wide fluctuations in the values of the parameter, no further effort is needed or justified.

Simulation is ideally suited for sensitivity analysis because of the experimenter's degree of control. Unlike experimentation with real world systems, the simulation modeler has absolute control over his model and can vary one parameter at a time if need be, observing the results upon the behavior of the model.

1.17 Implementation and documentation

The last two elements that must be included in any simulation project are implementation and documentation. No simulation project can be considered successfully completed until it has been accepted, understood, and *used*. Management scientists' greatest failure has been in gaining acceptance and use of their labors. In his survey, Gershefski [12] found that the median percentage of total time to develop a model was 25% for the formulation of the problem, 25% for collection and analysis of historical data, 40% for the development of the computer model, and 10% for implementation. It is not surprising, therefore, that Rubenstein [28] found one of the greatest causes of failure in operations research and management sciences projects to be the user's inadequate understanding of results, and thus a lack of implementation. The small percentage of time devoted to implementation is even more difficult to understand when one considers that refining the model, training the user, adjusting the model to changing conditions, and insuring that the results found are valid (which can only be done by implementation), tend to be the most difficult problems the analyst faces. From the author's personal experience, 25% to defining the problem, 20% to gathering and analyzing historical data, 30% to developing the model, and 25% to implementation, is a more realistic breakdown of total project time.

Documentation is closely linked to implementation. Careful and complete documentation of the development and operation of the model can greatly increase its useful life and chances of successful implementation. Good documentation facilitates modification and ensures that the model can be used even if the services of the original developers are no longer available. In addition, careful documentation can help the modeler to learn from his mistakes and perhaps provide a source of subprograms that can be reused in future projects.

In this chapter, we have tried to set the stage for a more intense consideration of the art of simulation. We emphasize that simulation is an art and not a science. There are no hard and fast rules concerning what should not be done or included in the model. The following chapters provide a more detailed treatment of the various aspects of dealing with the development and running of simulation models.

REFERENCES

1. ACKOFF, R. L., and M. W. SASIENI, *Fundamentals of Operations Research*, John Wiley & Sons, Inc., New York, 1968.
2. AMSTUTZ, A. E., *Computer Simulation of Competitive Market Response*, Massachusetts Institute of Technology Press, Cambridge, Mass., 1967.
3. ARMSTRONG, R. H., and J. L. TAYLOR (eds.), *Instructional Simulation Systems in Higher Education*, Cambridge Monographs on Teaching Methods, No. 2, 1970.
4. BARISH, N. N., *Economic Analysis for Engineering and Managerial Decision-Making*, McGraw-Hill Book Company, New York, 1962.
5. BARTEE, E. M., *Engineering Experimental Design Fundamentals*, Prentice-Hall, Inc., Englewood Cliffs, N. J., 1968.

6. BEKEY, G. A., and W. J. KARPLUS, *Hybrid Computation*, John Wiley & Sons, Inc., New York, 1968.

7. CHERRYHOLMES, C. H., and M. J. SHAPIRO, *Representatives and Roll-calls: A Computer Simulation of Voting in the Eighty-eighth Congress*, Bobbs-Merrill Company, Inc., New York, 1969.

8. DUTTON, J. M., and W. H. STARBUCH, *Computer Simulation of Human Behavior*, John Wiley & Sons, Inc., New York, 1971.

9. ELMAGHRABY, S. E., "The Role of Modeling in I.E. Design," *The Journal of Industrial Engineering*, Vol. XIX, No. 6, June, 1968.

10. FISHMAN, G. S., and P. J. KIVIAT, "The Analysis of Simulation-Generated Time Series," *Management Science*, Vol. 13, No. 7, Mar. 1967.

11. FORRESTER, J. W., *Industrial Dynamics*, Massachusetts Institute of Technology Press, Cambridge, Mass., 1961.

12. GERSHEFSKI, G. W., *Corporate Models, The State of the Art*, University of Washington, Seattle, Wash. 1970.

13. GUETZKOW, H. (ed.), *Simulation in Social Science*, Prentice-Hall, Inc., Englewood Cliffs, N. J., 1962.

14. GUETZKOW, H., *Simulation in International Relations: Developments for Research and Teaching*, Prentice-Hall, Inc., Englewood Cliffs, N. J., 1963.

15. HERMANN, C. F., *Crisis in Foreign Policy: A Simulation Analysis*, Bobbs-Merrill Company, Inc., New York, 1969.

16. HITCH, C. J., *Decision-Making for Defense*, University of California Press, Berkeley and Los Angeles, 1967.

17. HOGATT, A. C., and F. E. BALDERSTONE (eds.), *Symposium on Simulation Models: Methodology and Applications to Behavioral Sciences*, South-Western Publishing Co., Cincinnati, Ohio, 1963.

18. KRESGE, D. T., and P. O. ROBERTS, *Techniques of Transportation Planning: Systems Analysis and Simulation Models*, Brookings Institution, Washington, D. C., 1971.

19. LITTLE, J. D. C., "Models and Managers: The Concept of a Decision Calculus," *Management Science*, Vol. 16, No. 8, Apr. 1970.

20. McKENNEY, J. L., "A Clinical Study of the Use of a Simulation Model," *The Journal of Industrial Engineering*, Vol. XVIII, No. 1, Jan. 1967.

21. MEADOWS, D. L., *Dynamics of Commodity Production Cycles*, Wright-Allen Press, Cambridge, Mass., 1970.

22. MEIR, R. C., *et al.*, *Simulation in Business and Economics*, Prentice-Hall, Inc., Englewood Cliffs, N. J., 1969.

23. MORRIS, W. T., "On the Art of Modeling," *Management Science*, Vol. 13, No. 12, Aug. 1967.

24. NAYLOR, T. H., *Computer Simulation Experiments with Models of Economic Systems*, John Wiley & Sons, Inc., New York, 1971.

25. NAYLOR, T. H., and J. M. VERNON, *Microeconomics and Decision Models of the Firm*, Harcourt Brace Jovanovich, Inc., New York, 1970.

26. PACHER, A. H., *Models of Economic Systems*, Massachusetts Institute of Technology Press, Cambridge, Mass., 1972.

27. ROWE, A. J., "Simulation—A Decision-Aiding Tool," A.I.I.E. International Conference Proceedings, New York, 1963.

28. RUBENSTEIN, A. H., *et al.*, "Some Organization Factors Related to the Effectiveness of Management Science Groups in Industry," *Management Science*, Vol. 15, No. 8, Apr. 1967.

29. SHANNON, R. E., and W. E. BILES, "The Utility of Certain Curriculum Topics to Operations Research Practitioners," *Operations Research*, Vol. 18, No. 4, Jul.-Aug., 1970.

30. SIEGAL, A. I., and J. J. WOLF, *Man-machine Simulation Models*, Interscience Publishers, New York, 1969.

31. TURING, A. M., "Computing Machinery and Intelligence," *Mind*, Vol. 59, Oct. 1959.

32. WATSON, J. D., *The Double Helix—The Discovery of the Structure of DNA*, Atheneum Publishers, New York, 1968.

33. WESTON, F. C., "O.R. Techniques Relevant to Corporate Planning Function Practices, An Investigative Look," presented at 39th National Meeting, Operations Research Society of America, *Operations Research Bulletin*, Vol. 19, Suppl. 2, Spring, 1971.

EXERCISES

1. Read case study I in Appendix A.
   Could this system be analyzed by another technique?
   Discuss whether the author considered all the necessary steps.

2. The time between arrivals of customers is uniformly distributed from 1 to 20 minutes. For 50% of the customers the service time is 8 minutes, whereas for the other 50% it is 14 minutes. Devise a suitable random number generator (do not use a table of random digits) for the arrival times and another for the service times. Simulate 4 hours of activity. Keep track of total customer waiting time and service facility idle time.

3. T. A. Chance lives in Las Vegas. Each work day he walks to a nearby restaurant, eats his breakfast, and then plays "3 difference" while waiting for his bus to take him to work. In the game "3 difference," the player repeatedly flips an unbiased coin until the difference between the number of heads and number of tails tossed is equal to three. The player pays $1 for each flip of the coin, but he receives $8 at the end of each game he completes. Each morning, Mr. Chance sets aside $10 for playing this game and plays until either (a) he loses all the money he has set aside for that day or (b) one game is completed. Simulate one week's play (5 days) to determine if Mr. Chance comes out ahead.

chapter two

SYSTEMS INVESTIGATION

Order and simplification are the first steps toward the mastery of a subject. The actual enemy is the unknown.

Thomas Mann

2.1 Systems characteristics and behavior

The concept of systems plays a critical role in our modern viewpoint of the world. The fundamental idea of the systems approach to attacking problems has become so ingrained into contemporary technology that we tend to take it for granted and talk about it as if it were a precise methodology (which it is not). The systems approach attempts to study total-systems performance rather than to concentrate on the parts. It stems from the recognition that even if each element or subsystem is optimized from a design or operational viewpoint, the total-systems performance may be suboptimal owing to interactions between the parts. Because of the increasing complexity of man-organized systems and the need to cope with this complexity, the necessity for systems thinking has become more and more important.

All complex systems have certain characteristics that cause many of the failures and frustrations we experience in trying to improve their behavior. Among these are the following:

1 *Change.* The present condition or state of a system is the integrated result of the past and the foundation for the future. No real world system remains static over a long period of time. Elements enter and leave the system either through a birth-and-death process or by passing through the system boundary.

2 *Environment.* Each system has its own environment and is in fact a subsystem of some broader system. The environment of a system is a set of elements and their relevant properties, which, although not a part of the system, if modified can produce a change in the state of the system. Thus, a system's environment consists of all external variables that can affect its state.

3 *Counterintuitive behavior.* Cursory examination of complex systems will sometimes indicate needed corrective action, which is often ineffective or even adverse in its results. Cause and effect are often not closely related in time or space; symptoms may appear long after the primary causes. Obvious solutions may actually intensify a problem rather than solve it.

4 *Drift to low performance.* Complex systems generally tend toward a condition of reduced performance with time. Parts deteriorate and inefficiencies creep in, their counterintuitive nature causing detrimental design changes.

5 *Interdependency.* No activity in a complex system takes place in total isolation. Each event is influenced by its predecessors and affects its successors. In addition, real world activities generally parallel and ultimately influence each other.

6 *Organization.* Virtually all complex systems consist in highly organized elements or components. Parts combine into hierarchies of subsystems, which interact to carry out the function of the system.

Much of the behavior of complex systems is caused by the preceding characteristics. Although complex systems and their environments are objective (i.e., they exist), they are also subjective in that the particular selection of elements to be included or excluded and their configuration are dictated by the researcher. Different analysis of the same objective process or phenomenon may conceptualize it into very different systems and environments. For example, as Ackoff [1] points out, "An architect may consider a house together with its electrical, heating and water systems as one large system. But a mechanical engineer may consider the heating system as a system and the house as its environment. To a social psychologist a house may be an environment of a family, the system with which he is concerned. To him the relationship between the heating and electrical system may be irrelevant, but to the architect it may be very relevant."

Thus we must be very careful in defining the system to be studied and determining the boundary between the system and its environment.

A systems description for simulation purposes entails two steps: (1) a static representation and (2) a dynamic representation. The first deals with determining or defining the existence of the subsystems. What components of the system will be included in the model, what elements will be excluded or considered part of the environment, and what structural relationships will be recognized between them? The second phase deals with change. What state changes are possible in the system and its environment (as defined), and what are the sequence relationships among these changes? The *state of a system* at a particular moment is defined as the set of relevant properties which that system displays at that time.

The actual process of formulating the model is somewhat individualistic. Expert judgment and intuition are crucial to the whole process. This reliance upon judgment and expertise permeates every aspect of the modeling effort: in deciding the most fruitful approach; in deciding what the model includes; in designing the model; in determining the facts and their relative importance; and in interpreting the results.

The system elements are the components, parts, and subsystems that perform a function or process. The relationships among these elements and the manner in which they interact determine the behavior of the overall system and how well it fulfills its overall purpose. In defining a system, we realize the importance of relevance. The first step in creating any model is to specify its *purpose*. As we have stated several times, there is no such thing as *the* model of a system. We can model any system in numerous ways, depending upon what we wish to accomplish. Both the elements and the relationships must be selected to achieve a *specific purpose* and the choices made to that end. Moreover, our specification must take into account our ability to handle the types of analysis desired and should be as simple as our stated purpose will allow.

Unfortunately, although the need for relevant specification and simplification is readily apparent, the method of implementation is not. In the present state of the modeling art, a theory of system definition can neither guarantee that we have abstracted our system from the real world correctly nor assure that we have modeled it in an optimum manner for our purpose. In this all-important matter of system definition, we must rely on art, experience, intuition, advice, and hypothesis. Yet our procedure for systems definition need not be a random process or entrusted to fate. Following a generalized approach and framework based upon past experience increases our chances of success.

We wish to construct a model of the real system that neither oversimplifies to the point where the model becomes trivial nor carries so much detail that the model becomes clumsy and prohibitively expensive to operate. Therefore, our first problem is to gain an adequate knowledge of the system to be modeled. The model will continuously undergo changes during its development as our knowledge of the system increases. Therefore, the most effective approach is gradually starting from a general qualitative treatment and proceeding in time to a more detailed and exact representation as our knowledge of the process or system increases.

We should begin any investigation of an existing or proposed system by establishing a plan to show how the investigation will be made, what methods will be used, and what schedule will be followed. Gathering data and information will be necessary throughout the study. Data gathering is usually interpreted to mean gathering numbers, which is indeed important; however, the modeler must also gather other data that are not so easily expressed numerically. For example, he will require data on the inputs and outputs of the system, as well as information on the various components or elements of the system and the interconnections and relationships among them.

Students taking formal course work in quantitative methods seldom stop to contemplate where and how they will get the information, numbers, etc. that they will use to construct their models. Textbooks almost invariably give the student pertinent information and data without reference to how it was gathered. The student becomes schizophrenic when faced with his first unstructured problem in the real world, where he must determine what data and information are needed and how he must gather them. Each investigation has two stages: fact finding and analysis. Fact finding involves establishing exactly the nature and purpose of the system. Analysis concerns recording the facts to display the structure of the system prior to modeling.

So many different types of systems might concern the analyst that no attempt at classificiation can be completely satisfactory. However, the definitions proposed by Ackoff [1] might prove helpful in defining the system and its goals. We might consider systems from two viewpoints: What is a system trying to achieve? What brings about change in a system?

Some systems are *state maintaining*, in that their purpose or function is to try to react to changing conditions in such a manner that they maintain some predetermined or preset state. For example, the law enforcement system is devised to maintain a state of peaceful order, where both persons and property are safe from violence and unlawful seizure. Likewise, the purpose of the heating and cooling system of a house is to maintain room temperatures within a small preset range. Such a system reacts only to changes in its environment; as long as the desired state prevails, the system's only action is surveillance.

Other systems are *goal seeking*, i.e., they are devised to seek a state that does not presently prevail. For example, the educational system strives to educate and prepare students for later life; it does not seek to maintain the student's present state of knowledge but rather to expand and increase it to the maximum extent possible. Systems with automatic pilots are also goal seeking, as are most unmanned military weapons, such as antiaircraft missiles, etc. Most goal seeking systems can improve with experience and modify their own behavior as they pursue their goals. A system may also be multi-goal seeking. That is, it may seek two or more goals simultaneously, thus necessitating tradeoffs when the behavior required to achieve the different goals conflicts.

Finally, we have *purposeful* systems, which set their own goals. That is, they can devise both methods and ends under either constant or changing conditions. Human beings are the most familiar example of such systems—e.g., piloted aircraft and other systems with human operators, who can change the system's goals. In purposeful systems, as soon as one goal is met another is immediately set; or, if it becomes obvious that a certain goal cannot be met, an alternative goal is selected.

The changes within a system can be *reactive*, *responsive*, or *autonomous*. A reaction in a system is an event that is deterministically caused by another event. For instance, when we throw a light switch (event 1), the system reacts by the lights coming on (event 2), i.e., a direct cause and effect deterministic reaction. In responsive change, an event may create a cause but is insufficient to bring about an effect. For example, when it gets dark we usually turn on a light. We do so in response to the darkness, but darkness is not a sufficient event to cause the lights to come on (unless the switch is automatic). Finally, autonomous changes are self-determined events, in which no other predecessor event is necessary. Much of human behavior is of this nature.

2.2 Beginning a study

Experience indicates that beginning a simulation project properly may well make the critical difference between success and failure. Ideally, such studies are initiated by a decision maker or group of decision makers faced with a problem requiring a decision (the analyst may be the decision maker who instigates the study). Often the project is initiated or sponsored by an authority who cannot make the final decisions but who is responsible only for recommendations. In such a case, the results of the study may have to serve the dual purpose of (1) helping the sponsor formulate his recommendations and (2) justifying, supporting, and helping to sell the sponsor's recommendations.

We begin our analysis by collecting information and data adequate to provide both the initial framework for the study and a description of the system. We formulate the problem especially carefully to be sure that we properly understand the sponsor's needs and requirements. The usual simulation project begins by the sponsor describing the situation to the analyst in general terms and in a rather vague and imprecise way. This is not too surprising since, had he known precisely what the problem was and the best way to solve it, he would have solved it himself. We must therefore consider the sponsor's problem description as a set of symptoms requiring diagnosis.

The sponsor describes the problem situation in terms of his background and experience; therefore, it is often put in terms of profit and loss figures, inventory levels, delays, bottlenecks, or other operational data available to the sponsor. The flow of events will be: diagnosis of symptoms → problem definition → model formulation. Therefore, defining the problem converts the sponsor's description of symptoms into the precise language of our formal models. Our understanding of this process is minimal, and much research remains to be done to develop a logical strategy for problem formulation. The present state-of-the-art in problem formulation is very crude and imprecise compared with the power and precision of our formal modeling techniques. In trying to formulate the problem we must become thoroughly familiar with all aspects of the sponsoring organization's operations relevant to the problem. This includes forces or factors outside the organization that can have an impact, and an understanding of the subjective as well as objective aspects of the problem. Sensitivity, imagination, and meticulous attention to detail are required. Often, very important clues or hints are casually given and overlooked because their significance is not recognized.

The design, development, and implementation of a simulation model are expensive. The project cost for even the simplest model will easily run 5 to 6 man-months ($30,000) or more, whereas some of the more complex simulation models have cost as much as 5 million dollars. The average cost and time to develop simulation models are $100,000 and 6 to 12 months. Thus the organization, planning, and management of the project should not be treated casually. The first phase of the study involves overall planning, including budgeting to effect a reasonable match between the potential benefits of a solution and the cost of obtaining it.

Limits of the proposed study should be established early so that both analyst and sponsor have a clear idea of what will and will not be done. Initial estimates of the potential benefits and the cost of the work to be done should be made. Unless the short-term benefits are expected to exceed the cost by a factor of over ten to one, the project should probably not be undertaken. Obviously, a return of ten to one exceeds most investment opportunities. However, historically the costs are invariably underestimated by a factor of three to one, whereas the benefits to be derived are usually overestimated by a similar factor. Consider a hypothetical case with initial benefits estimated at $100,000 and project costs at $10,000. If we discount the benefits by a factor of three and triple the costs, we obtain the following results:

	Benefits	Cost	B/C Ratio
Initial estimates	$100,000	$10,000	10.0
Actual results	33,300	30,000	1.11

These results are approximately equivalent to many other business opportunities. Analysts often argue that in some cases it is desirable to perform the study anyway because of its side benefits, such as understanding the system better. Such arguments have some validity, but unless the modeling activity shows a substantial long-run return, it will not survive.

To formulate a problem properly, we must break it down into two phases. Phase one is the orientation period and formulation of the sponsor's problem; phase two is the formulation of the research problem. The steps proposed for phase one or the orientation period are as follows [2]:

1 Identify the primary decision maker or decision makers and the decision making process relative to the system being considered.

2 Determine the relevant objectives of each of those responsible for some aspect of the decision.

3 Identify other participants and the degree of influence they may have on a solution as well as the channels through which they may exercise this influence.

4 Determine the objectives and vested interests of these other participants.

5 Establish which aspects of the situation are subject to the control of the decision makers and the range of control they are able to exercise.

6 Identify those aspects of the environment or problem context that can affect the outcome of possible solutions which are not controlled by the decision makers.

7 Determine what objections or counteractions are likely to arise from other participants opposed to changes in the system.

Very seldom in our complex society are major decisions made by a single decision maker. The more important the decision the more likely it is that several people will play a significant part in arriving at a final solution and its implementation. Thus, although the sponsor may be a single person or group of persons, the final acceptance or rejection of the resulting recommendations usually involves other parties. It is important that these other participants and their objectives, fears, etc., be recognized from the very beginning of the study if we are to maximize the probability of success.

We must also recognize that very, very seldom is there a single objective to be achieved. For example, we want increased range for a missile system, but we also want to maintain light weight, high reliability, and an effective payload; we seek increased profits, but we also want to maintain stable employment levels and continued long-term growth. Thus we usually have both acquisitive goals (seeking properties that are not presently achieved) and retentive goals (maintaining present properties). In the process of formulating a problem, we seek alternatives that are considered to be appropriate means for reaching desired ends. However, these ends often are merely instrumental to achieving other objectives. We thus recognize a series or hierarchy of ends and objectives. The fact that certain goals may be desirable because of their importance to the achievement of more distant ends, leads to the arrangement of these goals in a hierarchy—each level to be considered as an end relative to the levels below it, and as a means relative to the levels above it. Unfortunately, these means-ends hierarchies are seldom integrated, completely connected chains. Often, they do not build nicely one on the other but actually conflict—i.e., the achievement of one goal reduces the achievement of another. Hence, tradeoffs and the relative weighting of objectives become prime necessities and themselves some of the goals of the study.

As Herbert A. Simon [17] points out, it is also important to recognize that every decision involves elements of two kinds: *factual* and *value*. Factual elements are statements about the observable world and the way in which it operates. We can test these elements to determine their veracity. Value elements, on the other hand, are ethical or moral statements whose validity we cannot test. The same dichotomy is true of objectives. Objectives such as weight, range, speed, profit, etc., are factual elements. Objectives such as deterrent effect, morale, good health care, excellent education, adequate housing, etc., are value elements. In other words, most decision problems consist of both objective and subjective elements. That a certain change to an automobile will give it greater speed is a factual element; whether greater speed is desirable is a value element. The analyst must therefore recognize clearly which are the subjective (value) and which the objective (factual) elements in a problem situation during the formulation phase.

The second part of problem formulation concerns laying the groundwork and planning the research phase of the study. At a minimum, this process entails the following steps:

1 Explicitly specifying the task to be accomplished and the system to be studied.

2 Determining the constraints imposed in terms of personnel, schedule, funds, and computer time.

3 Establishing a task management system to coordinate people, control resources, and transmit knowledge.

4 Securing the participation of all required personnel.

5 Obtaining access to all pertinent information and data that may be significant for the problem's solution.

6 Evolving appropriate criteria as bases by which the study's results can be evaluated.

7 Establishing the limits of the proposed study so that both analyst and sponsor have a clear idea of its scope.

The manner in which the study is planned, implemented, and controlled is significant in determining the ultimate success or failure of the project. Therefore, this aspect of the effort should not be taken lightly or given only cursory attention. Because of its great importance, we have devoted Chapter 7 to this subject, and will thus defer any detailed discussion until later. In the following sections we discuss means of gathering the information and data needed to define the system and to formulate management's problems and the research problem.

2.3 Fact finding

The most obvious source of data is through measurement and observation, but several other sources are equally important, among which are documents, interviews, and participation. Documents include manuals, blueprints, written procedures, reports, specifications, correspondence, and the like. In most organizations there usually exists an overabundance of paper work and documentation, but not always of the right type. The ability to uncover and interpret properly appropriate documentation is a tremendous aid to understanding and defining a system. The analyst must be very careful, however, regarding what data he accepts from documentation and interviews. Such data should be accepted only tentatively and later cross-checked for validity from other sources and against experience and observation. Data contained in documents may be obsolescent, and information given in correspondence and interviews may be biased by personality traits or perceived self-interests. However, despite these very real dangers, both sources may call attention to discrepancies, supply facts, or provide a different point of view.

Interviews are extremely useful for many reasons, but particularly when valid documentation is scarce. Many procedures and operating rules are never written down; even when they are, much valuable information can only be obtained through interviews. Successful interviewing is an art difficult to acquire. Interviewing in modeling work has two main objectives: (1) to discover and verify facts; and (2) to provide the opportunity to meet and overcome resistance to possible changes caused by the modeling investigation. The analyst must always remember that the people he interviews are fully aware that his efforts will very probably result in changes of some sort in the system. The person being interviewed is therefore concerned and wary regarding the nature and effect of those changes.

Because the analyst faces considerable actual or potential resistance from people suspicious of change, tact is essential. Resistance to change often produces a somewhat frigid climate for the interview. The analyst should remember that when he investigates and tries to model existing systems, he is usually concerned with well-established existing procedures and habits. The ineffective employee may fear exposure, and the effective employee may not welcome anyone who potentially may interfere with his methods. Interviewing is the process of obtaining information by being a good listener, facilitating conversation, and focusing the discussion on the appropriate subject.

If possible and appropriate, the analyst should ask the person being interviewed to go through the motions of his job in detail before he listens to the employee's description, owing to the fact that a person's viewpoint of a situation often contains a certain amount of subjective opinion. It is always important to distinguish between facts and opinions; both are necessary and valuable but must be handled differently. The analyst must also guard against his own biases and opinions coloring or distorting the fact-finding process. If he is experienced, he is sometimes inclined to jump to conclusions based upon problems he has had in the past. Since very few problems (if any) are exactly the same, he must be careful not to guess, not to rely too heavily on his past experience, and not to prejudge the issue.

Whenever possible, after gathering all the pertinent data he can from documentation and interviews, the analyst should try to participate in, or at least personally observe, the system in action. Morse and Kimball [13] state that the analyst should "fly in the bomber plane, travel in the bus, operate the early-warning radar, or make purchases in the store, as the case may be." When he cannot participate, he should at least observe the whole system in operation. He should follow the order through the system, watch the missile crew in operation, observe the maneuvers, or stand around the department and watch. By such participation he not only gets the feel of the system and validates prior data gathered through other sources, but he may also learn a great deal from the people whose day-to-day function is the operation of the system.

Hopefully, out of the processes of observing, participating, interviewing, and examining documentation, will come a clear, unbiased understanding of the system and how it operates or is supposed to operate. This fact-finding step needs to be well documented, and each piece of information must be checked and counterchecked to insure its validity. One often is asked, "Why do you need all those detailed reports (or, why should you waste the time observing the operation) when so-and-so can tell you about it?" The answer, of course, is that the need for unbiased, impersonal facts rather than opinions is so critical to successful research that each fact must be checked and rechecked, despite its source.

In the following sections, we discuss some possible methods of documenting the information disclosed during the fact-finding investigation so as to insure its greatest utility. As stated earlier, we first need a static representation of the system to be modeled. A group of very useful techniques for this purpose might appear under the heading *schematic models*, by which we mean all those methods of analysis involving the graphical representation of a system's operation. They present an abstraction of the way in which events occur or may occur in the real world, and traditionally are the basic tools of the engineer. Their purpose is to aid the analyst in understanding and documenting how a process or system functions. Although numerous schematic methods are available, we limit our presentation to only flow process charts, flow diagrams, multiple activity charts, block diagrams, and logic flow charts. Others may be found in textbooks on methods improvement [3, 14] and analyses of production management [6, 7].

2.4 Flow charts

The *flow process chart* is a comparatively easy tool to use and is an excellent technique for determining how the step-by-step details of a process are actually performed or anticipated to be performed. It presents the operation in a logical and condensed form by showing the step-by-step progress of the process or procedure being charted. One limitation, however, somewhat limits its generality: it is most easily used to follow a single product, material, form, or person through a series of steps; multiple forms, materials, and people can be charted, but the procedure becomes more complex.

The process chart was originated by Frank B. Gilbreth and later refined and further developed by others, such as Ralph M. Barnes and Allan H. Mogensen. Gilbreth developed a sort of symbolic shorthand to indicate each of several types of activity or lack thereof in a process. His 39 symbols have been simplified over the years. The American Society of Mechanical Engineers established five symbols that are now widely accepted and cover the vast majority of activities encountered in a study:

Symbol	Name	Results
○	Operation	Produces, prepares, and accomplishes
⇨	Transportation	Moves
□	Inspection	Verifies
D	Delay	Interferes, waits
▽	Storage	Keeps, retains

The circle stands for an *operation* in which something is being done or intentionally changed physically or chemically; something is being created, assembled, or taken apart. These operations are productive, such as typing a form, drilling a hole, welding a part, etc. Calculating, planning, and receiving information are also considered operations.

The arrowhead represents movement or *transportation.* It depicts the movement of the item being charted from one work station to another, or from one department to another. On the process chart, any movements within the immediate area of an operation—such as reaching for a tool, placing a report in the outgoing basket at the edge of the desk, etc.—are considered part of the operation and not a transport movement. However, if the person or thing must move to another location, then a transportation should be indicated. Sometimes it is advantageous to place letters inside the symbols to indicate the mode of transportation, such as "T" for truck, "C" for conveyor, or "E" for elevator.

The square indicates an *inspection* or comparison against predetermined plans or standards, including examination of the item for quality, quantity, and proofreading; reading printed information prior to making some decision; etc. Counting parts, checking a form for error, comparing information received to prior directives, are all inspections and should be represented by squares. In many processes, an inspection is frequently combined with an operation—for example, when a worker counts and inspects parts as they are packed for shipping. When this occurs, we can combine the two symbols by drawing the circle inside the square.

The big D is used for temporary *delays*, such as when a person is interrupted in an activity, a letter is in an outgoing box waiting to be picked up, or data are in a buffer waiting to be telemetered to earth. Items in a queue awaiting service would be shown as a delay, as would items awaiting pickup at a loading dock.

The triangle stands for *storage*, which occurs when parts or items are kept in a formal and deliberate inactive status. Examples are components in a storeroom, information in computer memory or on tape for later reference, records on file, spare parts in a warehouse, and plasma in the blood bank. The difference between storage and delay is that the first is planned and deliberate, whereas the latter is beyond control or unplanned.

With this or a similar set of symbols we may portray any procedure or process. A process chart is constructed relatively simply and can take several different forms. The simplest is the flow process chart, as is shown in Fig. 2.1. The flow process chart depicts a single item. It is a running commentary on what happens to a specific part or person while it or he accomplishes a specific purpose. Other pertinent data may also be included, such as distances traveled, times for different operations or delays, etc.

Another way to visualize a process is a *flow diagram*, such as Fig. 2.2, used in conjunction with the flow process chart. The flow diagram allows us to display the process flow superimposed upon a to-scale floor plan of the physical area where the process takes place. The operations and other events are numbered so that we can compare them with the flow process chart. Such a chart helps us visualize the piling up of activities, possible interferences, and inefficiencies in the system. Special variations of flow process charts and flow diagrams may be devised to study more complex operations involving several items or multiple copies.

Another very useful device for schematic representation is the *multiple activity chart*, which combines some of the details of the process with the analysis of the time balance between items. Its function is to display the coordination needed between two or more people, machines, or combinations of both against a scale of time. Again, this concept is essentially simple. The chart records the times required to do something where men, machines, or both

Flow chart

Summary

	Present		Proposed		Difference	
	No.	Time	No.	Time	No.	Time
○ Operations	2					
⇨ Transportations	5					
□ Inspections	0					
D Delays	6					
▽ Storages	2					
Distance traveled	7	Ft		Ft		Ft

Process Cut and terminate # 62 jumper cables

☐ Man or ☒ Material # 10 cable wire

Chart begins Inventory (raw material)

Chart ends Inventory (in process)

Charted by T.P.C. Date May 1, 1973

	Details of (proposed) method	Operation	Transport	Inspection	Delay	Storage	Distance in feet	Quantity	Time	Possibilities: Eliminate	Combine	Change: Sequence	Place	Person	Improve	Notes
1	Stockroom (raw materials)	○	⇨	□	D	▽										
2	Materials handler	○	⇨	□	D	▽	20									
3	Work bench	○	⇨	□	D	▽										
4	Cutting machine operator	○	⇨	□	D	▽	5									
5	At cable cutter	○	⇨	□	D	▽										
6	Cut to length	○	⇨	□	D	▽										
7	At cable cutter	○	⇨	□	D	▽										
8	Materials handler	○	⇨	□	D	▽	10									
9	Work bench	○	⇨	□	D	▽										
10	Terminate	○	⇨	□	D	▽										
11	Work bench	○	⇨	□	D	▽										
12	By hand (tote bin)	○	⇨	□	D	▽	28									
13	Inventory (check in)	○	⇨	□	D	▽										
14	By hand	○	⇨	□	D	▽	8									
15	Stockroom (in process inventory)	○	⇨	□	D	▽										
16		○	⇨	□	D	▽										
17		○	⇨	□	D	▽										
18		○	⇨	□	D	▽										
19		○	⇨	□	D	▽										
20		○	⇨	□	D	▽										

Figure 2.1 Example of a flow chart [8].

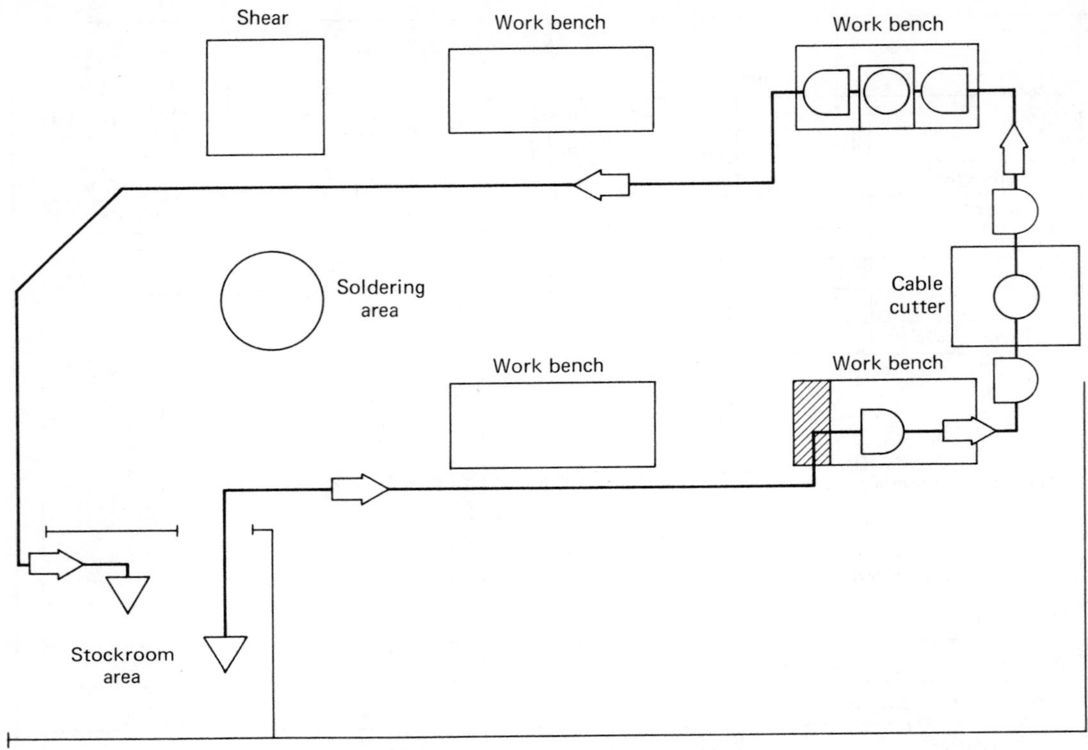

Figure 2.2 Example of a flow diagram.

are involved. The basic symbols used in this kind of chart are as follows:

	Man	Machine
(dotted box)	Any work or activity which is independent of either machines or other people	Any work or operation which is done by the machine without operator help
(blank box)	Work done in conjunction with a machine or other person; also called cooperation time	Operating time where the machine and operator are working together
(hatched box)	Waiting for the machine or another person	Waiting for the operator

Figure 2.3 shows what a multiple activity chart might look like. Such charts have been used in the space program to study checkout and countdown procedures, to study astronaut time

utilization, and to calculate when certain experiments should be run. In hospitals, they are used to study operating room procedures and clinical laboratory situations. Since in most complex systems multiple activities occur simultaneously, such charts can help us to see where combined and independent activity should take place in our simulation.

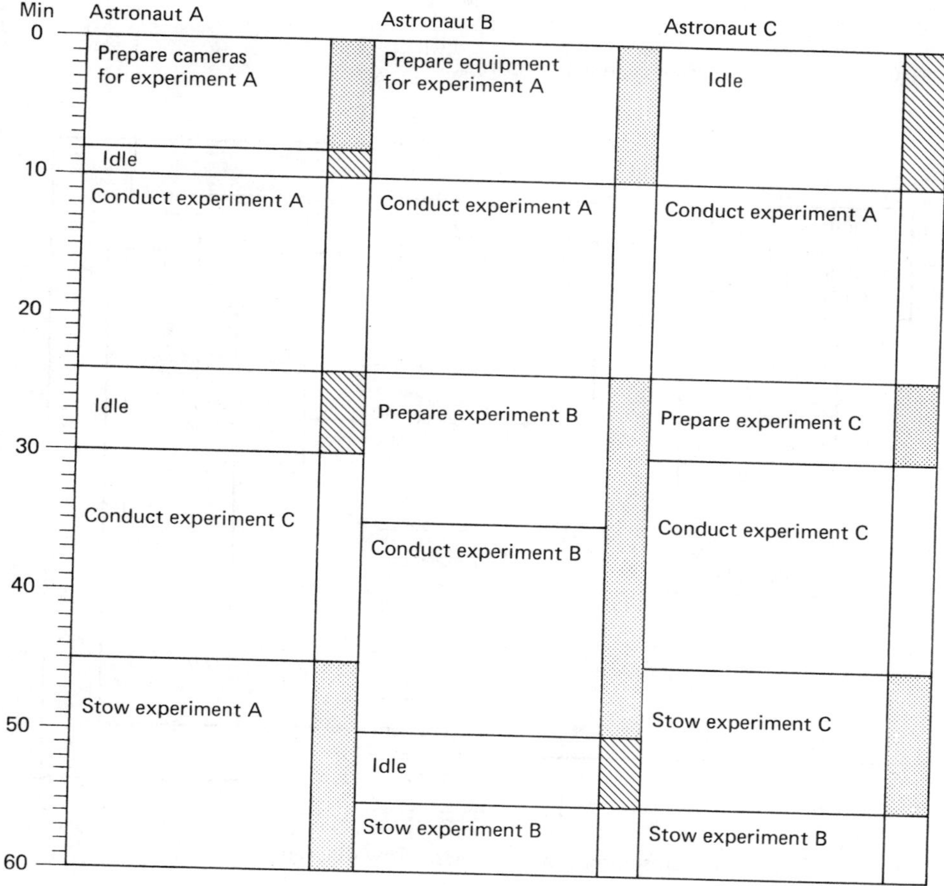

Figure 2.3 Multiple activity chart.

2.5 Block and logic diagrams

Simple in conception, block diagrams are perhaps the most useful device available to the modeler. The simplest form of block diagram is portrayed in Figs. 2.4 and 2.5. Block diagrams identify major elements or subsystems and show the existence of relationships among them. Arrows depict the flow of transactions between

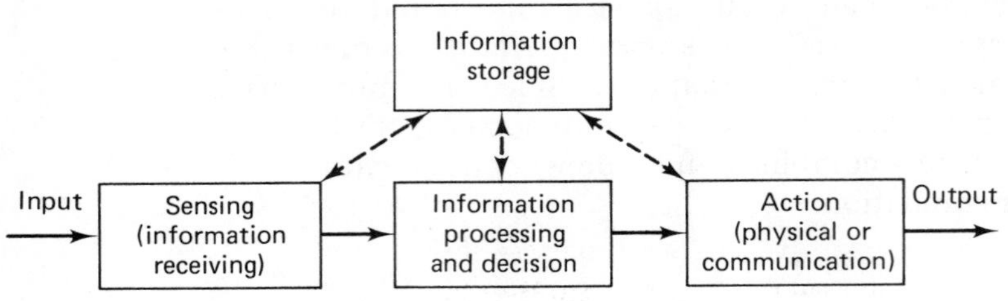

Figure 2.4 Man-machine systems functions.

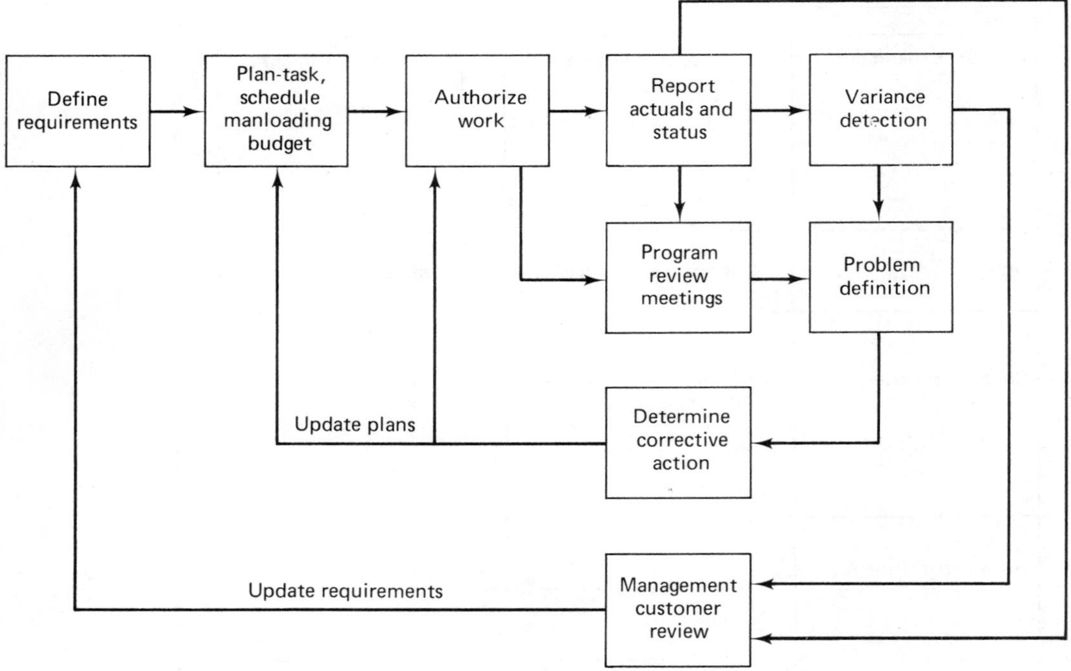

Figure 2.5 Closed-loop management flow diagram.

elements, and boxes containing descriptive or identifying words represent the elements or subsystems.

A slightly more sophisticated block diagram appeared in Chapter 1 (see Fig. 1.3). In this type of block diagram (actually a system logic flow chart), instead of rectangular blocks we introduce the systems flow chart symbols used by computer programmers and systems analysts. For the time being, we need only six symbols:

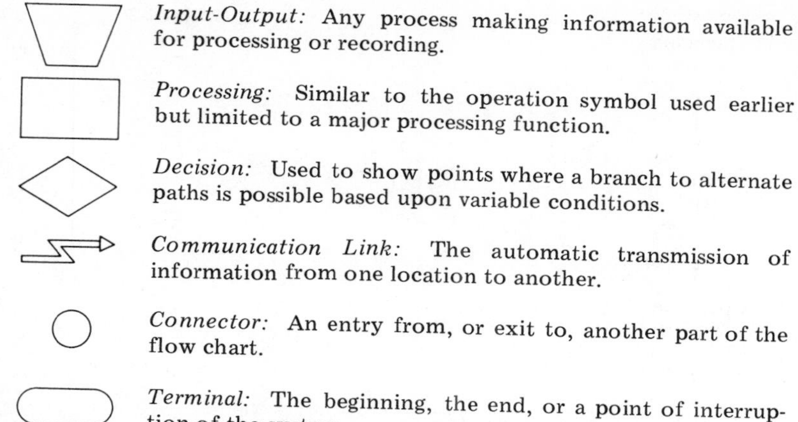

A flow chart consists basically in a schematic diagram of lines connecting symbols or boxes that contain some abbreviated statements. The symbols represent groups of elementary steps or operations portraying the behavior of the system being analyzed; the lines indicate the relationships among the groups. Arrows indicate the direction of flow or order of execution. The statements within the symbols identify what is going on at each step, and the process is read in a top-to-bottom order (see Fig. 2.6).

One feature of the flow chart that varies from the traditional block diagram and gives greater flexibility is its branching capability. The flow chart can depict decision or choice points in the sequence. The various lines leaving the decision box are labeled (usually "yes" or "no") and the line for the correct answer to the question is followed. Consider the flow chart for the all-too-familiar process shown in Fig. 2.6—changing a flat tire. Several branching opportunities exist, one of which entails the flat tire. Different action is required depending upon whether it is a front or rear tire that is flat. Likewise, the answer to the question of whether the spare is flat calls for different sequences of events. It might be necessary to iterate several times through the loop on the right-hand side if the first several service stations called will not send help.

This demonstrates one of the advantages of describing a system by a flow chart rather than by a verbal description. Spoken or written language is serial in nature; thus, when a decision or question is reached, one branch must be followed through to completion, and then we must return and follow or describe the other branch or branches. Therefore, we say language gives a sequential, linear, and one-dimensional description of a system. The flow

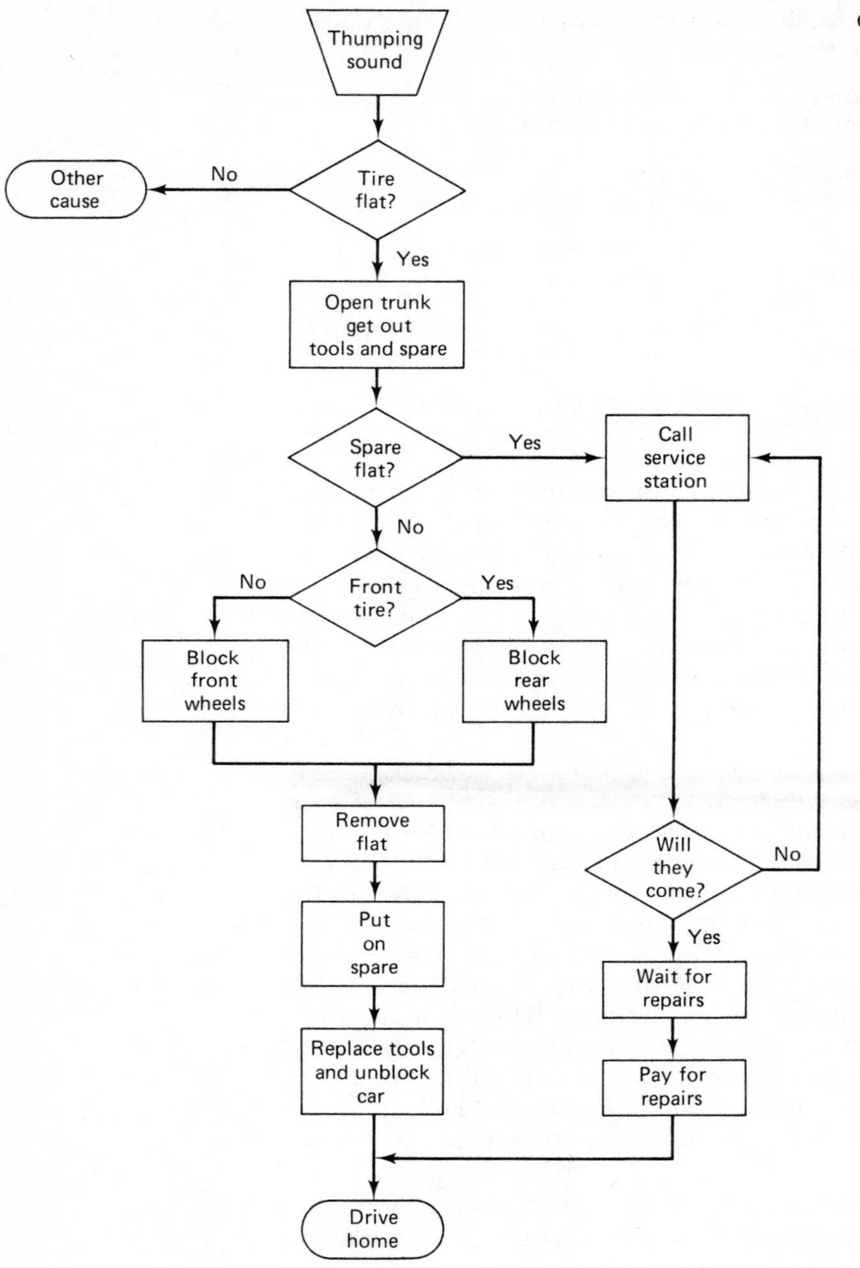

Figure 2.6 Changing a flat tire.

chart, on the other hand, presents two or more branches simultaneously, thereby allowing a multidimensional, parallel description. Bohl gives an excellent discussion of flow charting procedures [5].

The main purpose of the flow chart and/or block diagram is to aid in the analysis and communication of the logical structure of the system and the relationships among its elements. This is especially important during the definition of the system, since at that time we often do not know how the system should or does operate. We can represent very complex systems in block diagram or flow chart form. Doing so often exposes ignorance about a system and enables the analyst to be reasonably sure that he has covered all aspects of the system. In addition, we now have our first static model of the system from which we may begin to work toward the dynamic model.

2.6 Example of system definition

To demonstrate the process of using block and logic diagrams, we shall consider the system definition phase of an actual study [16]. The purpose of the study was to evaluate certain planned

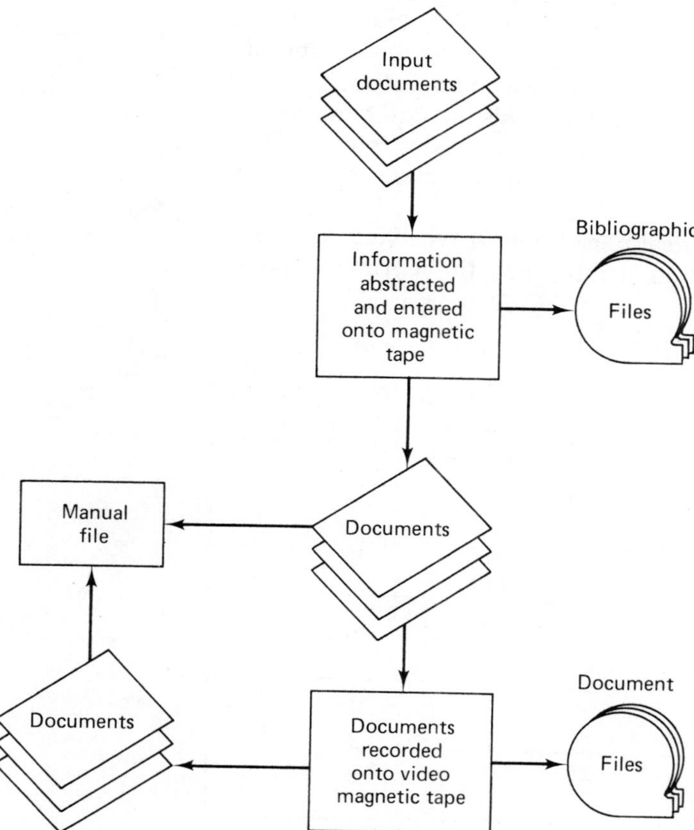

Figure 2.7 General information-storage procedures [16].

additions to NASA's Apollo Parts Information Center (APIC). The APIC was an information center established to collect, store and disseminate parts and materials information that was important to the Apollo moon exploration program. Specifically, the center was concerned with parts and materials information such as item identification, requirements information, qualification data, test data, failure data, environment and environmental extremes data, usage data, and inspection data.

Figure 2.7 depicts the general procedures for storing information. Information required for retrieval is abstracted from the various input documents and stored on magnetic tapes, which comprise the bibliographical files. After the abstracted information has been coded, the documents themselves are recorded onto video magnetic tape to comprise the document files. Original documents are then stored in standard filing cabinets.

All incoming inquiries are received at inquiry control by telephone. At inquiry control there are two rotary lines, each line consisting of two telephones. Inquiries received at the center can be generated either by users who have access to an output terminal or by those who do not. The output terminals include facilities for directly receiving responses from the center via teletype and facsimile. Figure 2.8 shows a block diagram representation of the center operations.

The center employs technical personnel for receiving, standardizing, and coding inquiries to assure that the correct type of

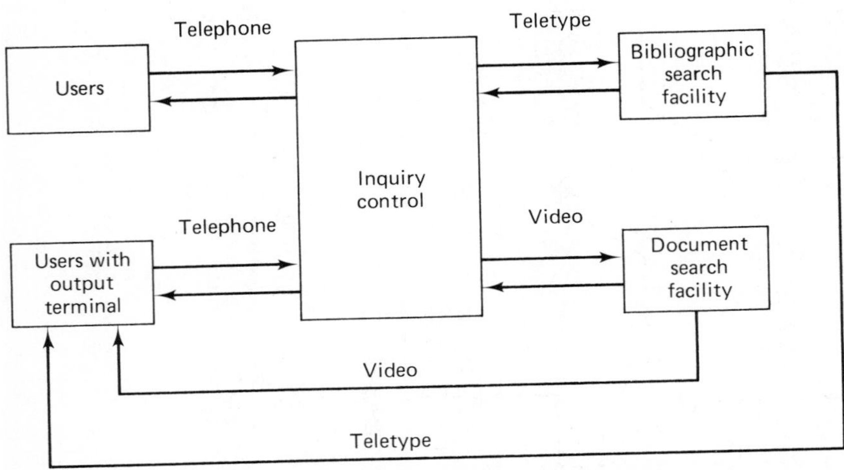

Figure 2.8 Generalized inquiry response block diagram [16].

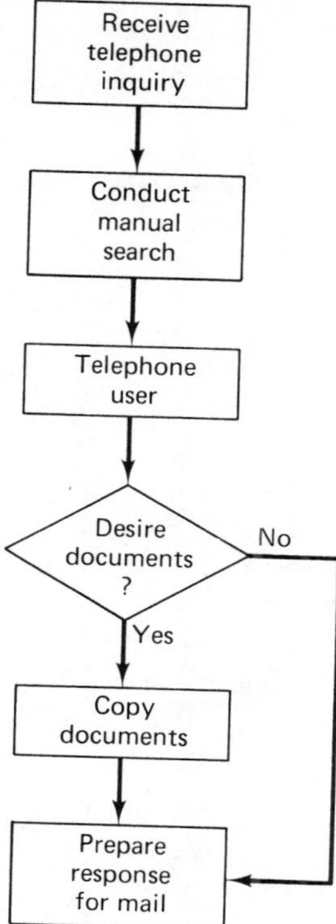

Figure 2.9

information is sent from the center. Three classes of inquiries or processing routes are possible: special manual search, bibliographical search, and document search. Logic flow charts for each are shown in Figs. 2.9, 2.10, and 2.11, respectively. Hopefully, these logic flow charts are self-explanatory, but a more detailed discussion is found in case study I, Appendix A.

2.7 Constructing the model

In the process of converting our knowledge of the system into a mathematical model, we must accomplish four tasks:

1 Specification of the purpose of the model.

2 Specification of the components to be included in the model.

3 Specification of the parameters and variables associated with the components.

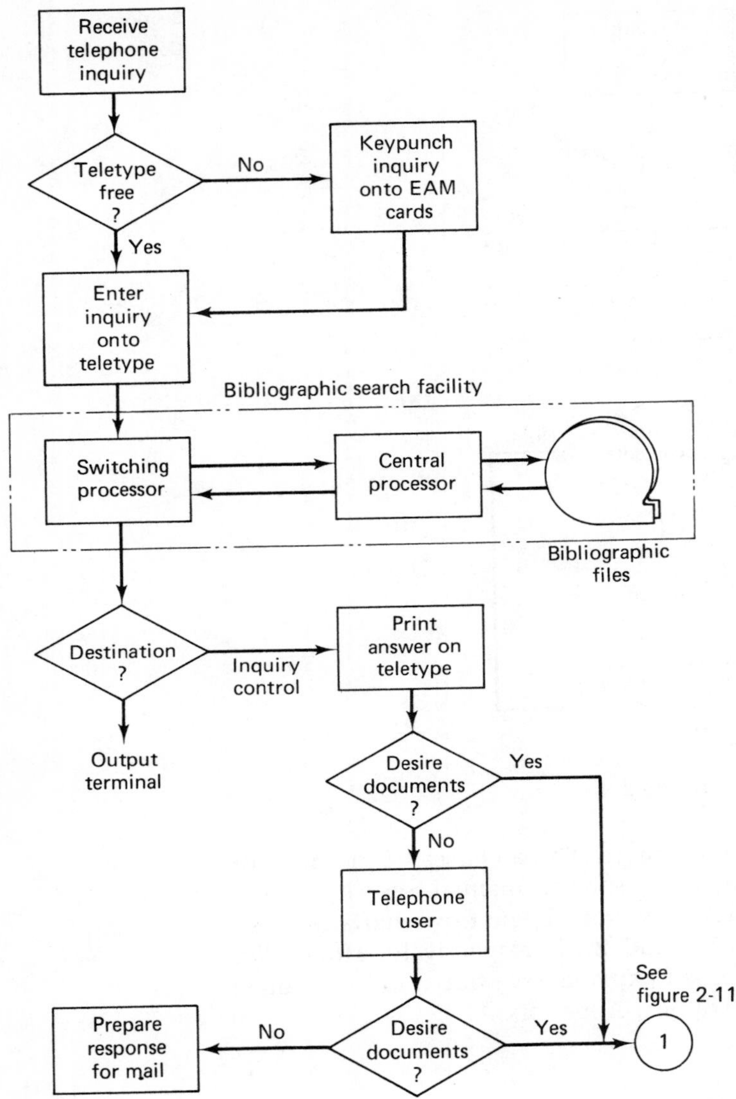

Figure 2.10 Processing the bilbliographic search class of inquiries [16].

4 Specification of the functional relationships among the components, parameters, and variables.

Hopefully, one outcome of the orientation period will be the determination of the explicit goals or purpose of the simulation project. Simulation experiments are conducted for a wide variety of purposes, some of which are as follows:

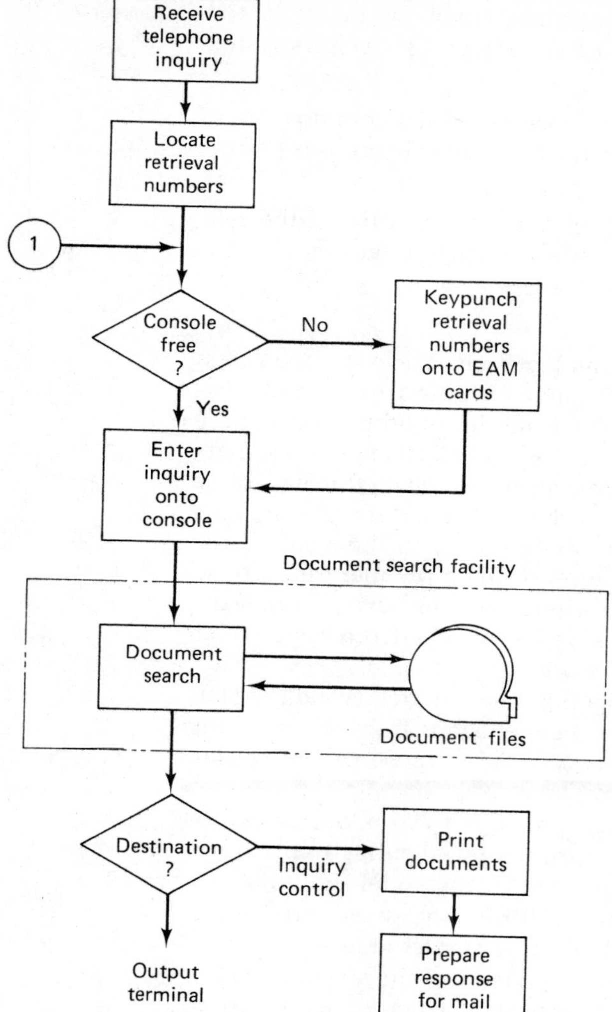

Figure 2.11 Processing the document search class of inquiries [16].

1 Evaluation: determining how good a proposed system design performs in an absolute sense when evaluated against specific criteria.

2 Comparison: comparing competitive systems designed to carry out a specified function, or comparing several proposed operating policies or procedures.

3 Prediction: estimating the performance of the system under some projected set of conditions.

4 Sensitivity analysis: determining which of many factors are the most significant in affecting overall system performance.

5 Optimization: determining exactly which combination of factor levels will produce the best overall response of the system.

6 Functional relations: establishing the nature of the relationship among one or more significant factors and the system's response.

This list is not exhaustive and merely suggests the most common goals or purposes; the explicit purpose of the model has significant implications for the whole model building and experimentation process. For example, if the model's goal is to evaluate a proposed or existing system in an absolute sense, this imposes a heavy burden upon the accuracy of the model and demands a high degree of isomorphism. On the other hand, if the goal is the relative comparison of two or more systems or operating procedures, the model may be valid in a relative sense even though the absolute magnitude of responses varies widely from that which would be encountered in the real world.

The whole process of designing the model, validating it, designing experiments, and drawing conclusions from the resulting experiments is closely tied to the specific purpose of the model. No one should build a model without having an explicit experimental goal in mind. Unfortunately, the builder does not always understand the real world problem well enough initially to ask the right questions in advance. Thus, the model should have a more or less open structure so that additional questions arising as a result of early experimental results can be answered later.

After we have specified (at least tentatively) the specific goal or purpose for which the model is to be constructed, we begin to identify the pertinent components. This process entails itemizing all the components of the system that contribute to the effectiveness or ineffectiveness of its operation. Once a complete list of the components of a system is specified, we next determine whether each component should be included in our model. This is easier said than done, since at this stage of model development it is not always clear whether a component is significant to the overall goal. One of the key questions to be answered is whether a particular component is to be considered part of the model or part of the environment.

One pertinent consideration in deciding which components are to be included and which excluded is the question of how many variables are to be included in the model. In general, we have little difficulty in deciding upon the endogenous or output variables. If we have done a good job in specifying the goals or purposes of the study, the required output variables are fairly obvious. The real difficulty arises in determining which input and status variables produce the effects observed and which can be manipulated to produce the effects desired. We are also faced with conflicting objectives: on one hand, we try to make the model as simple as possible for ease of understanding, ease of formulation, and computational efficiency; on the other hand, we try to make the model as accurate as possible. Consequently, we need to simplify reality but only to the point where there is no significant loss of accuracy.

It is not easy to find the proper balance. Clearly we do not wish to omit any component or variable that has a large effect upon the system's performance. However, at this early stage of development, whether the effect is significant may not be at all clear. Here we must rely on sound judgment, insight, and the experience of those familiar with the system. It may be necessary and desirable to test these judgments by experiments and/or statistical analysis of available data. We should always attempt to verify our assumptions and hypotheses about how the system operates as rigorously as possible.

Usually, the tendency on the part of the analyst is to include initially too many components and variables. We agree with Ackoff and Sasieni when they state, "In general, the extent to which a phenomenon is understood is inversely proportional to the number of variables required to explain it" [2]. Unfortunately, if we follow our natural tendencies, the model will end up so complex and intractable that we can neither explain nor experiment efficiently with it. Certainly we should omit no variable without as rigorous a justification as possible. On the other hand, if we build the model in a modular or building block form, we can add additional variables or components later if it appears desirable. In most cases, we can aggregate or combine components and variables at the early stages with the idea of separating them later if necessary, after we better understand the system and how it operates.

Once we have decided which components and variables we shall include in our model, we must then determine the functional relationships among them and the values of the parameters to be used. Again, formidable problems confront us. First, it may be difficult (if not impossible) to quantify or measure certain variables that are important to the behavior of the system. Second, the relationships between components and variables may not be clear. Third, the data and information we need may not be available, or perhaps not exist in the form we need. Each of these factors are considered in more detail in the following sections.

2.8 Component modeling

When attempting to model individual components or elements (subsystems) of a complex system, we can find ourselves faced with several different types of problems, which we broadly categorize as either *direct* or *inverse problems.* Consider the system shown in Fig. 2.12.

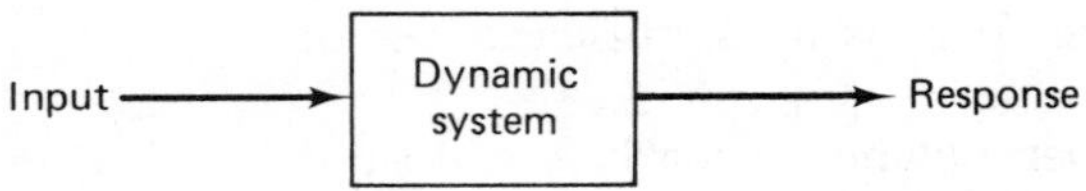

Figure 2.12 Simple system.

Here we are concerned with three distinct entities—the input, the system, and the response. To model this component, we must have available or assume knowledge about two of the three entities. If we know the equations describing the behavior of the dynamic system, then the direct problem is finding the response of the system to a given input. This is the most ideal situation and relatively easy to model. The equations may have been derived in the course of designing the system or may be based upon previous research on similar systems. It might also take the following form: given a system description and a response, find the input that caused it. In this latter case, if the equations describing the system are known and a desired response is specified or desired, the problem of finding the necessary inputs to produce such a response comes under the category of *control problems.*

The much more difficult inverse problem arises when we are given a set of inputs and corresponding outputs from a system and must derive or find a mathematical description of the system. Such a problem is known as the *identification* or *structure recovery problem.* The difficulty is that several mathematical expressions may appear to explain the input-output relationship. For example, suppose we have fed the integer numbers 1, 2, 3, 4, 5, 6 into the system in that order and received an output response of 2, 4, 6, 8, 10, 12. Reasonable hypotheses would be that if the input is n, then the output is $2n$, and to predict that the next output in the sequence would be the even numbers 14, 16, ..., $2n$. However, the original data could be fitted equally well to the formula

$$2n + (n-1)(n-2)(n-3)(n-4)(n-5)(n-6)$$

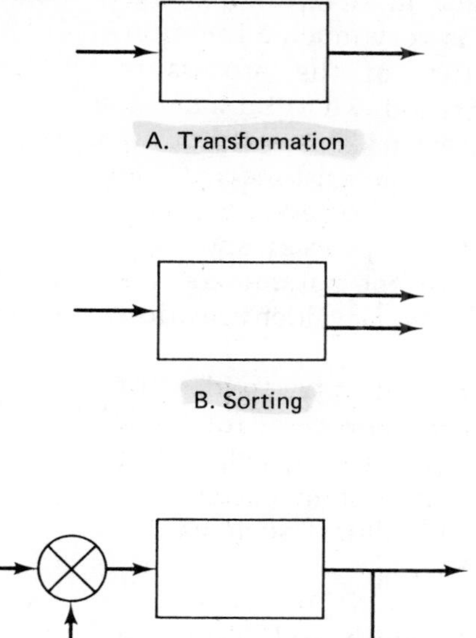

Figure 2.13 Basic system components.

But this equation would predict that the next few numbers in the sequence would be 734, 5056, etc., a very different prediction from the first one.*

In general, system components convert inputs into outputs. They are of three different types, which constitute the basic building blocks of complex systems: (1) *transformation* elements, where one or more inputs are operated upon in some prescribed manner and transformed into one or more outputs; (2) *sorting* elements, where one or more inputs are separated or sorted into two or more different outputs; and (3) *feedback* elements, where the input is modified in some manner as a function of the output. These three building blocks are shown schematically in Fig. 2.13.

Depending upon the purpose of our model, we may be interested in the specific nature of the operation of the component or we may only be interested in the amount of time the component takes to perform the operation. The conversion process may be either *deterministic* (output uniquely determined by a given input) or *stochastic* (uncertainty of output for a given

*This example is from E. B. Wilson, Jr., *An Introduction to Scientific Research*, McGraw-Hill Book Company, New York, 1952, p. 29.

input). Either way, the difficulty we face in determining and defining the structure of the components is very much a function of our a priori knowledge. If the nature of the process is completely unknown or very poorly understood, we call it a *black box* identification problem. In this case, we usually attempt to describe it by means of overall transfer characteristics, such as linear or curvilinear equations. In other cases, however, we may know a great deal about the nature of the process and lack knowledge only about the specific values of the parameters. We sometimes refer to such a problem as a *gray box* identification problem.

As mentioned, the way in which we proceed to try to identify the structure or nature of the conversion process for each component depends upon our degree of a priori knowledge. Past experience in model building indicates that certain patterns or approaches are often used by the model builder, such as the following [2]:

1 The structure of the system is sufficiently well known or is so simple and transparent that it can be understood by inspection and/or discussion with those who designed or operate the subsystem.

2 The structure is not readily apparent but appears to be analogous to some existing theoretical description.

3 The structure of the system is not apparent but can be extracted by statistical analysis of the data that describe the operation of the system. A knowledge of which variables play an important part in the performance of the system is usually required.

4 It is not possible to isolate the effects of individual variables by analysis of existing data and we must resort to experimentation.

5 No data exist and it is not possible to experiment directly on the system.

In the following sections we discuss some of the statistical techniques and other methodologies that may be useful in both determining and describing relationships between components and variables.

2.9 Monte-Carlo sampling

Monte-Carlo sampling is basic to the concept of simulating systems containing stochastic or probabilistic elements. It traces its modern origin and name to the work of von Neumann and Ulan in the late 1940's when they coined the term and applied the technique to the solution of certain nuclear shielding problems. The mathematical technique had been known for many years but was revived for the secret work at Los Alamos and given the code name "Monte Carlo." The Monte-Carlo method was so successful that its popularity spread to various fields, and the term has almost become synonymous with the term simulation in the minds of many. Although our primary interest in the Monte-Carlo sampling technique is its utility in simulating probabilistic situations, it may also be used in certain completely deterministic problems that cannot be solved analytically [12].

In the Monte-Carlo technique, artificial experience or data are generated by the use of some random number generator and the cumulative probability distribution of interest. The random number generator may be a table of random digits, a roulette wheel, a computer subroutine, or any other source of uniformly distributed random digits. The probability distribution to be sampled may be based upon empirical data derived from past records, may result from a recent experiment, or may be a known theoretical distribution. The random numbers are used to produce a randomized stream of variates that will duplicate the expected experience, which would be produced by the probability distribution being sampled.

The process or technique is relatively simple in concept. To draw an artificial sample at random from a population described by some probability function:

1 Plot or tabulate the data of interest (not the random numbers) as a cumulative probability distribution function with the values of the variate on the x axis or abscissa and the probabilities from 0 to 1 plotted on the y axis or ordinate.

2 Choose a random decimal number (RN) between 0 and 1 (to as many places as desired) by means of a random number generator.

3 Project horizontally the point on the y axis (ordinate) corresponding to this random decimal number until the projection line intersects the cumulative curve.

4 Project down from this point of intersection on the curve to the x axis (abscissa).

5 Write down the value of x corresponding to this point of intersection. This value of x is then taken as the sample value.

6 Repeat steps 2 through 6 until as many random variates as are desired have been generated, keeping track of the sequence in which they were drawn.

We can easily see the rationale behind this technique if we look at a very simple example using empirical data. Suppose we have a system where in any 10-minute period customers will show up for servicing with the following probability:

Customers	Probability	Cumulative Probability
0	0.40	0.40
1	0.25	0.65
2	0.20	0.85
3	0.15	1.00

Assume we want to generate artificial experience for five time periods. We first could draw the cumulative probability distribution shown in Fig. 2.14. Then, from the table of random numbers in Appendix C-I, we would draw five two-digit numbers and place a decimal point in front of each. Each of these five numbers will be used to generate the number of customers arriving during a particular period. If the random numbers drawn are 09, 54, 42, 80, 20, respectively, then the result is as shown in Table 2.1.

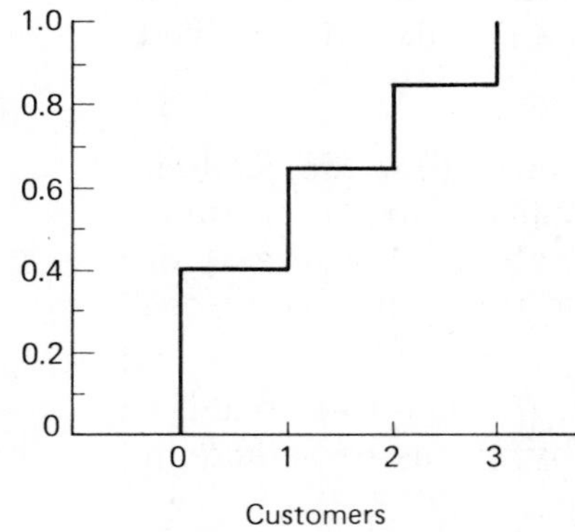

Figure 2.14 Cumulative distribution.

By running a few more samples the reader can easily convince himself of the obvious rationale behind the method. If the random numbers used are indeed uniformly distributed and random (i.e., each of the numbers from 00 to 99 has an equal chance of coming up or being selected), then each number from the data of interest will come up with the same relative frequency in this synthetic

Table 2.1 *Monte-Carlo sample from Fig. 2.14.*

Time Period	Random Number	Number of Customers
1	0.09	0
2	0.54	1
3	0.42	1
4	0.80	2
5	0.20	0

process as in the "real world" but their arrangement will be randomized. Thus, we have generated artificial experience (the number of customers arriving during each 10-minute period) that is typical of what we have been experiencing with the real system in the past.

The preceding example utilized discrete data. In the case of continuous data, the process is identical except that the curve is smooth rather than stepped (see Fig. 2.15). Since the normal or Gaussian distribution is so commonly encountered in the real world, this process has been carried out ahead of time and the result tabulated in Appendix C-II. This table of random normal numbers is for a normal distribution with mean (μ) = 0 and standard deviation (σ) = 1. To draw a normally distributed variate of a population with known parameters (μ, σ), we draw a random normal number (RNN) from the table, multiply it times the standard deviation, and add the result to the mean:

$$\text{Random normal variate} = \mu + (\text{RNN})\ \sigma$$

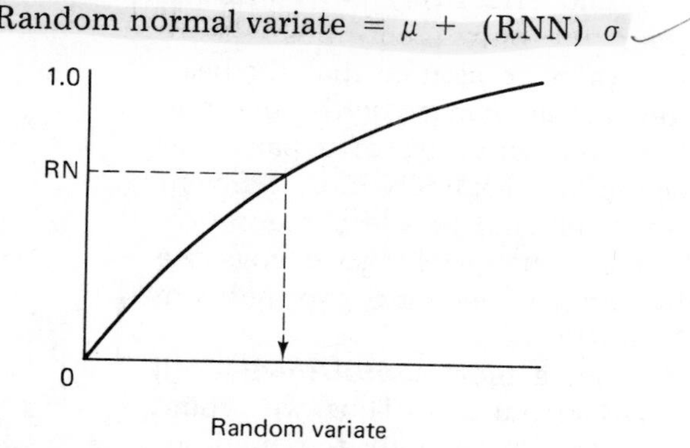

Figure 2.15 Monte-Carlo from continuous distribution.

In using the tables in Appendices C-I and C-II, we enter the table at any point of our choosing and then follow some sort of regular pattern in reading off the numbers. Suppose we wanted to draw a sample of five (5) random values from a normally distributed population with mean = 100, standard deviation = 10. We would go to Appendix C-II, draw out five (5) random normal numbers (RNN), and plug them into the equation: $x = 100 + (\text{RNN})10$ (see Table 2.2).

Table 2.2 *Monte-Carlo of normal distribution* $\mu = 100$, $\sigma = 10$.

RNN	Calculation	X
0.906	100 + (0.906)10	109.06
1.179	100 + (1.179)10	111.79
-1.501	100 + (- 1.501)10	84.99
-0.690	100 + (- 0.690)10	93.10
1.372	100 + (1.372)10	113.72

These examples are for hand-generated random variates. In a computerized simulation, these random variates are generated by computer subroutines. Most computer centers have random variate generators for most of the better-known probability distributions as a part of their library of user routines. Appendix B provides several of these in FORTRAN if they are not already available to the reader.

The design of a simulation model containing stochastic or probabilistic elements always raises the question of whether to use empirical data directly in the Monte-Carlo sampling or whether to use one of the theoretical distributions. This question is very important and fundamental for three reasons. First, the use of raw empirical data implies that all one is doing is simulating the past. The use of data from last year would effectively be replicating only the performance of that year; the only events possible are those that have occurred. It is one thing to assume that the basic form of the distribution will remain unchanged with time and quite another to assume that the idiosyncracies of a particular time period will be repeated. Second, it is generally more efficient of computer time and storage requirements to use a theoretical probability distribution. Third, it is much easier to change the parameters of a theoretical distribution generator to perform sensitivity tests or ask "what if" questions.

As a general rule, we believe that a more useful model will result if we can use theoretical distributions. Thus we would suggest that empirical data be tested by one of the methods discussed in Sections 2.12 or 2.13 to determine if it fits a known distribution to a statistically accepted level of confidence. If it does, then the theoretical distribution should be used.

2.10 The staggering drunk

For illustrative purposes, let us consider the classical staggering drunk or random walk problem. Suppose a drunk is standing on a street corner when he decides to go for a walk to sober up. Assume that there is an equal probability of his going north, south, east, or west as he reaches each street corner. If he walks 10 blocks, what is the probability that he will end up within 2 blocks of where he started?

We could approach this problem by designating his location at each corner by a two-dimensional vector (X, Y), where X is the east-west axis or direction and Y is the north-south axis or direction. Each time he moves a block to the east we could increase X by 1, and each time he moves west we could decrease X by 1. Likewise, each time he goes north we could add 1 to Y, and each time he moves south decrease Y by 1. Now, if we designate his starting position as (0,0), then at any stage we will know exactly where he is in relation to his starting point. At the end of the 10-block walk, if the sum of the absolute values of X and Y are greater than 2, then our drunk has ended up more than 2 blocks away from his starting point.

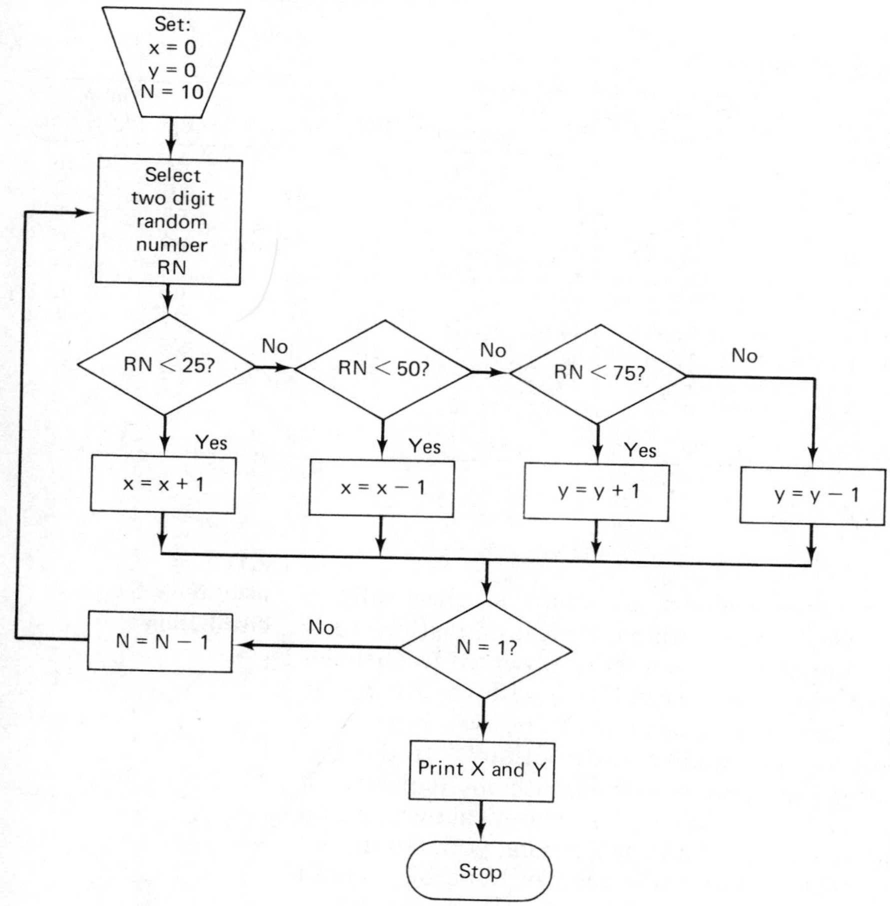

Figure 2.16 Logic flow, staggering drunk simlation.

Since we said that at any corner (including the start) there is an equal probability that he will go in any direction, we want there to be a probability of ¼ for his choosing any particular direction. We could therefore draw a series of two-digit random numbers (one for each corner) to decide in what direction he moves next. Thus, if the number drawn is 00 through 24, he goes east and we add 1 to X; if the random number is 25 through 49, he goes west and we subtract 1 from X; if the number is 50 through 74, he goes north and we add 1 to Y; or if the random number is 75 through 99, he goes south and we subtract 1 from Y. Figure 2.16 shows a logic flow chart for our simulation and Table 2.3 shows the results of five trials. Obviously five trials are insufficient to obtain a good estimate of the true probability of his ending up within two blocks of his starting point; a discussion of adequate sample sizes appears in Chapter 5.

Table 2.3 *Staggering drunk simulation results.*

End of Block	Trial 1 RN	Trial 1 Location	Trial 2 RN	Trial 2 Location	Trial 3 RN	Trial 3 Location	Trial 4 RN	Trial 4 Location	Trial 5 RN	Trial 5 Location
1	73	0, 1	10	1, 0	05	1, 0	06	1, 0	64	0, 1
2	21	1, 1	89	1, −1	88	1, −1	95	1, −1	76	0, 0
3	45	0, 1	14	2, −1	10	2, −1	04	2, −1	79	0, −1
4	76	0, 0	81	2, −2	04	3, −1	67	2, 0	54	0, 0
5	96	0, −1	30	1, −2	48	2, −1	51	2, 1	28	−1, 0
6	94	0, −2	91	1, −3	19	3, −1	95	2, 0	05	0, 0
7	53	0, −1	06	2, −3	44	2, −1	73	2, 1	71	0, 1
8	57	0, 0	38	1, −3	21	3, −1	10	3, 1	75	0, 0
9	96	0, −1	79	1, −4	95	3, −2	76	3, 0	53	0, 1
10	43	−1, −1	43	0, −4	11	4, −2	30	2, 0	29	−1, 1
SUCCESS?		YES		NO		NO		YES		YES

2.11 Identifying the distribution

If some of the elements of the system behave stochastically, a problem that arises several times during a typical simulation study is that of testing the compatability of a set of observed frequencies with some theoretical frequency. That is, we wish to answer the question, "Could the data or sample values I observed have come from a certain specified probability distribution?" If the frequency of the observed data compares well with the expected or theoretical frequency, we can then use the theoretical distribution in our model to represent the parent or underlying population.

Usually we cannot make a reasonable guess or hypothesis about

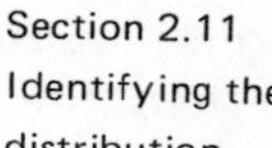

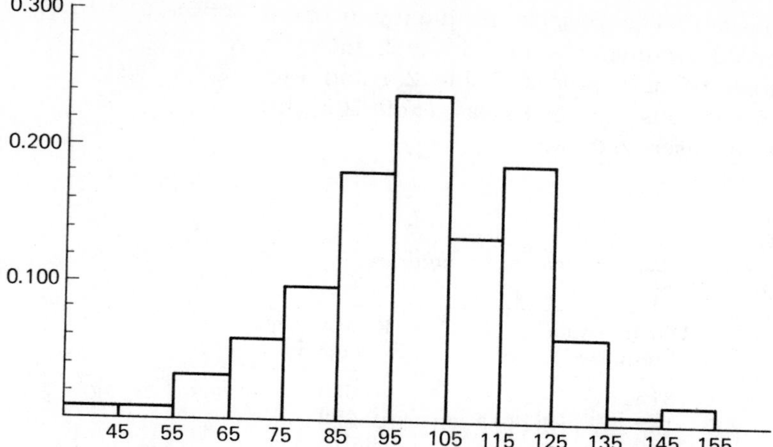

Figure 2.17 Frequency distribution of data in Table 2.4.

the distribution of a random variable until we have collected and analyzed data (either historical or experimental). The collected data are usually summarized in a frequency distribution, such as Fig. 2.17. If we are dealing with a discrete variable, we record the frequency with which each individual value occurs. If the variable is continuous, we break the range of values into equal intervals or classes and record the frequency occurring within each interval or class. The number of class intervals is usually taken between 5 and

Table 2.4 *Weekly production.*

Weekly Production (X)	Frequency	$P(x)$
Below 46	1	0.008
46—55	1	0.008
56—65	3	0.025
66—75	7	0.058
76—85	11	0.092
86—95	21	0.175
96—105	28	0.234
106—115	16	0.134
116—125	22	0.183
126—135	7	0.058
136—145	1	0.008
146 and up	2	0.017
	120	1.000

20, depending upon the data. The relative frequency in each interval is then the observed frequency count in each interval, divided by the total number of data points. Table 2.4 and Fig. 2.17 represent a set of continuous data, whereas Table 2.5 and Fig. 2.18 demonstrate a set of discrete data.

Table 2.5 *Frequency distribution of telephone inquiries per one-hour interval.*

N No. of Inquiries	No. of 1-hr Intervals with N Inquiries	Relative Frequency
0	315	0.619
1	142	0.279
2	40	0.078
3	9	0.018
4	2	0.004
5	1	0.002
	509	1.000

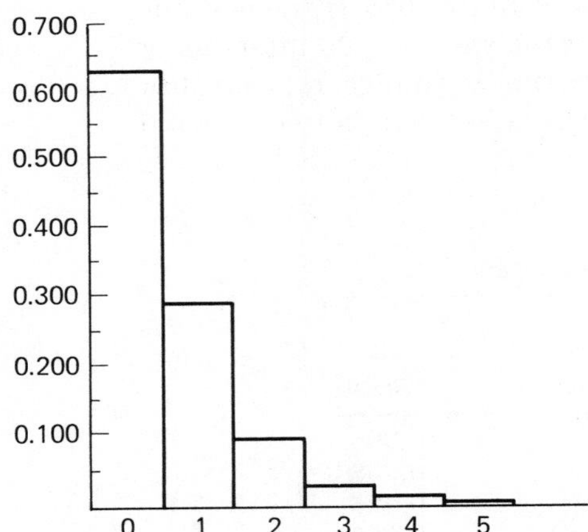

Figure 2.18 Relative frequency diagram of data in Table 2.5.

After the analyst has obtained a relative frequency distribution, selection of the possible probability distributions from which it may have derived becomes a matter of judgment and experience. One aid is to compare visually the observed frequency distribution with those of several theoretical distributions. Thus, if we compare Fig. 2.18 with Fig. 2.19, we see that it might have derived from a

Poisson distribution. On the other hand, the data portrayed in Fig. 2.17 appear to be normally distributed. Visual comparisons of distributions serve only to suggest what distributions we want to try; they are never sufficient justification to accept a certain hypothesis or theoretical distribution.

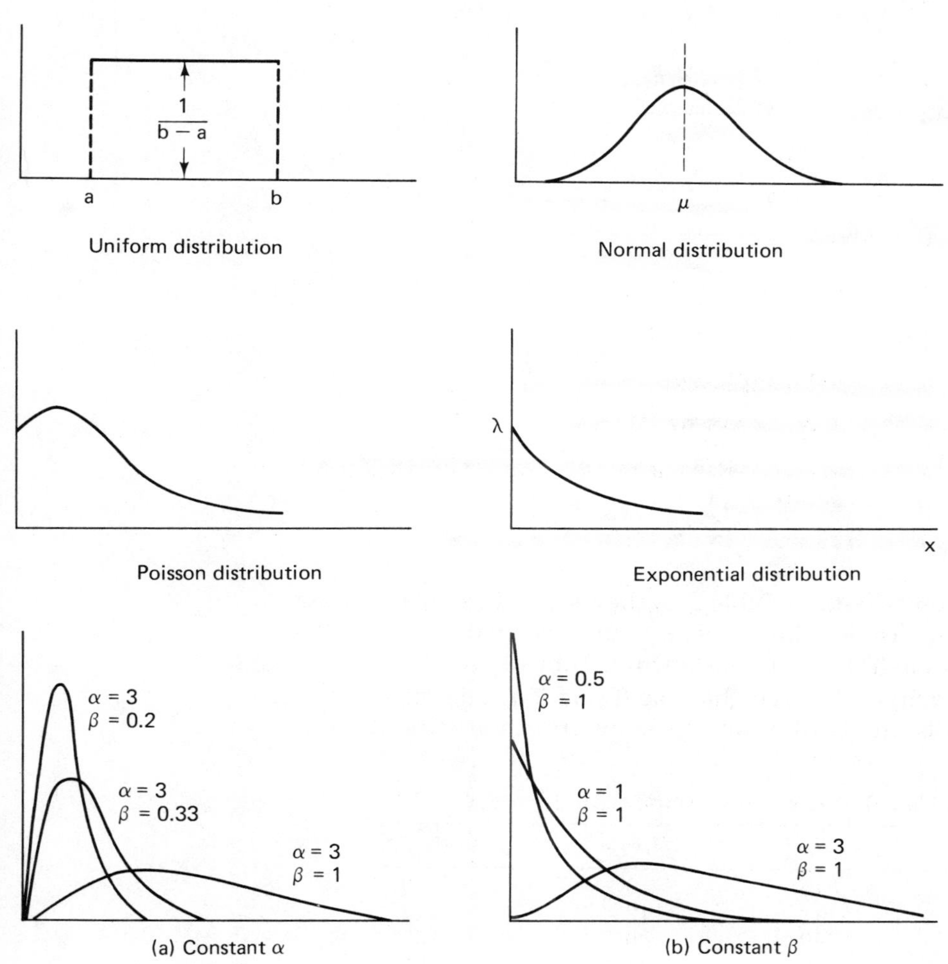

Figure 2.19 Probability curves.

Once the analyst has identified one or more theoretical distributions—such as normal, Poisson, binomial, gamma, etc.—

that he believes might fit his empirical data, he must then determine the distribution parameters so as to proceed with statistical testing. When the hypothesized distribution is a function of two parameters, he can usually estimate these parameters from the sample mean and sample variance.

For grouped data, we can calculate the mean and variance as

$$\text{Mean} = \bar{X} = \frac{\sum_{i=1}^{k} M_i F_i}{n}$$

$$\text{Variance} = s^2 = \frac{\sum_{i=1}^{k} M_i^2 F_i - n\bar{X}^2}{n-1}$$

where

n = total sample size = $\sum_{i=1}^{k} F_i$

k = no. of classes or intervals

M_i = midpoint of ith interval or value of ith class if discrete data

F_i = frequency in ith interval or ith class

For the discrete data in Table 2.5, the required calculations are summarized in Table 2.6, whereas the calculations for the continuous data in Table 2.4 are shown in Table 2.7.

We originally hypothesized that the Table 2.5 data might have come from a Poisson distribution. We know from statistics that for

Table 2.6 *Parameter calculations for Table 2.5.*

M_i	F_i	M_iF_i	$M_i^2F_i$
0	315	0	0
1	142	142	142
2	40	80	160
3	9	27	81
4	2	8	32
5	1	5	25
	509	262	440

$$\bar{X} = \frac{262}{509} = 0.5147$$

$$s^2 = \frac{440 - 509\,(0.5147)^2}{509 - 1} = 0.6007$$

Table 2.7 *Calculation of parameters of continuous data.*

M_i	F_i	M_iF_i	$M_i^2 F_i$
40.5	1	40.5	1,640.25
50.5	1	50.5	2,550.25
60.5	3	181.5	10,980.75
70.5	7	493.5	34,791.75
80.5	11	885.5	71,282.75
90.5	21	1900.5	171,995.25
100.5	28	2814.0	282,807.00
110.5	16	1768.0	195,364.00
120.5	22	2651.0	319,445.50
130.5	7	913.5	119,211.75
140.5	1	140.5	19,740.25
150.5	2	301.0	45,300.50
	120	12,140.0	1,275,110.00

$$\bar{X} = \frac{12,140}{120} = 101.17$$

$$s^2 = \frac{1,275,110 - 120(101.17)^2}{120 - 1} = 393.83$$

$$s = \sqrt{393.83} = 19.86$$

a Poisson distribution the mean = λ and also that the variance = λ—i.e., the mean and variance are equal. As we see from Table 2.6, the mean and variance are not equal for our observed data (i.e., 0.5147 < 0.6007), which might lead us to reject the hypothesis of a Poisson distribution. However, in this particular case we have both practical and theoretical reasons to continue to suspect a Poisson. When it is as likely that this event will occur in one time interval as in any other and the occurrence of an event has no effect on whether another occurs, we can strongly suspect a Poisson distribution. In addition, if there is a high probability of zero occurrences in any time interval and the mean or average number of occurrences per time interval is small, we have further reason to expect a Poisson. In our example, the data in Table 2.5 represent the number of inquiries received by a technical information service, and they fit all these criteria. If we wish to continue to consider the Poisson as a possibility, we could do so be letting λ equal the average of the sample mean and sample variance, or

$$\lambda = \frac{0.5147 + 0.6007}{2} = 0.5577$$

2.12 Chi-square goodness of fit test

To test statistically the hypothesis that a set of empirical or sample data does not differ significantly from that which would be expected from some specified theoretical distribution, we can consider two "goodness of fit" tests. One measure or test of the discrepancy existing between an observed and an expected frequency is supplied by the statistic χ^2 (read chi-square). The chi-square test was proposed by Karl Pearson in 1903, although Sir Ronald Fisher fully developed the test and published the table of critical values used today in 1924. The chi-square statistic is given by

$$\chi^2 = \sum^{k} \frac{(f_o - f_e)^2}{f_e}$$

where

f_o = observed frequency for each class or interval

f_e = expected frequency for each class or interval predicted by the theoretical distribution

$\sum^{k}$ = sum over all k classes or intervals

If $\chi^2 = 0$, then the observed and theoretical frequencies agree exactly, whereas if $\chi^2 > 0$ they do not. The larger the value of χ^2, the greater is the discrepancy between the observed and expected. If $\chi^2 > 0$, we must compare our calculated value against the tabulated values of χ^2 (Appendix C, Table C-III) to determine if we could expect that much variation from chance causes alone. The χ^2 statistic is tabulated by degrees of freedom vs. $(1 - \alpha)$ or significance level. In practice, the null hypothesis, H_0, is that there is no significant difference between the observed frequency distribution and that of a specified theoretical distribution with the same parameters. If under this hypothesis the computed value of χ^2 is greater than the critical or tabulated value of χ^2 (at a given level of significance and appropriate degrees of freedom), we would conclude that the observed frequencies differ significantly from the expected frequencies at that level of confidence, and we would reject the H_0.

In using the χ^2 goodness of fit test, several points should be remembered:

1 Relative frequencies or percentages cannot be used—i.e., we must use actual counts or actual frequencies.

2 The *expected* frequencies for each class or interval should equal 5 or more. If they do not, then adjacent classes or intervals should be grouped or lumped together.

3 The degrees of freedom are given by $\nu = k - 1 - m$, where ν = degrees of freedom, k = number of classes or intervals, and m = number of empirical or sample data population parameters necessary to calculate the expected frequencies.

EXAMPLE 2.12.1

Assume we want to test the data in Table 2.5 for fitting a Poisson distribution at a 0.95 significance level. We know that the Poisson distribution is given as

$$P\{x = n\} = Px(n) = \frac{\lambda^n e^{-\lambda}}{n!}$$

where

$P\{x = n\}$ = probability of getting n occurrences

e = 2.71828

λ = a positive constant (which is both the mean and variance)

In the preceding section, we calculated $\lambda = 0.5577$ for the sample data of this example; therefore, the H_0 is, "There is no significant difference between the observed data and those which would be given by a Poisson distribution with mean = 0.5577." Substituting this value for λ and then in turn $n = 0$, $n = 1$, $n = 2$, etc., into the equation for a Poisson distribution, we get the calculations shown in Table 2.8.

Table 2.8 χ^2 *calculations for Table 2.5.*

n	$Px(n)$	f_e		f_o		$\frac{(f_o - f_e)^2}{f_e}$
0	0.571	291		315		1.98
1	0.319	162		142		2.47
2	0.089	45		40		0.56
3	0.017	9		9		
4	0.003	1	11	2	12	0.09
5	0.001	1		1		
	1.000	509		509		5.10

To obtain f_e, we multiply the appropriate $Px(n)$ by 509. Therefore, our calculated $\chi^2 = 5.10$. Looking up the critical value for χ^2 from Appendix C, Table C-III for a confidence level of 0.95 and degrees of freedom = $4 - 1 - 1 = 2$, we find $\chi^2 = 5.99$. Thus, since our calculated χ^2 is less than the tabulated critical value of χ^2, we do not reject the H_0 of no difference. The last three classes were grouped in the calculations to get a frequency of at least 5 in each class; thus, we have 4 classes instead of the original 6. Also, in figuring the degrees of freedom, we took off one additional degree of freedom because we used λ from the observed data to calculate the expected frequency.

Assume that we have a random number generator that has generated 500 random digits, and we have observed the frequency shown in Table 2.9. If the generator is truly random, we would expect each digit to have shown up about 50 times. Using a significance level of 0.99, we test the results to see if they differ significantly from a uniform distribution. The calculations are shown in Table 2.9.

Table 2.9 χ^2 *calculations for Example 2.12.2.*

Digit	0	1	2	3	4	5	6	7	8	9	Totals
Observed	62	58	36	28	40	70	60	40	72	34	500
Expected	50	50	50	50	50	50	50	50	50	50	500
χ^2	2.88	1.28	3.92	9.68	2.00	8.00	2.00	2.00	9.68	5.12	46.56

Tabulated $\chi^2 = 21.7$ for ($\nu = 10 - 1 = 9$ and $\alpha = 0.01$)

$\therefore \chi^2$ cal $> \chi^2$ tab, reject H_0 of no difference

2.13 Kolmogorov-Smirnov test

Another widely used goodness of fit test is the Kolmogorov-Smirnov test. This test was suggested by Smirnov in 1939 and developed by Kolmogorov and Smirnov, who assumed the probability distribution under test to be continuous and the population mean and variance to be known. Smirnov's table of the critical values was published in 1948 [18]. Like the chi-square test, the Kolmogorov-Smirnov test can be used to test the degree of agreement between the distribution of a set of empirical or sample data and some specified theoretical distribution. The test is conducted by developing or specifying the cumulative probability distribution that would result from the theoretical distribution and comparing it to the cumulative probability distribution of the empirical or sample data. The process is based upon the class in which the theoretical and observed distributions have the largest absolute deviation. This deviation is then compared to the critical values in Table C-VII, Appendix C, to determine if a deviation of this magnitude could occur owing to random variation.

EXAMPLE 2.13.1

For demonstration purposes, let us use the data from Table 2.5, which we have already tested against the Poisson distribution by the chi-square method in Example 2.12.1. Previously, we calculated $\lambda = 0.5577$ for this set of data.

The H_0 is, "There is no significant difference between the observed data and those which would be given by a Poisson distribution with mean = 0.5577 and $n =$ 509." We first must obtain the cumulative distributions for both the observed data and those which would be given by the theoretical distribution, and then calculate the absolute difference for each class. This is done in Table 2.10.

Table 2.10 *Kolmogorov-Smirnov calculations.*

No. Inquiries	I Observed Frequency	II Observed Probability	III Theoretical Probability	IV Cumulative II	V Cumulative III	VI $\|D\|$ (IV-V)
0	315	0.619	0.571	0.619	0.571	0.048
1	142	0.279	0.319	0.898	0.890	0.008
2	40	0.078	0.089	0.976	0.979	0.003
3	9	0.018	0.017	0.994	0.996	0.002
4	2	0.004	0.003	0.998	0.999	0.001
5	1	0.002	0.001	1.000	1.000	0.000

From Table 2.10, the largest absolute deviation is 0.048 for zero inquiries. It is this deviation we wish to test against the critical values given in Table C-VII, Appendix C. From this table we can see that for $n =$ 509 and $\alpha =$ 0.05, the critical value is given by $\frac{1.36}{\sqrt{n}}$. Thus,

$$D \text{ critical} = \frac{1.36}{\sqrt{509}} = \frac{1.36}{22.56} = 0.0603$$

Since our largest deviation was 0.048, we do not reject the null hypothesis.

The question of when to use the chi-square test and when to use the Kolmogorov-Smirnov test naturally arises. In general, for very small samples the chi-square test is not applicable at all and the Kolmogorov-Smirnov should be used. Also, when the sample size is small enough that we have to combine adjacent categories in order to use the χ^2 test, some power is lost. On the other hand, if the sample size is large, the χ^2 test is probably preferable.

Each test has certain strengths and weaknesses, and we can provide only general guidelines for choosing between them. The chi-square test is very powerful for large samples ($n \geqslant 100$); although some authors indicate that they have obtained good results to $n \geqslant 30$, there is little reason not to use the Kolmogorov-Smirnov test in the range of $99 \geqslant n \geqslant 10$. For sample sizes under 10, the Cramer-Von Mises test [15] appears to be the most appropriate to use. In both the chi-square and Kolmogorov-Smirnov goodness of fit tests, the analyst can specify the number of classes or intervals. This choice is important, since it determines the degrees of freedom for each test, and, in general, the more degrees of freedom one can use the more discriminating the test will be. In the chi-square test, the number of classes or intervals is often determined by the necessity of a minimum of five data points in each class. The Kolmogorov-Smirnov test, on the other hand, can either use group data or employ each observation as an individual class, thus allowing the effective analysis of smaller samples. For anyone deeply involved in simulation, I would suggest obtaining Don T. Phillips' monograph [15]. This publication presents a machine-independent computer code in FORTRAN IV, which allows the user to test quickly a set of n observations against 10 common theoretical probability density functions using chi-square, Kolmogorov-Smirnov, Cramer-Von Mises, or Moments (normality) goodness of fit tests. In addition to running the specified tests, the algorithm will, if desired, compute the best estimate of the parameters and print out a histogram. This very excellent computer algorithm allows the analyst to find the best fitting distribution very quickly and efficiently with a minimum of work.

2.14 Curve fitting

In many subsystems, two or more variables are functionally related; we would like to determine this relationship. Sometimes it is a straightforward, simple one that is easily discovered or known; however, more often the functional relationship is extremely complex or even completely unknown. In the latter case we may be forced to hypothesize or approximate the true functional relationship by using some relatively simple mathematical relationship, such as a linear or polynomial equation. In trying to establish such a mathematical equation to describe the functional or structural relationship between two or more variables when historical data exist, the techniques of *regression* and *correlation analysis* are very useful. Regression analysis takes a set of data and fits it to an equation whose form is preselected by the analyst. Correlation analysis gives us some indication of how well the data points fit or cluster around the equation so derived.

The first step in deriving an equation to approximate the desired relationship is to collect data showing the corresponding values of the variables under consideration. For example, suppose we suspect that the yield of a chemical process is a function of the amount of catalyst added to the reaction. We might then let y equal the yield and x equal the amount of catalyst added. Then a sample of N runs of the process taken either from historical data or obtained through a special experiment would reveal the catalyst contents $x_1, x_2, \ldots, x_N$ and the corresponding yields $y_1, y_2, \ldots, y_N$.

The next step would be to plot these points (x_1, y_1), (x_2, y_2), $\ldots$, (x_N, y_N) on a rectangular coordinate system. The result is called a *scatter diagram*, from which it is often possible to visualize a smooth curve approximating the data. In Fig. 2.20, for example, the data would appear to fit a straight line fairly well. On the other hand, the data in Fig. 2.21 appear to fit a nonlinear curve much better.

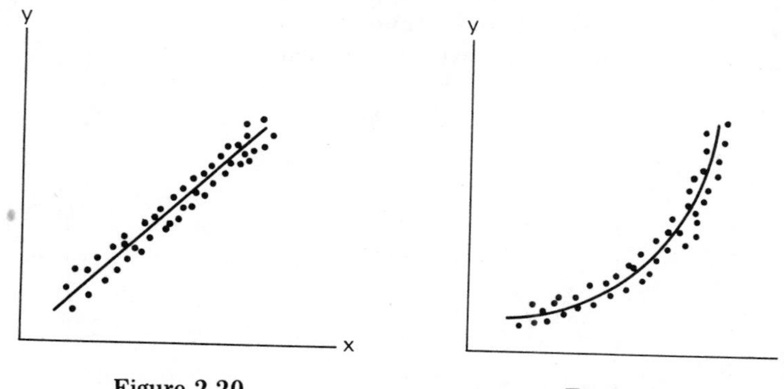

Figure 2.20 **Figure 2.21**

The problem of finding equations for the approximating curves that best fit the given sets of data is called *curve fitting.* The analyst must select the curve to be fitted. For reference, we list several common types of approximating curves and their equations. All letters other than x and y are constants or parameters and x is the independent variable.

1 $y = a_0 + a_1 x$	straight line
2 $y = a_1 x + a_2 x^2$	parabola or quadratic curve
3 $y = a_0 + a_1 x + a_2 x^2 + a_3 x^3$	cubic curve
4 $y = a_0 + a_1 x + a_2 x^2 + a_3 x^3 + a_4 x^4$	quartic curve
5 $y = a_0 + a_1 x + a_2 x^2 + a_3 x^3 + a_4 x^4 + \dots + a_n x^n$	nth degree curve
6 $y = 1 / (a_0 + a_1 x)$ or $1 / y = a_0 + a_1 x$	hyperbola
7 $y = ab^x$ or $\log y = a_0 + a_1 x$	exponential curve
8 $y = a_0 + a_1 \log x$	logrithmic
9 $\log y = a_0 + a_1 \log x$	cubic logrithmic

Many, many others, of course, could be used. To decide which to use, we can examine scatter diagrams and compare results with the general shapes of curves given by the different equations; Fig. 2.22 shows the general shapes of the curves given by some of the different equations. It is also sometimes useful to examine scatter diagrams of transformed variables. To do this, we can use special graph paper for which one or both of the scales are calibrated logarithmically. These are called semilog or log-log graph paper, respectively. Thus, for example, if a scatter diagram of log y vs. x shows a linear relationship, we would use Equation 7. Likewise, if a plot of log y vs. log x appears to be linear, we would use Equation 9.

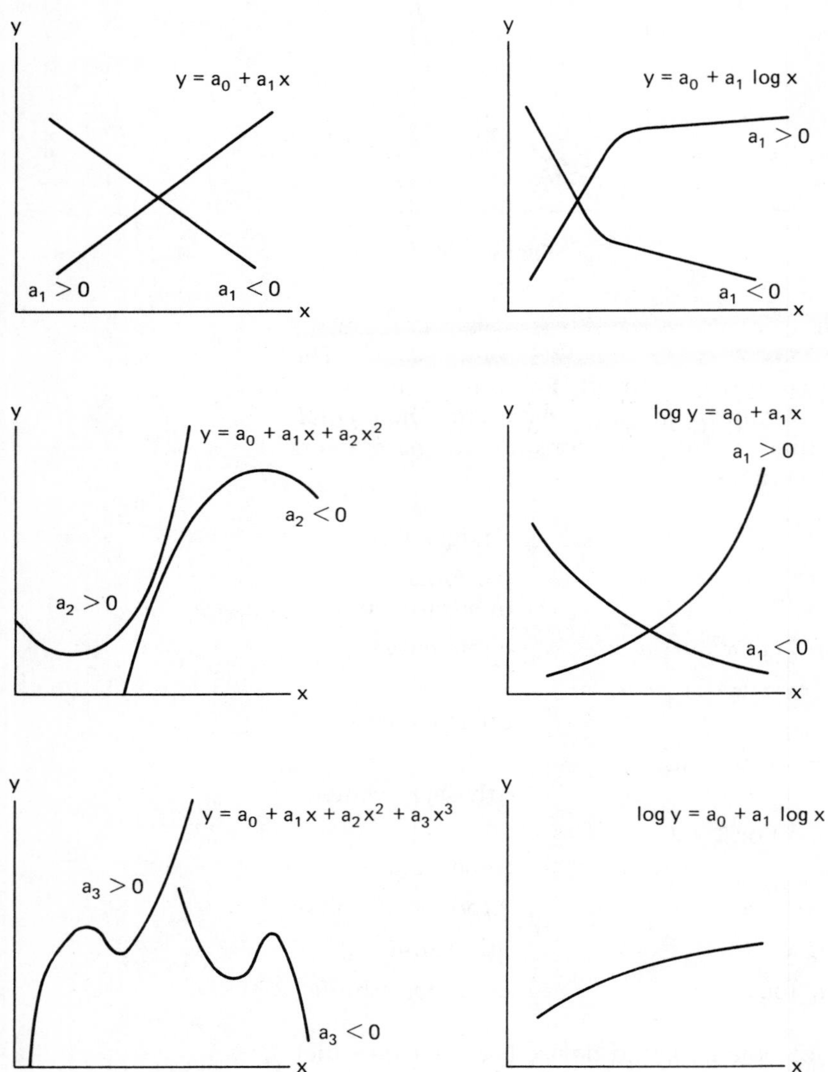

Figure 2.22 General shape of regression curves.

Before going further, we must define what we mean by the curve of best fit. One method would be to use individual judgment. We could place a ruler or French curve on the scatter diagram and move it about until it appears to pass through the center of the points. That is, we could move the ruler or curve until we think we have minimized the deviations of the points from the prospective line. The disadvantage in this method is that different analysts will obtain different curves and different equations. Thus, we would rather have a criterion of "best fit" that is objective, intuitively reasonable, and relatively easy to implement mathematically.

The method most often used is one which is in many respects a formalization of the procedure for fitting a curve by eye; that is, we want to minimize the deviations of the points from the prospective line. This method is called the *method of least squares*. If we consider Fig. 2.23, in which the data points are given by (x_1, y_1), (x_2, y_2),..., (x_n, y_n), for any given x, say x_1, there will be a difference between y_1 and the corresponding value given by the curve. We could call this difference D_1 and refer to it as the deviation. This deviation could be positive, negative, or zero. Correspondingly, for each of the data points, we could calculate deviation values D_2, ..., D_N. One measure of the goodness of fit of the curve to the data would then be the sum of the absolute deviations, or $|D_1| + |D_2| + \cdots + |D_N|$. Because the deviations can be either positive or negative, it is easier mathematically to square the numbers and deal with the deviations squared. Thus an equally good measure of goodness of fit would be the sum of the deviations squared. We therefore define the best fitting curve to be the one curve from all the curves approximating a given set of data

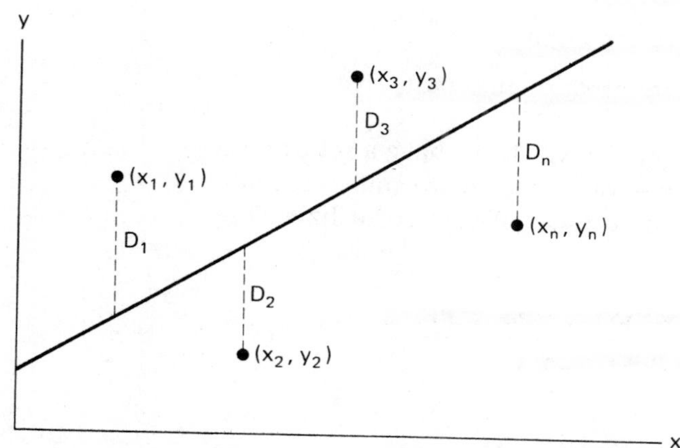

Figure 2.23

points that has the property in which $D_1^2 + D_2^2 + \cdots + D_N^2$ is minimum. This definition is for the least squares criterion for the best fitting curve.

2.15 Regression analysis

The mathematical technique of fitting a set of data points to a chosen curve according to the least squares criterion for best fit is called *regression analysis.* The general form of the curve to be fitted must be chosen by the analyst after examination of the data plotted on a scatter diagram. The mathematics involved are indifferent from the viewpoint that they will find the best fit of the data to the curve chosen whether the choice was good or not. By fitting data to a curve we mean only the process of calculating the values of the constants or parameters in such a manner as to minimize the sum of the deviations squared. The analyst must have chosen the correct equation to be fitted.

A detailed dicussion of regression analysis is beyond the scope of this book. A comprehensive treatment of regression analysis is available in Ezekiel and Fox [11], Draper and Smith [10], and many other excellent texts. We thus restrict our discussion to the simplest case of predicting y as a linear function of a single variable x.

The basic model for a linear relationship between a dependent variable y and an independent variable x is given as

$$y = a_0 + a_1 x + \epsilon \tag{2.1}$$

where

a_0 = the intercept

a_1 = the slope of the line

ϵ = is a random error component

The quantities a_0, a_1, and ϵ are unknown. If we have a set of data consisting of corresponding values of x and y, we must estimate the values of a_0 and a_1 from these data. We do so utilizing the following equations:

$$a_0 = \frac{(\Sigma\, y)(\Sigma\, x^2) - (\Sigma\, x)(\Sigma\, xy)}{n\Sigma x^2 - (\Sigma x)^2} \tag{2.2}$$

$$a_1 = \frac{n\, \Sigma\, xy - (\Sigma\, x)(\Sigma\, y)}{n\Sigma x^2 - (\Sigma x)^2} \tag{2.3}$$

EXAMPLE 2.15.1

Suppose we have 4 data points (n) and we wish to fit them to a linear equation. The data and calculations are shown in Table 2.11.

Table 2.11 *Fitting data to a linear equation.*

	x	y	x^2	xy
	0	1	0	0
	2	4	4	8
	2	3	4	6
	3	5	9	15
Sums	7	13	17	29

$$a_0 = \frac{(13)(17) - (7)(29)}{4(17) - (7)^2} = \frac{221 - 203}{68 - 49} = 0.947$$

$$a_1 = \frac{4(29) - (7)(13)}{4(17) - (7)^2} = \frac{116 - 91}{68 - 49} = 1.316$$

$$\therefore\ y = 0.947 + 1.316x$$

2.16 Correlation

We should restress the fact that because we have fitted a straight line (or any other curve) to the data does not mean that the physical data or real relationship are best described by that equation. Mathematical operations always establish the value of the parameters that gives the best least squares fit to the equation form chosen. Taking an obvious example, we see that the data in Fig. 2.24 are obviously not from a linear relationship, even though we can calculate a best fit to a straight line.

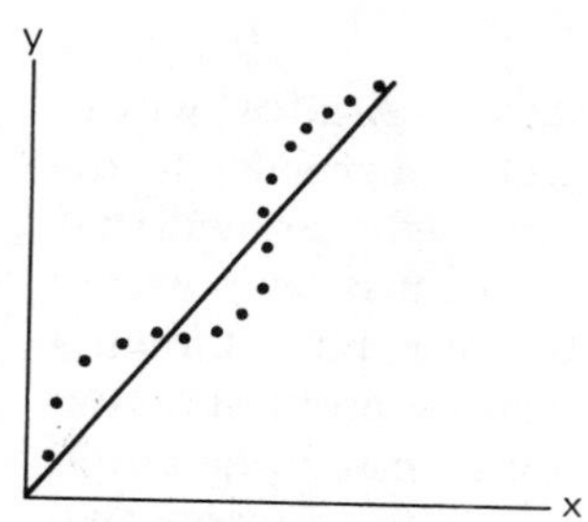

Figure 2.24

To evaluate whether the data are a "good fit" to our line or equation, we need the concept of *correlation.* Correlation tells us how close the data points cluster around the curve or line. While regression defines a proposed relationship between the variables, correlation tells us how good that relationship is. A high correlation between variables shows that they change their values in a related manner (Fig. 2.25), but we must realize that this does not prove or imply a cause and effect relationship. Regression analysis assumes that there is a cause and effect relationship between the dependent and independent variables; correlation studies make no such assumption.

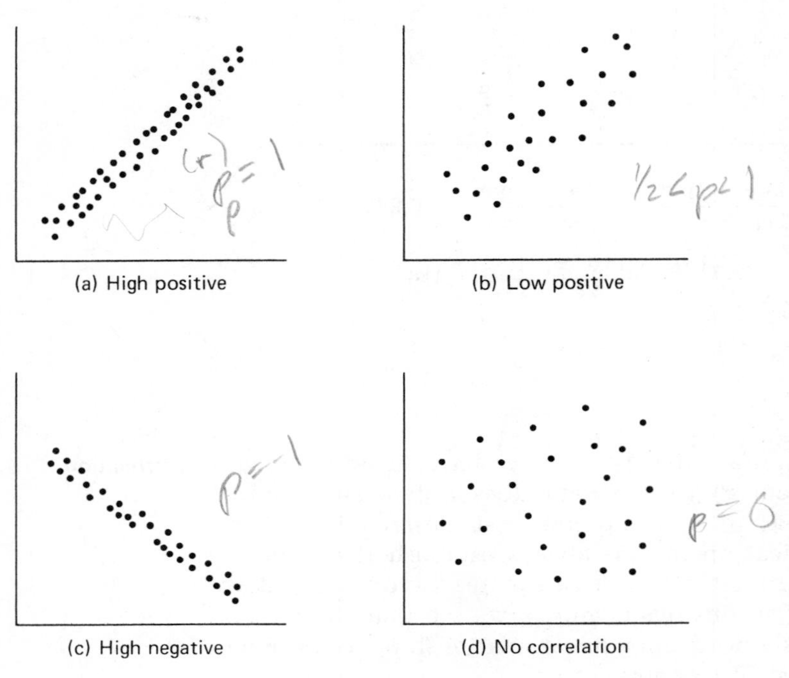

Figure 2.25

An often-quoted story is that of the state legislator whose analysis showed a very high correlation between beer sales in his state and increases in teacher incomes. He supposedly argued that it would do no good to raise teachers' salaries because they would just spend it on more beer. He could probably also have shown a high correlation between rising teachers' salaries and increased crime, increased illegitimate births, and inflation. In fact, he could show a high correlation and regression curve between any two phenomena that are rising more or less in proportion to each other.

The important point is that the analyst makes an assumption or hypothesis about a cause and effect relationship between the variables when he does a regression analysis. That assumption may or may not be true. Correlation tells him how closely his data points cluster around the curve he has hypothesized. It still does not tell him whether the basic assumption of cause and effect is valid. Correlation analysis does tell him whether the data are consistent with his hypothesis.

Correlation coefficients will range from −1 to +1. A coefficient of −1 means perfect negative correlation—i.e., as x goes up, y goes down and all the data points lie exactly on the curve. A coefficient of 0 means absolutely no correlation, and a coefficient of +1 means perfect positive correlation. Very, very seldom do we get any of these three values; usually, we get some fraction that we must then test for statistical significance.

For a simple linear regression problem (i.e., one dependent and one independent variable, linearly related), the calculation of the coefficient of correlation r is

$$r = \frac{n\Sigma XY - (\Sigma X)(\Sigma Y)}{\sqrt{[n \Sigma X^2 - (\Sigma X)^2][n \Sigma Y^2 - (\Sigma Y)^2]}} \tag{2.4}$$

EXAMPLE 2.16.1

Using the same data from Example 2.15.1, we can calculate the coefficient of correlation, as in Table 2.12.

Table 2.12

	X	Y	XY	X^2	Y^2
	0	1	0	0	1
	2	4	8	4	16
	2	3	6	4	9
	3	5	15	9	25
Sums	7	13	29	17	51

$$r = \frac{4(29) - (7)(13)}{\sqrt{[4(17) - (7)^2][4(51) - (13)^2]}}$$

$$= \frac{25}{\sqrt{(19)(35)}} = \frac{25}{25.8} = 0.969$$

The total variation of Y is defined as $\Sigma (Y - \bar{Y})^2$—i.e., the sum of the squares of the deviations of Y from the mean $\bar{Y}$. The ratio of the amount of this variation, which can be explained by our regression equation, to the total variation observed is called the *coefficient of determination* and is given by the square of the correlation coefficient. Thus, in the previous example where we calculated an $r = 0.969$, the coefficient of determination $= r^2 = (0.969)^2 = 0.939$. That is to say, 93.9% of the variation of Y as X changes can be explained by our relationship that $Y = 0.947 + 1.316X$.

2.17 Nonlinear and multivariable problems

Because our purpose is to discuss the art of simulation and not statistics, we have limited our discussion and examples to the simplest cases. The same basic ideas and philosophy carry over to situations where the hypothesized relationships are nonlinear and/or where the number of independent variables is greater than one. Obviously the mathematics of nonlinear and/or multivariable regression and correlation analysis are much more complex and outside the scope of our major concern in this book. Most computer centers have very good subroutines in their user library for performing these more complex calculations. The reader can find computer codes written in FORTRAN for multiple linear regression and curvilinear regression in Bartee [4]. Other programs may be obtained from COSMIC* at nominal cost. Detailed treatments of the theory and application of regression and correlation analysis can be found in the two previously referenced books [10, 11] as well as in most good texts on applied statistics.

We postpone the discussion of establishing confidence limits and statistical testing of the various coefficients until Chapter 6, where we consider validation procedures in more depth.

2.18 Panel of experts

When we cannot determine the value of parameters empirically from experiments or past history, we must resort to subjective evaluations. In such a case, it is often advantageous and desirable to use the collective judgment of a panel of experts rather than that of a single individual. Such a panel should consist of experts or individuals especially knowledgeable about the process being modeled and, if feasible, responsible decision makers. Several techniques can be used for eliciting and amalgamating the group's opinion, but perhaps the most useful is the Delphi procedure.

The Delphi procedure was developed at the RAND Corporation by Olaf Helmer and Norman Dalkey [9]. It is an iterative procedure that subjects each participant's views to the criticism of each of the others without face-to-face confrontation. The idea is to provide a mechanism that preserves the anonymity of personal viewpoints so as to minimize the influence of vocal and persuasive individuals on group behavior. All interactions among the participants are handled through a steering group or coordinator. The steering group controls the deliberations and preserves anonymity. Group response is computed and given back to the participants by the steering group via some form of averaging (usually either the mean or the median).

*COSMIC is Computer Software Management and Information Center, University of Georgia, Athens, Georgia, 30601.

Section 2.18 Panel of experts

Let us consider an example. Consider the common problem of determining the value of a particular number N. Suppose we have 12 participants on our panel. The Delphi procedure has several variations, but we might proceed as follows. (1) Ask each participant independently to give us his estimate of N. (2) Arrange the responses on a scale in the order of their magnitude and determine the quartiles, Q_1, M, and Q_3 so that each of the four intervals contained one-quarter of the estimates. The result for the 12 participants might look like Fig. 2.26. (3) Communicate the values of Q_1, M, and Q_3 to each participant and ask him to reevaluate his estimate, and, if his new estimate is either above Q_3 or below Q_1, to state briefly why he believes his opinion is correct. (4) Compute the results of round two, and send the new values of Q_1, M, and Q_3 (which will usually be less dispersed than round one) to the participants along with the written justifications for the extreme points (always preserving anonymity). Ask each respondent to consider the new information and arguments and revise his previous estimate, if desired. If a participant's revised estimate in round three is above Q_3 or below Q_1, ask him to state briefly why he considered unconvincing the arguments that might have drawn his estimate toward the median. (5) Repeat this procedure for as many rounds as deemed desirable by the steering committee, or until the range between Q_1 and Q_3 has narrowed to some predetermined value. This usually entails only 3 or 4 rounds, since the arguments will repeat in short order. Then take the median value as representing the group position as to what the value of N should be.

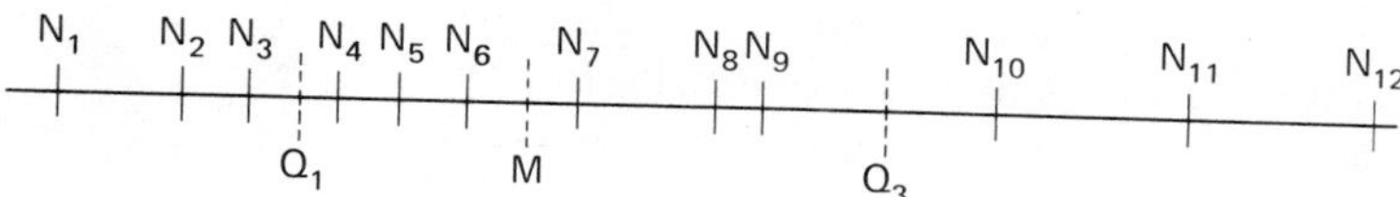

Figure 2.26 Results of Delphi estimates.

As mentioned earlier, several variations of the Delphi procedure are possible. For example, instead of using the median and quartiles we might use the mean and standard deviation. In this case, the steering group might provide the participants with the mean and standard deviation in each round and ask for brief justifications of all estimates outside of ±1 standard deviation. Of course, we assume that the responses will be normally distributed about the mean, which may or may not be a good assumption. The rounds could be continued until the standard deviation is reduced to a predetermined size or until no further reduction in the dispersion of estimates is evident. We are not necessarily trying to force complete conformity, and a spread of opinions on the final round is the normal outcome.

The purpose of the Delphi procedure is to reduce the psychological pressures of face-to-face confrontation and the possibility of an especially vocal or powerful individual dominating the final outcome; however, it is not a foolproof method. For example, it is not known how much of the convergence that takes place is induced by a desire to conform to the group opinion or how much by the elimination of the basic causes of disagreement. Placing the responsibility for justifying their responses on the participants clearly tends to cause those without strong convictions to move their estimates closer to the median. In addition, those who originally believed they had a good argument might tend readily to abandon their positions when they see that others were not immediately convinced. This may increase the bandwagon effect instead of reducing it, as intended.

The Delphi procedure, consisting of anonymous response, iteration, controlled feedback, numerical estimates, and statistical group response, imperfect as it is, shows great promise as a valuable tool for simulation modelers. As E. S. Quade [19] reports, experiments indicate the following:

1 Face-to-face confrontation is not as efficient as the Delphi procedure.
2 Improvement in accuracy of estimates increases with the number of respondents and with increasing iterations.
3 The accuracy of the estimates decreases as the time between responses increases.
4 There is greater acceptance of the group opinion by the individual participants using Delphi than there is with face-to-face procedures. This aspect is obviously important if some of the participants are the decision makers responsible for implementing the results of the simulation.

2.19 Conclusions

Undoubtedly one of the most difficult and critical parts of any study using simulation is the systems definition and model formulation phase. Our success or failure at later stages will be largely determined by how well we have:

1 defined the problem;
2 established the goals of the study;
3 defined the boundaries of the system;
4 determined the relevant components and variables;
5 hypothesized and abstracted the relationships between the components and variables;

6 estimated the values of the pertinent parameters.

If we are skillful and careful in the initial design of our model and develop it in a modular fashion, we will be able to improve and correct our model at later stages as we gain experience with the system. Nevertheless, the more mistakes we make initially the more difficult and costly will be any later modifications required. The real world can be overwhelmingly complex, and we have to make simplifying assumptions about how it works. But we must never forget that we have made these simplifying assumptions and begin to believe that the output of our model is reality.

The capabilities of modern computers continue to expand at a fantastic rate. Advances in processing speed, memory storage capacity, and time sharing provide the analyst with an increasingly powerful tool. But the present and future advances in simulation modeling depend not only on the development of new computer capabilities but also upon our ability to make efficient use of these capabilities. Since the efficient utilization of the computer rests greatly with the software of the system, many special programming languages have been designed to help make the modeler's work easier, less time consuming, and less subject to error. In general, these languages provide the user with a set of modeling concepts used to describe the system, a systematic way to name the variables, and a programming system which will convert the model description into a machine language that will allow the computer to perform the simulation. In Chapter 3 we consider some of the programming languages that might be helpful.

REFERENCES

1. ACKOFF, R. L., "Towards a System of Systems Concept," *Management Science*, Vol. 17, No. 11, Jul., 1971.
2. ACKOFF, R. L., and M. W. SASIENI, *Fundamentals of Operations Research*, John Wiley & Sons, New York, 1968.
3. BARNES, R. M., *Motion and Time Study: Design and Measurement of Work*, 5th ed., John Wiley & Sons, New York, 1963.
4. BARTEE, E. M., *Statistical Methods in Engineering Experiments*, Charles E. Merrill Publishing Co., Columbus, Ohio, 1966.
5. BOHL, M., *Flowcharting Techniques*, Scientific Research Associates, Inc., Chicago, 1972.
6. BOWMAN, E. H., and R. B. FETTER, *Analysis for Production Management*, Richard D. Irwin, Inc., Homewood, Ill., 1961.
7. BUFFA, E. S., *Modern Production Management*, John Wiley & Sons, Inc., New York, 1965.

8. CULLINANE, T. P., "Modeling Material Handling Systems" *Industrial Engineering*, Vol. 6, No. 1, Jan. 1974.

9. DALKEY, N., and O. HELMER, "An Experimental Application of the Delphi Method to the Use of Experts," *Management Science*, Vol. 9, 1963, p. 458.

10. DRAPER, N. R., and H. SMITH, *Applied Regression Analysis*, John Wiley & Sons, Inc., New York, 1966.

11. EZEKIEL, M., and K. A. FOX, *Methods of Correlation and Regression Analysis*, John Wiley & Sons, Inc., New York, 1959.

12. HAMMERSLY, J. M., and D. C. HANDSCOMB, *Monte-Carlo Methods*, Methuen & Co., Ltd., London, 1964.

13. MORSE, P. M., and G. E. KIMBALL, *Methods of Operations Research*, John Wiley & Sons, Inc., New York, 1951.

14. NADLER, G., *Work Design*, Richard D. Irwin, Inc., Homewood, Ill., 1963.

15. PHILLIPS, DON T., *Applied Goodness of Fit Testing*, O. R. Monograph Series, No. 1, AIIE-OR-72-1, American Institute of Industrial Engineers, Atlanta, Ga., 1972.

16. SCHROER, B. J., "A Simulation Model for Evaluating the Projected Expansion of the Apollo Parts Information Centers Inquiry Answering Service," unpublished Master's Thesis, University of Alabama, Huntsville, 1967.

17. SIMON, H. A., *Administrative Behavior*, 2nd ed., The Free Press, New York, 1957.

18. SMIRNOV, N., "Table for Estimating the Goodness of Fit of Empirical Distributions," *Annals of Mathematical Statistics*, Vol. 19, 1948.

19. QUADE, E. S., "An Extended Concept of Model," *Proceedings of the 5th International O.R. Conference*, J. R. Lawrence (ed.), Tavistock Publications, Ltd., London, 1970.

EXERCISES

1. Read case study II, Appendix A.
   a) In the model the aircraft reverses its field instantaneously as soon as it fires its last missile. Does this obvious shortcoming significantly affect the results?
   b) If T_t is the time required for the airplane to make a 180° turn after firing its last missile, how would this change the equations in the model?

2. Draw a flow process chart and flow diagram of your morning activities from the time you get up until you leave the house on a typical day.

3. Assume you are designing a candy vending machine. All purchases cost ten cents. The machine will accept nickels, dimes, or quarters but returns all change in nickels. It will be restocked with candy and 50 nickels each morning. Draw a logic flow diagram that will provide the necessary control mechanism.

4. Given A, B, and C, which are random, uncorrelated variables. Variable A is normally distributed with $\mu = 100$ and $\sigma = 20$. Variable B is also normally distributed with $\mu = 20$ and $\sigma = 5$. Variable C is distributed in accordance with the following table. Use Monte-Carlo sampling to estimate the mean of a new variable Z, which is defined as $Z = (A + B) / C$, using a sample size of 10.

Value of C	Probability
10	0.10
20	0.25
30	0.50
40	0.15

5. Assume that variables A and B in Problem 4 have a correlation coefficient of +1 (i.e., as one goes up, so does the other). Use a Monte-Carlo sample of size 10 to estimate the mean of Z.

6. Reconsider the staggering drunk problem discussed in Section 2.10. Assume that instead of equal probabilities of going in each direction at any given corner, the probabilities are as follows: 50%, going straight ahead; 20%, going right or left; and 10% reversing direction. Devise a simulation of this situation, and, using 10 trials, calculate the probability that he ends up within two blocks of his starting point after he walks five blocks.

7. Simulate a fighter-bomber that is to be sent to attack an important factory with air-to-ground missiles. Each missile is individually guided. The dimensions of the factory are 200 by 500 feet. The attack run is to be made parallel to the long dimension of the target, and the aiming point is the geometric center of the factory. The actual impact point of each missile can be defined by a horizontal deviation, x, and a vertical deviation, y (see sketch). At the distance from which the missiles are to be launched, the two deviations are independent and are normally

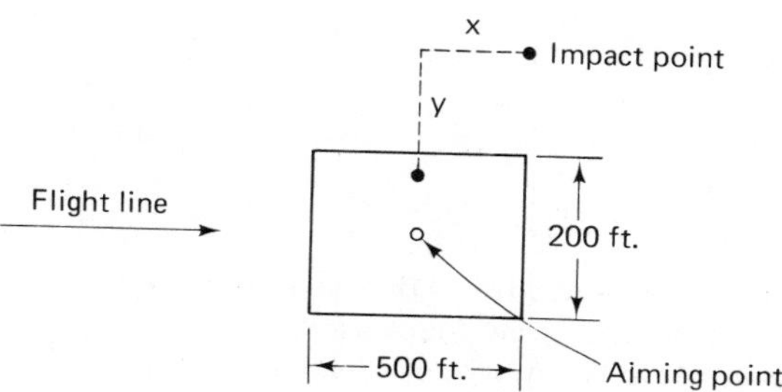

distributed about the aiming point with a mean of zero. The standard deviation is 200 feet in the x dimension and 100 feet in the y dimension. The airplane is to fire six missiles on each run. Using a sample size of 10 runs, estimate the mean number of hits per attack.

8. Use of a component follows a Poisson distribution as shown in the table. Stock is examined at the end of each 5-day week and if it has declined to a level of six parts or less, an order is placed for 10 units. Each order is delivered in accordance with the probability distribution shown in the table (days till delivery are working days). When the simulation begins, it is the beginning of a week, there are 10 parts on hand, and no orders are outstanding. Simulate 5 weeks of operation. Plot the inventory level at the end of each day and calculate the number of days during which shortages exist.

Daily Demand	Probability	Days till Delivery	Probability
0	0.30	2	0.10
1	0.36	3	0.50
2	0.22	4	0.30
3	0.09	5	0.10
4	0.03		

9. The Fire Chief in Diablo, Alabama, has found that the number of fires occurring during a 24-hour day follows a Poisson distribution with a mean of four fires per day (see table). In examining past fires, the Chief finds that ¾ of all fires require only one truck, and the amount of time required per fire is normally distributed with $\mu = 3$ hr and $\sigma = ½$ hr. The other ¼ of fires reported require 2 trucks, and the time to put them out is normally distributed with $\mu = 4$ hr and $\sigma = 1$ hr. Assume the necessary fire trucks are always available to respond; what is the average daily demand in hours per day? Use a simulation with a sample of 10 days.

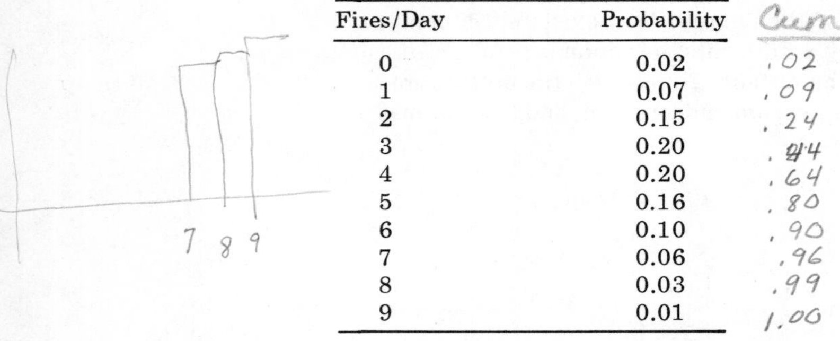

Fires/Day	Probability
0	0.02
1	0.07
2	0.15
3	0.20
4	0.20
5	0.16
6	0.10
7	0.06
8	0.03
9	0.01

10. Trucks arrive randomly at a terminal to be unloaded. After analyzing historical data, we conclude that the daily arrival rate follows a Poisson distribution with a mean of 3 trucks per day. The weight of each truck's cargo is an important factor in unloading time. Past records show that cargo weights are normally distributed with mean equal to 30,000 lb and

a standard deviation of 5000 lb. The number of pounds per hour that a crew can unload also varies and is a function of the type of cargo. The probability of each type of cargo and the unloading rate for each are shown in the table. A crew contains three people: a fork lift operator, paid $4.00 per hour, and two laborers, paid $2.50 per hour each. It is company policy to unload all trucks on the day after they arrive. All trucks arriving the previous day must be unloaded regardless of any overtime costs incurred. The union contract requires a 50% pay premium for any hours over 8 worked on any day. Based upon a simulation of 10 days, determine how many crews should be employed to minimize total unloading costs.

No. Trucks Arriving	Probability
0	0.05
1	0.15
2	0.22
3	0.22
4	0.17
5	0.11
6	0.05
7	0.03

Type of Load	Probability	Unloading Rate/Hr (lb)
A	0.40	8000
B	0.35	7000
C	0.25	5000

11.* Reconsider Problem 9. If the number of fires per day is Poisson distributed, the time between fires will be distributed as a negative exponential with mean equal to ⅓ days. The city of Diablo presently has 3 fire trucks. A new fire truck costs $50,000, and the yearly cost for the necessary crew per truck is $84,000. If a fire occurs and no truck responds, it cost $5,000 in additional damages for each 15 minutes (or part thereof) to a maximum of $30,000 until a truck does respond. The time required to respond when a truck is available is considered negligible. If only one truck responds to a fire requiring 2 trucks, the additional loss is half that incurred by no response (for the period until another truck arrives). Develop a computer simulation model, and, based upon a simulation of 365 days, determine how many fire trucks the city should have.

12.* Assume that in Problem 10 the trucks must be unloaded the same day they arrive instead of the next day. Assume the trucks arrive randomly during an 8-hour period starting at 8 A.M. If arrivals are at a mean rate of 4 per day, the time between arrivals is distributed as a negative exponential with mean equal to ¼ day. Base your answer on a 3-month (60 working days) simulation of activity. Assume no more than 4 hours overtime can be worked on any given day.

13. The following table shows the number of daily burglaries reported in Paradise, Texas, during a typical month. Using $\alpha = 0.05$, test the data for goodness of fit to a Poisson distribution, using a chi-square test.

* Problems 11 and 12 require a computer model to be programmed. Appendix B gives random variate generators.

M	T	W	T	F	S	S
	2	3	4	6	4	3
5	4	2	1	4	5	3
4	5	3	5	8	2	2
3	1	3	6	2	1	3
2	7	1				

14. Retest the data in Problem 13 using a Kolmogorov-Smirnov test for goodness of fit, still with $\alpha = 0.05$.

15. The Purple Taxicab Company of Boston keeps daily track of the number of accidents involving its cabs. The results for one year are shown. An operations researcher from M.I.T. has told them that accident statistics are usually Poisson distributed. Test this assertion, using $\alpha = 0.01$.

No. Accidents	No. Days
0	217
1	113
2	29
3	4
4	2

16. As a part of a simulation model they are building, the Lights-Out Municipal Utility has examined the January usage of electricity by its 1200 customers. The distribution of usage is as shown. Test to determine if the usage by family is normally distributed, using $\alpha = 0.05$.

KW Used	No. Households
1000—1499	8
1500—1999	24
2000—2499	96
2500—2999	224
3000—3499	417
3500—3999	240
4000—4499	114
4500—4999	57
5000—5499	16
5500—5999	4

17. Case study A-IV in Appendix A presents a simulation study of refuse collection in Cleveland. The authors present historical data and data from their simulation model (Table 1), which they state demonstrates the validity of their model. Test their assertion by a Kolmogorov-Smirnov test with $\alpha = 0.05$.

18. The following are 200 random digits generated on an IBM-360/30 using the library subroutine RANDU. Use a chi-square goodness of fit test with $\alpha = 0.01$ to determine if they are uniformly distributed.

4144	4864	1889	7555	8328	1071	6878	3522	9233	3697
9082	1215	5548	2353	4186	3033	5025	0149	7588	4064
6273	1062	9912	9910	0252	2319	1647	9066	9215	4233
2461	6669	7862	7151	2140	8530	1840	4205	9032	5807
3548	9025	2213	2053	2400	5817	3901	0148	5781	3350

19. For eight responses of an independent variable x we obtained the following responses for a dependent variable y:

x	4	1	12	6	3	9	7	4
y	2	1	7	3	2	6	5	3

Calculate (a) the line of regression, and (b) the regression estimate for y when $x = 7$.

20. The yield of a chemical process is thought to be a function of the amount of catalyst added to the reaction. The result of an experiment is shown. Compute (a) the line of regression, (b) the coefficient of correlation, and (c) the coefficient of determination.

Yield (%)	67.3	69.5	70.4	69.6	73.8	76.8	75.8
Catalyst (lb)	17	18	19	20	21	22	23

21. Attempting to establish an equation used in a model developed for the study of urban problems, we hypothesize that there exists a linear relationship between vocational training and increased earnings. The following results were computed by a statistician from data on years of training (X) and weekly increase in earnings (Y) for a group of 25 trainees who have gone through an Office of Economic Opportunity (OEO) training program.

$$n = 25 \qquad \Sigma X^2 = 318$$
$$\Sigma X = 80 \text{ (years)} \qquad \Sigma Y^2 = 67{,}650$$
$$\Sigma Y = 1240 \text{ (dollars)} \qquad \Sigma XY = 4490$$

Compute (a) the line of regression, (b) the coefficient of correlation, and (c) the coefficient of determination.

chapter three

MODEL TRANSLATION

Man need not bend to linquistic circumstance but may easily bend language to his needs.

Joshua Whatmough

3.1 The computer and modeling

In the most general sense, computer modeling may be performed by one of three techniques. General purpose digital computers, electronic analog computers, and, more recently, hybrid computers are all being used extensively for simulation modeling. The advantage of one technique over another lies in the basic features of digital and analog computers. The analog machine represents the variables of a problem by physical quantities that are easily generated or controlled, such as electrical voltages. It develops a solution by performing simultaneous (or parallel) operations, whereas conventional digital computers carry out sequential (or serial) processing. This results in a considerable computational speed advantage for the analog machine, particularly for solving systems of differential equations.

On the other hand, the digital computer is capable of much greater precision and dynamic range because of its ability to count, obey logic rules, perform floating point arithmetic, and utilize long word lengths. Thus, one of the chief distinctions between the analog and digital computers is the manner in which the dependent variables are handled within the computer. In analog computers, the dependent variables (although not necessarily the independent variables) appear in continuous form. In the digital computer, all variables, both dependent and independent, appear in discrete form. In analog machines the accuracy of the numbers (i.e., the number of significant digits) is limited by the quality of the circuit components, whereas the accuracy of the digital machine depends upon the number of digits carried and is limited only by the size or capacity of the memory registers.

The attractive possibility of combining the speed of the analog computer with the precision and logic control of the digital computer provided the initial motivation for hybrid computational methods, starting in the late 1950's. The hybrid computer may be one of several possible systems, but the most common is the combination of a general-purpose digital computer, an analog computer, and an interface or linkage. Within this class of hybrid techniques is a considerable range of systems types and hybrid operational modes, depending upon the relative amounts of "computing power" in the analog and digital parts of the system. In the following paragraphs, we try to point out and summarize the characteristic strengths and weaknesses of all three techniques as they apply to simulation modeling.

In general, a large-scale digital computer can do any job that can be accomplished on an analog computer. But many problems can be handled faster, easier, and at less cost on the analog machine. The following characteristics of *analog computers* are enumerated by Bekey and Karplus [2]:

1 Dependent variables within the machine are treated in continuous form.

2 Accuracy is limited by the quality of the computer components.

3 Parallel operation is possible with several computations occurring simultaneously.

4 Real-time or compressed time operation is possible with computing speeds limited primarily by the bandwidth characteristics of the computer elements and not by the complexity of the problem.

5 Such operations as addition, subtraction, multiplication, differentiation, integration, and nonlinear function generation are performed efficiently. On the other hand, the analog computer has very limited ability to make logical decisions, store numerical data, provide extended time delays, and handle non-numerical information.

6 Programming techniques consist largely in substituting analog computing elements (such as amplifiers, resistors, capacitors, etc.) for corresponding elements in the physical system being simulated.

7 Actual hardware from a system being studied in the simulation can be included.

8 The user can experiment by adjusting coefficient settings on the computer, thereby gaining insight into system operation.

The *digital computer* is what the layman usually visualizes when he hears the term "electronic computer." Like the analog computer, it possesses both strengths and weaknesses. The following are attributes associated with the digital computer, as listed by Bekey and Karplus [2]:

1 All data within the computer are handled in discrete form.

2 Operation is serial, involving the time sharing of all operational and memory units. Only one, or a limited number, of operations are carried out simultaneously.

3 Accuracy is relatively independent of the quality of computer components and determined primarily by the size of the memory registers, the number of significant digits carried, and the numerical technique selected.

4 Solution times can be rather long and are determined by the complexity (i.e., the number of arithmetic operations required for the solution) of the problem.

5 There is an ability to trade-off solution time and accuracy, i.e., to reduce errors by increasing the length of computation time and vice versa.

6 A limited number of arithmetic operations are possible, including addition, subtraction, multiplication, and division, but more complex operations, such as differentiation and integration, must be performed by approximation (numerical) techniques.

7 Numerical and nonnumerical data are memorized indefinitely.

8 Logical operations are performed and decisions are made utilizing either numerical or nonnumerical data.

9 Floating decimal point operation is possible, thereby eliminating scale-factor problems.

10 Programming techniques include special computer languages, often bearing little direct relationship to the problem under study.

Hybrid computer techniques are an effort to combine the best characteristics of the analog and digital computers and to avoid their shortcomings. The speed capability of analog equipment suggests the need to augment a digital computer for certain problems, and the computational accuracy, flexibility, data storage capability, and logical control functions of a digital computer are highly desirable additions to an analog system. The digital computer portion of a hybrid system adds three important characteristics to the combination: (1) the possibility of controlling the analog computer at high speed and in complex mode-switching operations; (2) the addition of a data storage and memory system; and (3) the possibility of more precise computations and logical operations. Advantages of hybrid operation are summarized by Bekey and Karplus [2] as follows:

1 To combine the speed of an analog computer with the accuracy of a digital machine.

2 To permit the use of actual hardware from the system being studied in a digital simulation.

3 To increase the flexibility of an analog simulation by using digital memory and logic control.

4 To increase the speed of digital computations by using analog subroutines.

5 To permit the processing of incoming data that are partially discrete and partially continuous.

Hybrid computers are usually justified because they overcome certain shortcomings in present-day analog and digital computers. However, as these limitations are reduced or eliminated, we can anticipate that the rationalization for many hybrid computer simulations will disappear. For example, the increasing speed, reduced cost, and on-line time sharing of large-scale digital computers has already narrowed the range of justifiable hybrid applications.

The preceding discussion of the relative strengths and weaknesses of analog, digital, and hybrid computers serves to alert the reader to the fact that all computerized simulations need not be restricted to the use of the digital computer. Under certain sets of circumstances, advantages can be gained by using analog and hybrid machines. Ideally, selection of a computer is the analyst's decision. Unfortunately, not everyone is blessed with access to all three types of computers, and, in actual practice, the user will probably have to make do with the type of hardware configuration available.

3.2 Programming languages

Early effort in a simulation study is concerned with defining the system to be modeled and describing it in terms of logic flow diagrams and functional relationships. But eventually one is faced with the problem of describing the model in a language acceptable to the computer. Most digital computers operate in a binary language of data representation or in some multiple of binary such as octal or hexadecimal. Since these languages are awkward for communication, programming languages have evolved for users to converse with the computer. Unfortunately, so many general and special purpose programming languages have been developed over the years that it is nearly impossible to decide which language best fits or is even a near best fit to any particular application. Over 170 languages were in use in the United States alone in 1972 [25]. Consequently, the usual procedure is to adopt a language known by the analyst, not because it is best but because it is known. Any general algorithmic language can express the desired model; however, one of the specialized simulation languages may have very distinct advantages in terms of ease, efficiency, and effectiveness.

This chapter does not teach the reader to program in any of the languages described, nor does it discuss implementation techniques. Hopefully it does make the reader aware of the characteristics, strengths, and weaknesses of some of the more popular languages. The major differences between special purpose simulation languages are, generally, (1) the organization of time and activities, (2) the naming and structuring of entities within the model, (3) the testing of activities and conditions between elements, (4) the types of statistical tests possible on the data, and (5) the ease of changing model structure. In the following sections, we compare several languages, after first showing the various philosophies of language design and describing a number of key factors involved in choosing a language.

Kiviat [12] has concisely defined some of the terms necessary in the study of programming languages. His definitions and those of Sammet [24] have been merged to form the basis of the following discussion. A computer programming language is a set of symbols recognizable by a computer by which a programmer instructs the computer regarding operations he wishes it to perform. At the lowest or most basic level, we have *machine language* programming, in which the instructions are written in binary notation (or other form of data representation used by the machine), which corresponds directly to machine functions. Coding a program in machine language is a difficult task. Beside the tedious, time-consumming process of writing instructions in binary notation (zeros and ones), the programmer must designate storage locations for both data and instructions. In addition, he must know the locations of and keep track of every switch, indicator, and register that he uses. He must control their functions by the instructions in his program. This task is impractical for most computer users.

Because of the obvious difficulties of writing and testing a machine language program, several symbolic programming languages have been developed. Symbolic languages permit the programmer to use convenient symbols or mnemonics (memory aids) to write a computer program. For example, an *assembly language* program is a string of mnemonic symbols that correspond to machine language functions and are translatable by an *assembly program* or *assembler* into a basic machine language program. The gamut of assemblers runs from the least sophisticated, which simply substitute basic machine language codes for mnemonics and assign computer addresses to variable names and labels, to the more sophisticated, which can recognize additional symbols (known as macros) and construct complicated basic machine language programs from them. A program that uses meaningful labels is much easier to read and understand than one containing numerical addresses of locations in storage. When using a symbolic language, the programmer need not know or remember the exact location in which data and instructions are placed in storage. He can simply select a label for an instruction or data item and use that label each time he wishes to refer to the instruction or data item. Thus we might use A for add, D for divide, B for branch, NAME for employee name, RATE for hourly wage, and so on.

Assembler languages are machine-oriented (i.e., designed for a specific machine). Generally, one assembler language is available for use with each type of computer. On the other hand, *compiler languages* are procedure-oriented or problem-oriented languages. Compiler languages are sometimes referred to as high-level programming languages because the programmer who uses them is removed a considerable distance from concerns related to machine operations. Such a programmer is concerned with the steps necessary to solve the problem; he is not concerned with how the machine will perform the required operations. Compiler languages are macrolanguages—i.e., one compiler language statement is used to create several machine language instructions. This macroinstruction capability is important for at least two reasons. First, it reduces the time and effort required to program a complex problem, owing to reducing the number of instructions to designate a sequence of processing or machine steps. Second, it reduces the chances of programming error. The sequence of instructions inserted for each macroinstruction is pretested for correctness, whereas the programmer may easily err if he has to specify each detailed step for machine execution.

A compiler language is machine-independent. The compiler language statements are accepted by a *compiler program* and translated into either assembly language or basic machine language. If it is into assembly language, it must in turn be reduced to a basic machine language program for the particular machine to be used. This feature of machine-independence is significant. Having learned a compiler language, a programmer can use any machine by any manufacturer (as long as a compiler program is available) without having to learn a new language.

Finally, we must distinquish between general purpose languages and special purpose languages. Both are problem-oriented compiler languages. General-purpose languages, such as FORTRAN, COBOL, BASIC, PL/I, and ALGOL, are designed to solve a broad class of problems. Special-purpose languages, on the other hand, such as the various simulation languages, are designed to satisfy or solve a particular class or type of problem.

3.3 Advantages of simulation languages

The evolutionary development of simulation languages began in the late 1950's. At first, the languages used in simulation were specific-machine or general-purpose languages. Emshoff and Sisson [4] trace the historical background of simulation languages as a process of analysts going through a sequence of steps much like the following:

1 The analyst has a problem situation in mind.

2 He can state the situation in some native language, such as English augmented by technical terminology.

3 He describes the situation as precisely as possible in this language.

4 The description is used by a programmer (which may be the same analyst) to prepare a computer program in a general-purpose language.

Having repeated this sequence many times, analysts recognized that many situations being simulated could be categorized broadly as systems involving the flow of items through processes. Since many of the programs had functionally similar processes, the idea to develop special-purpose languages evolved almost simultaneously within several groups of researchers in the early 1960's. These languages changed gradually from assembly language programs with special features, through extended commercially available problem-oriented languages, to sophisticated special-purpose simulation languages. Any algorithmic programming language can be used for simulation modeling, but those languages designed specifically for the purpose of computer simulation provide certain useful features:

1 reducing the programming task;
2 providing conceptual guidance;
3 aiding in defining the classes of entities within the system;
4 providing flexibility for change;
5 providing a means of differentiating between entities of the same class by characteristics or properties;
6 describing the relationship of the entities to one another and to their common environment;
7 adjusting the number of entities as conditions vary within the system.

Emshoff and Sisson [4] believe that all simulations require certain common functions that differentiate a simulation language from a general algebraic or business programming language. Among these are the need to

1 create random numbers;
2 create random variates;
3 advance time, either by one unit or to the next event;
4 record data for output;
5 perform statistical analyses on recorded data;
6 arrange outputs in specified formats;
7 detect and report logical inconsistencies and other error conditions.

Furthermore, they state that for simulations in which discrete items are processed by specific operations, the following common processes are additionally present:

1 determining type of event (after retrieval from an event list);
2 calling subroutines to adjust the state variables as a result of the event;
3 identifying specific state conditions;
4 storing and retrieving data from lists (tables or arrays), including the event list and those that represent the state.

Some of the simulation languages are *languages* in the more general sense that, beyond linking the user with the computer as a means of conversing, they afford the user an aid to problem formulation. Having a vocabulary and a syntax, they are descriptive; consequently, users tend, after some utilization (as with other languages), to think in them. Thus Kiviat [12] believes the two most important reasons for utlizing simulation languages as opposed to general-purpose languages, to be programming convenience and concept articulation. Concept articulation is important in the modeling phase and in the overall approach taken to system experimentation. Program convenience becomes important during the actual writing of the computer program. Another advantage of the simulation languages is their use as communication and documentation devices. When written in English-like languages, simulations can be explained more easily to project managers and other non-programming-oriented users. A major cited disadvantage of using simulation languages is that because most were developed by individual organizations for their own purpose and released to the public more as a convenience and intellectual gesture than as a marketed commodity, most users, accustomed to having computer manufacturers do the compiler support work as a service, are not

Table 3.1 *Advantages and disadvantages.*

General Purpose Languages	
Advantages	*Disadvantages*
1) Minimum number of restrictions imposed on format of output. 2) Often already knowledgeable in language.	1) Longer programming time. 2) Debugging of simulation language terms not a feature.
Special Purpose Languages	
Advantages	*Disadvantages*
1) Requires less programming time. 2) Provides error checking techniques superior to those provided in general-purpose languages. 3) Provides a brief, direct vehicle for expressing the concepts arising in a simulation study. 4) Possesses ability to construct and furnish the user subroutines required as a part of any simulation routine. 5) Automatically generates certain data needed in simulation runs. 6) Facilitates collection and display of data produced. 7) Controls management and allocation of computer storage during the simulation run.	1) Must adhere to output format requirements of the language. 2) Reduced flexibility in models and increased computer running time.

set up to do this work themselves. However, more and more well-documented simulation languages are becoming available.

Generally speaking, the advantage of using a special-purpose language vs. a general-purpose is that those features which are assets in one are liabilities in the other. Table 3.1 compares the advantages and disadvantages of general-purpose and special-purpose languages compiled from various sources, primarily Naylor [22] and Evans, Wallace, and Sutherland [5].

3.4 Factors pertinent to language selection

Before a programming language is selected, the computer to be used (type and model) must be determined. Ideally, selection of the computer is one of the decision options open to the analyst. In actual practice, the user probably has available a particular hardware configuration and little latitude as to modification or choice. Once the computer to be used is known, we are ready to select the language. This selection should be a two-phased screening process. The initial phase can be accomplished at any time, even before a particular problem arises. In this phase, language possibilities are examined for their operational characteristics relative to the user's environment and capabilities. The second phase, which is related to the specific problem, must be accomplished after subsystem modeling and computer selection.

In the first phase of the selection process, we are concerned with the availability of references, documentation, and software compatability. Basically, we try to screen the multiplicity of available languages to find those that make good sense for us to consider later when we have a specific problem. The type of question to be answered deals with the general environment in which the analyst finds himself. Since this phase of the screening can be done at leisure and without regard to specific problems, it should be accomplished during slack periods or between projects. The following are among the questions we need to explore:

1 Are intelligibly written user's manuals available?
2 Is the language compiler compatible with available computer systems?
3 Is this language available on other computer systems where the user's problem might be run?
4 Does the language translator provide documentation and extensive error diagnostics?
5 When the organizing, programming, and debugging times are combined with the compiling and execution times, does the efficiency appear attractive?
6 What is the cost of installing, maintaining, and updating

the software for the language? (Since some languages are proprietary, there may be an explicit charge for these services.)

7 Is the language already known or easily learned?

8 Are a sufficient number of simulation studies anticipated for the future to justify the cost of learning and installing the new language?

In the second phase, we must deal with the characteristics of the specific problem at hand. Several different languages have probably survived the phase I screening and are now available for possible use. Phase II, therefore, deals with choosing the specific language to be used on the specific problem at hand, with the specific computer to be used. The following are among the issues to be dealt with in this phase:

1 What is the range and applicability of the world view of the language?

   a What are its time advance methods?

   b Is it event-, activity-, or process-oriented?

   c What is its random number and random variate generation capability?

2 How easily can state and entity variable data be stored and retrieved?

3 What is the flexibility and power provided by the language to modify the state of the system?

4 How easily can it be used to specify dynamic behavior?

5 What are the forms of output available, what are their utilities, and what statistical analyses can be performed on the data?

6 How easy is it to insert user-written subroutines?

The term "world view" appears in many publications describing simulation languages. It describes the way the language designer conceptualized the systems to be modeled using that language. Each simulation language has such an implicit view of the world that must be invoked when we use it. The world view of a typical discrete-change simulation language might be expressed as follows:

1 The world is viewed as a set of *entities* that may be modified or qualified by their characteristics, called *attributes.*

2 The entities interact with specific *activities* of the world consistent with certain *conditions*, which determine the sequence of interactions.

3 These interactions are regarded as *events* in the system, which result in changes in the *state of the system*.

We might also represent this typical world view sequentially as follows:

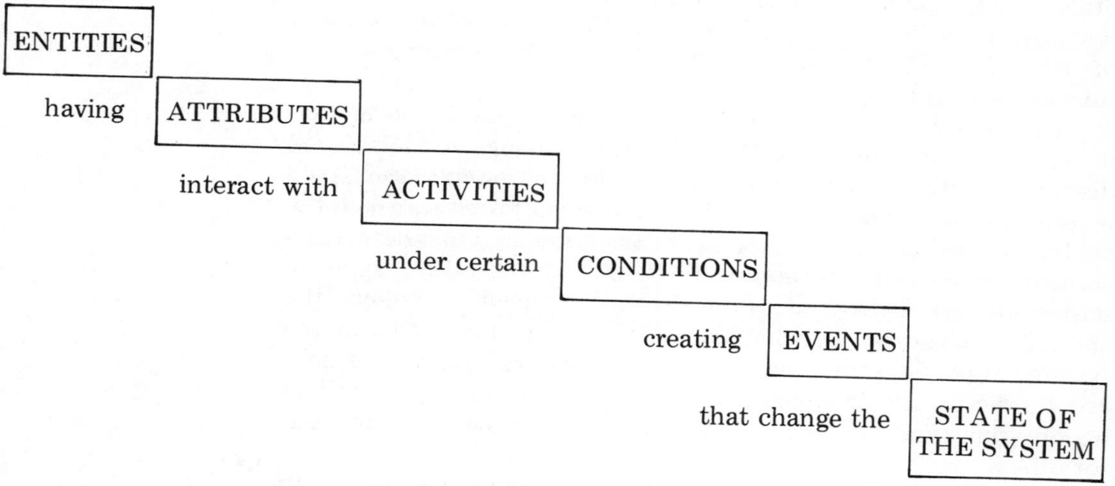

3.5 Time flow mechanisms

Since most simulation studies are concerned with a system's performance over a period of time, one of the most important considerations in designing the model and choosing the language in which to program it is the method used for timekeeping. Timekeeping in a simulation has two aspects or functions: that of advancing time or updating the time status of the system and that of providing synchronization of the various elements and occurrence of events. Since the actions of each element depend upon the state and actions of other elements, they must be coordinated or synchronized in time. Thus, the model must be designed to move through simulated time, causing events to occur in the proper order and with the proper time intervals between successive events. This poses a less than trivial problem for the designer. Although components of a real system function simultaneously, components in a digital simulation model function sequentially, owing to the fact that a digital computer executes its instructions one at a time (i.e., operation is serial) and thus can only consider the system components one at a time. Since events often occur simultaneously in different parts of the real world system, it is necessary to construct a timekeeping system that synchronizes the performance of the system components in the time domain.

Two basic timekeeping mechanisms are available for use: the fixed time increment and the variable time increment methods. They are also some times referred to as *fixed time step* and *next event step*, respectively. The fixed increment method updates the time in the system at predetermined, fixed-length time intervals (the simulation walks through time with a fixed stride). On the other hand, the next event or variable time increment method updates at the occurrence of each significant event, independent of the time elapsing between events (the simulation walks through time on events) [17].

Many writers find it convenient to classify simulation models into two major categories: (1) continuous change models or (2) discrete change models. *Continuous change models* use fixed increment time advance mechanisms and are appropriate when the analyst considers the system he is studying as consisting of a continuous flow of information or items counted in the aggregate rather than as individual items. In *discrete change models*, the analyst is interested in what happens to individual items in the system. Most discrete change models, therefore, utilize the next event type of timekeeping. Some problems are clearly described best by one type or the other, whereas either type might be used for others.

Figure 3.1 illustrates how time is represented and managed in each of the two formulations. The same sequence of events (e_i) are depicted as occurring on each axis in relation to time. The two events e_4 and e_5 are shown occurring simultaneously. The arrows point to the time values at which time is updated and the events appear to occur in each model. For the next event model,

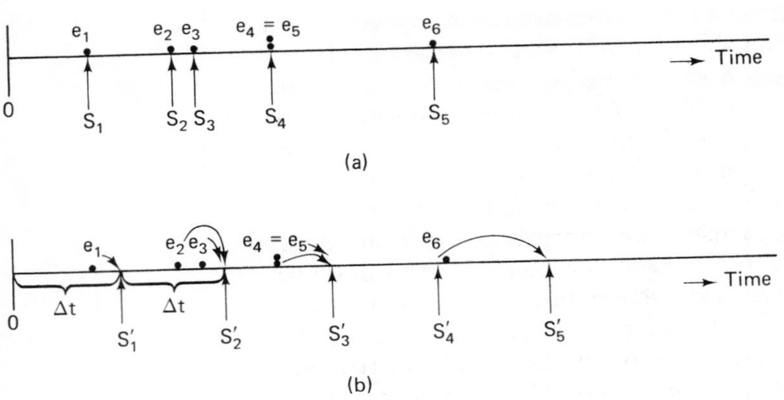

Figure 3.1 Progression of simulation time: (a) next-event model, (b) fixed time increment model.

whenever simulation time is changed it is advanced to the exact time of the earliest of all future event occurrences. The simulation time s_i would in sequence equal

$$
\begin{aligned}
s_1 &= e_1 \\
s_2 &= e_2 \\
s_3 &= e_3 \\
s_4 &= e_4 = e_5 = s_5 \\
s_6 &= e_6
\end{aligned}
$$

where the simulation time values are precisely the values e_1, e_2, ... , at which events occur. By contrast, in the fixed time increment model, the simulation time would successively assume the values

$$
\begin{aligned}
s'_1 &= \Delta t \\
s'_2 &= 2 \Delta t \\
s'_3 &= 3 \Delta t \\
s'_4 &= 4 \Delta t \\
s'_5 &= 5 \Delta t
\end{aligned}
$$

The values of simulation time are not at all dependent on the actual event occurrence times e_1, e_2, ... , which are being simulated. Simulation time is always advanced the same amount, a chosen time quanta Δt. Each method has its advantages. With the next event model, processing of events occurs in sequence, simulation time advancing to the time of the next earliest occurrence, when each occurrence in turn is processed. With a fixed time increment model, the processing of event occurrences does not proceed this way. It proceeds in groups, batches, or sets of event occurrences, as follows.

Given that time is set at some value, say s'_k, then all event occurrences having times e_p, e_q, e_r, ... , such that

$$s'_{k-1} < e_p, e_q, e_r, \ldots \leqslant s'_k$$

are processed before simulation time is advanced to the next value s'_{k+1}. The size of Δt has a critical influence on the simulation, since a positive increment of any size will permit the model to go through the motion of simulating. Erroneous results may be obtained if the size of Δt selected is not correct, because all event occurrences appear to have happened at the point of the upper end of the interval. Figure 3.2 illustrates the radically different results obtained by altering Δt. The next event model does not

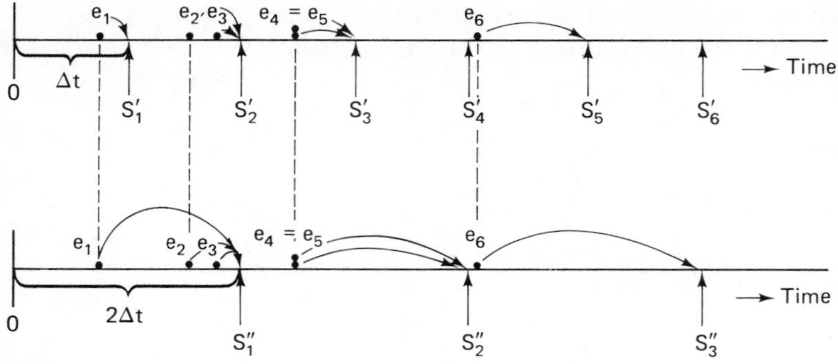

Figure 3.2 Alteration of Δt.

require selection of an artibrary and artificial time increment. Therefore, the possible pitfall of picking an increment that, unknown to the user, would alter his results is avoided. The next event model also has the advantage of processing events as simultaneously occurring only when they bear identical occurrence times. Simultaneity appears to exist only where it should. This avoids having to insert some arbitrary rule in the system for deciding which event that appeared to have occurred simultaneously, occurred first for the purpose of the simulation. This feature must be included in the fixed time increment concept. Should it be desirable for some reason to make a next event model look like a fixed time increment model, this can be done. Each occurrence time e_k can be replaced by the time

$$\left(\text{Greatest integer in } \frac{e_k}{\Delta t} + 1\right) \Delta t$$

The fixed time increment model cannot mimic the next event processing exactly. However, an approximation can be achieved by making the increments increasingly smaller. The time increment must be made small enough so that no two nonsimultaneous event occurrences occupy the same interval and give a false impression of simultaneity. Regardless of how much Δt is squeezed to obtain extremely small increment size, a false impression is still possible owing to the difference, however slight, of registered as opposed to actual occurrence times.

Since a significant portion of the computer run time is consumed by the tracking of events and updating the clock, an improper choice of the timekeeping mechanism can significantly increase the cost of runs. The proper or preferred time flow mechanism to use in a particular model is not easily determined. The following factors must be considered:

1 purpose or goal of the model;
2 precision of results required (usually a function of the first factor);
3 computer run time required for the different methods;
4 computer storage required;
5 programming effort required.

Unfortunately, the comparative answers to the last three factors cannot be determined easily without actually constructing the model both ways. The precision or accuracy required and the purpose of the model are undoubtedly the most important considerations. As we have tried to point out in the preceding discussion, the next event timekeeping procedure is generally the most accurate and precise, in that it keeps track of the exact time and sequence in which events occur. We must say this is true in general only, because the fixed increment method is equally precise and accurate if the increment is made sufficiently small. Here, as in other aspects of model design, a tradeoff exists between computer run time and precision of results.

Likewise, there is no clear cut universal advantage of one method over the other when we consider the computer run times required. Which method is preferable from this viewpoint is a function of the following:

1 the number of system components or entities capable of generating significant events;
2 the number of events per time period generated by each entity (density of events);
3 the mean or average time length of an event (the time from initial commitment to the time the change is completed).

As Conway, Johnson, and Maxwell [3] point out, if a system consists of m components, then in a run of T time units, using the fixed increment method, there will be $T \cdot m$ examinations of individual components to determine if updating is required. If the average length of an event is t time units, there will be $T \cdot m/t$ updatings. This number of updatings is required no matter which method is used. The next event method requires that we find the minimum of a set of m values for each of the Tm/t undatings that would involve $m-1$ comparisons. Thus, we are comparing $(Tm/t)(m-1)$ to Tm, and would prefer the fixed increment method if $t < (m-1)$.

In the early days of computer simulation studies, machine storage capacities presented a severe limitation upon what could and could not be done. But with the advent of large random access memories and rapid access auxiliary storage, this problem has become less important, although still significant. The storage requirements for each of the two methods is intimately intertwined with the listing and recordkeeping strategy used. This aspect is discussed in an excellent paper by R. E. Lave, Jr. [17].

Programming effort required should probably not even be listed as a factor to consider when choosing a timekeeping mechanism. The skills and experience of those doing the programming and the programming language used are probably much more important than the timekeeping mechanisms chosen. In addition, the programming effort required to do something one way vs. that for another is such as individual choice and so subjective that no guidance in this respect can be offered.

Therefore, we can offer no hard and fast rules as to when fixed increment vs. next event timekeeping is preferred. Under certain sets of circumstances each shows distinct strengths and advantages. The final decision depends upon the nature of the particular system being modeled. In general, we should consider a fixed time increment method when:

1 Events occur in a regular and fairly equally spaced manner.

2 A large number of events occur during some simulated time T and the mean length of events is short.

3 The exact nature of the significant events are not well known, such as in the early part of a study.

On the other hand, the next event timekeeping method:

1 saves computer time when the system is static i.e., no significant events occur for long periods of time;

2 requires no decision as to the size of time increment to use (which affects both computation time and accuracy);

3 is advantageous when events occur unevenly in time and/or the mean length of events is long.

Some special simulation languages restrict the user to either fixed increment or next event time flow mechanisms, whereas others allow the use of either.

3.6 Classification of languages

In the most general sense, three computer techniques are available for simulation. As mentioned earlier, these are digital, analog, and hybrid. In this section, we discuss and compare various computer languages that can be used for simulation modeling. Figure 3.3 depicts our classification scheme. Many of these languages have several versions and dialects. We have chosen, therefore, to use generic or family names only instead of listing all the various versions.

We do not discuss the pure analog simulation techniques because they are programmed by the wiring together of electrical components, such as amplifiers, potentiometers, resistors, capacitors, etc., into appropriate networks. Thus, no software or computer programming language is involved. To discuss the programming of analog computers is beyond our scope at present, but several very good books are available [10, 11]. All analog models are continuous time flow models, although the time scale can be speeded up or slowed down by using appropriate scaling factors. Some digital computer programs, such as APACHE and APSE,* can be used to help program and check out analog simulations. Using either of these languages, we can reduce the programming and check-out time on an analog computer. The imput language is equation-oriented in both cases and is an extension of a general-purpose language. Using either of these languages, we can reduce considerably the programming time on an analog computer. They provide a service but do not perform the entire job. Some programmers do use them, however, because they provide the convenience of setup calculation by digital programming yet retain the high speed-low cost problem solution capability of the analog system.

The spectrum of techniques labeled *hybrid simulation* stretches from true hybrid systems, which involve interconnection of analog and digital hardware, to hybridization, in which hybrid simulation occurs at a conceptualization level. As we shall see later, it is possible to cause the digital computer to mimic the analog, thus allowing an all-digital simulation, which performs like a hybrid simulation. Some hybrid techniques are mostly analog in nature and some mostly digital; in between are those capable of providing simulations containing both continuous and discrete events. The early attempts at true hybrid simulation involved general-purpose analog and general-purpose digital computers connected by completely separate interface equipment. This interface equipment linking the general-purpose computers consisted of analog-to-digital and digital-to-analog converters. It contained only the buffering capability that was absolutely necessary and had minimal synchronizing capability. The state-of-the-art has now progressed to the point where true hybrid equipment and

* Appropriate references for each language mentioned are given later.

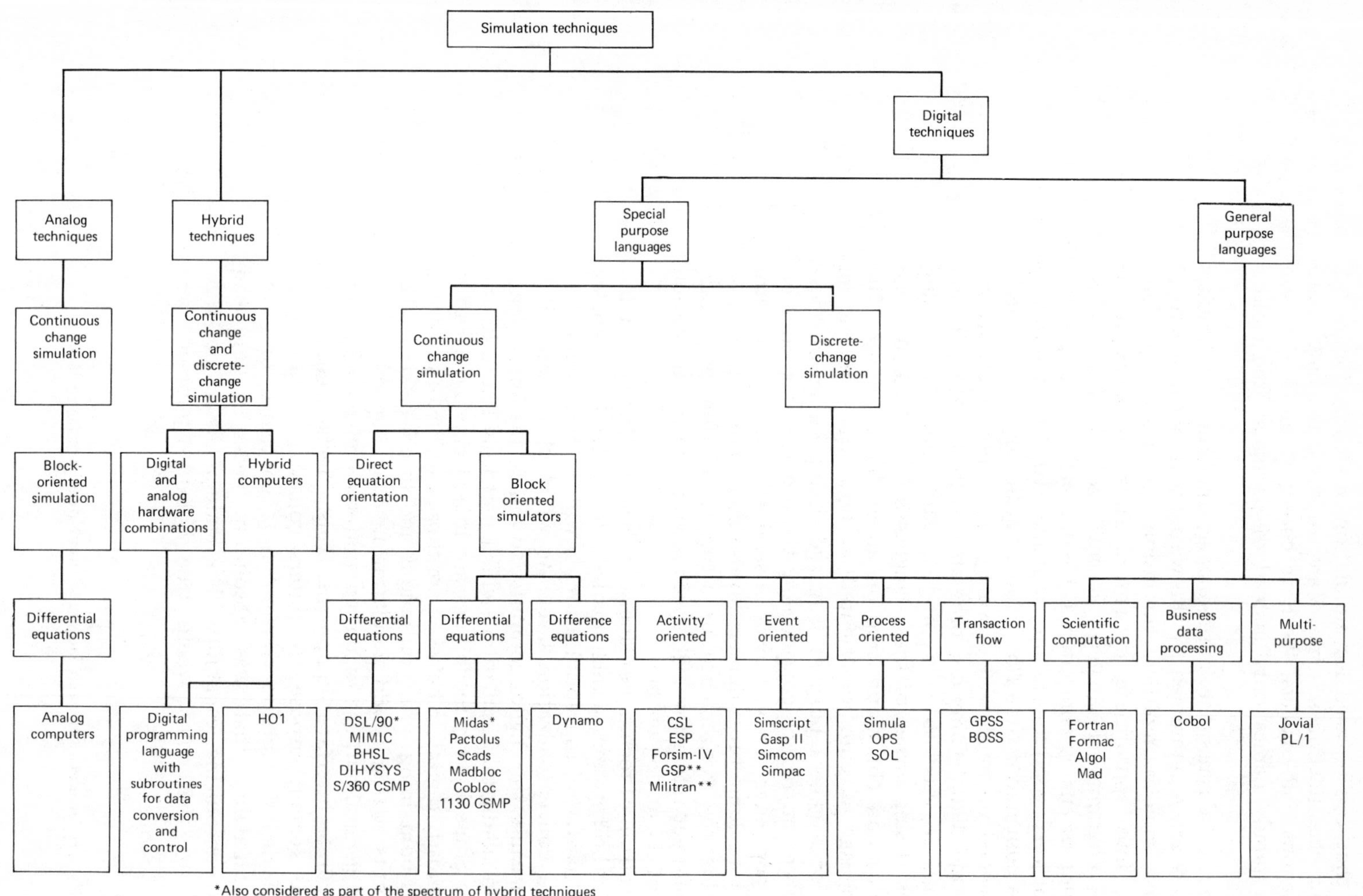

Figure 3.3 Language classification chart.

techniques are practical. Unfortunately, the development of software for these systems has progressed slowly because:

1 Initially, users associated with true hybrid systems were intimately familiar with the hardware and could program their simulations using machine language.

2 The primary users of hybrid systems have been analog programmers, who learned just enough about digital programming to solve their particular problems.

3 The systems have been largely one-of-a-kind, for which major investment in software design, development, and maintenance was not justified.

For these reasons, the majority of true hybrid simulations are accomplished via a procedure-oriented, general-purpose digital language as a supervisory program, often in combination with user-designed subroutines for checkout of the analog portion, data conversions, and control and diagnostics. Some attempts have been made to produce true hybrid simulation languages. HYTRAN was developed to automate the setup and checkout of the first commercially available hybrid system, called HYDAC. Both HYTRAN and HYDAC were developed by Electronic Associated, Inc., in about 1962. Since few systems were sold, HYTRAN was not widely used, but it did lay the foundation for the development of HOI (*H*YTRAN *O*perations *I*nterpreter). This language is used for automation of the setup and checkout of the analog subsystem and for complete executive control of the hybrid program.

Digital simulation techniques include the use of both general and special purpose languages. The relative advantages and disadvantages of each of these classes were presented in Table 3.1. Unquestionably, the use of a general-purpose language offers the programmer maximum flexibility in the design, implementation, and use of his model. However, this flexibility is purchased at the price of greater difficulty in programming. Most of this difficulty derives from the problems of sequencing, timekeeping, and control. Nevertheless, using a general-purpose language may be the only alternative available to the analyst who does not have access to a large-scale computer, since most of the existing special purpose languages were written for use with large-scale machines. In addition, many skilled and experienced programmers prefer general-purpose languages for the flexibility of output report format and the ability to write their own subroutines. Since a great deal of material has already been published on the general-purpose languages, we limit ourselves here merely to including them in our classification scheme.

The special purpose digital simulation languages include two distinct groups or types of simulation, which developed separately. These groups (the simulation of discrete-change and continuous-change processes) will be discussed separately, along with their historical development. We take up continuous-change simulation techniques first. One technique included in this category represents the end of the hybrid spectrum, in which hybridization occurs at the conceptualization level (i.e., the digital computer mimics a hybrid computer). The category of continuous-change simulation languages can be divided into three types of languages: the analog simulator languages; the language geared to the solution of difference equations representing a deterministic, closed-loop, continuous system; and the equation simulator languages. The first two types are block-oriented languages and the third is constructed directly from equations.

The continuous-change simulation languages are all descendants of the work of Selfridge in 1955. His unnamed program has proved to be the inspiration for the large field of analoglike simulation languages for digital computers. The most useful analog simulator languages include MIDAS, PACTOLUS, SCADS, MADBLOC, COBLOC, and 1130 CSMP. These languages emulate the behavior of analog and hybrid computers on a component-by-component basis. For example, a summer is replaced by a summation operational code, an integrator by an integration operational code, etc. These complements of functional components have their interconnections specified by a block-oriented language just as the analog computer patchboard electrically links analog computing components. The organization as well as the elements of the analog or hybrid computer are emulated. The analog simulator languages all draw their inspiration and motivation from the analog block diagram as a simple and convenient means for describing continuous systems.

Three languages fit the niche for a language helpful in modeling variables continuous in value but discrete in time. DYNAMO (*D*ynamic *M*odels) was developed at M.I.T., using first-order difference equations to approximate the continuous process because they are more practical than differential equations in representing time sequences of system input and output. The essential variables of a dynamic systems study, state variables and output variables, are described in DYNAMO by level equations and rate equations, respectively. The state variables (levels) describe the state or condition of the system at a given time; the rate variables (rates) describe how the states change with the passage of time. Auxiliary equations, components of the rate equations, completely describe the function of the rate equations. Although the system is described by the level and rate equations, the auxiliary equations are very important in representing feedback control, since they form the basis for control of the rates. If DYNAMO is not available, the MIMIC or CSMP can also be used.

The equation-based languages break away from the restrictions imposed by the complete block construction of the analog simulator languages and deal directly with the equations. In addition to the block modeling of the analog simulator languages, these add the power and convenience of algebraic and logical statements. The equation-oriented languages are a more recently developed family than the analog simulator languages. In 1966, the Simulation Software Committee of Simulation Councils, Inc., presented preliminary specifications for a Continuous System Simulation Language. The purpose was to standardize the language format and structure of digital analog simulator programs. It was hoped that future benefits would be comparable to those achieved by the American Standards Association, Standards Committee on FORTRAN. Just as there are considered to be first- and second-generation discrete-change languages, the continuous-change languages that emerged after these specifications were published showed a different orientation. Most of the languages developed after the publication of the specifications have a direct equation orientation. Some of the most useful of these include DSL/90, MIMIC, BHSL, DIHYSYS, and S/360 CSMP.

Kiviat [12] traces the separation of early discrete-change simulation theory into two schools: those schools introduced by IBM with their GPSS language that used flow chart symbols as basic model descriptors; and statement-oriented schools. In general, flow chart languages are easier to learn, but the statement-oriented languages are usually more flexible. Most of the newer languages are statement languages even though flow chart language is appealing and, in addition to GPSS, has been used in SIMCOM and BOSS. In our classification scheme, we have used four subcategories: activity, event, process, and transaction flow orientations. Transaction flow languages are actually process languages, since they take a synoptic view of systems, but we have established them as a separate category because of their flow chart orientation. Event, activity, and process languages (except SIMCOM) use programming statements to describe cause and effect relationships between the system elements.

The activity-oriented languages represent time-dependent acts as instantaneous occurrences in simulated time. In using these languages, we do not schedule the occurrences within a program, but specify under which conditions they can happen. No "activity scheduling" statements appear in these languages, but they contain executive programs that scan sets of conditions before each simulation time advance to determine whether any activities can take place. In this type of language, the program is composed of a test section and an action section. Whenever simulation time is advanced, all activity programs are scanned for possible performance. All test conditions must be met for the state-changing and time-setting instructions in the action section to be carried out. Should one of the test conditions not be met, the action instructions are passed over. By cyclic scanning of activity programs, we ensure that all possibilities have their opportunity to take place and that all interactions are accounted for. Included as activity-oriented languages are CSL, ESP, FORSIM-IV, GSP, and MILITRAN.

It may be that the problem at hand can be more efficiently written in an event-oriented language. Each event must be represented as an instantaneous occurrence in simulated time, scheduled to occur when it is known by the dynamics of the model that the proper conditions exist for its occurrence. Separate programs are required for, say, a man, a machine, and a part that interacts. An executive program automatically sequences the scheduled events so that they occur properly in simulated time. Event-oriented languages include SIMSCRIPT, GASP, SIMCOM, and SIMPAC.

Process-oriented languages attempt to combine the concise notation of the activity-oriented languages and the efficiencies of the event-oriented languages. A set of events associated with the behavior description of a system is a process. A process exists over time and can have a dynamic behavior. Processes are flexible, can be scheduled to occur, can be interrupted, have subprocesses that obey them, can be so programmed that they can delay themselves and other processes until certain conditions are met, etc. A key feature is that the process-oriented languages make a single program act as though it were several programs, independently controlled by either activity-oriented scans or event-oriented scheduling. The programming feature making this concept possible is the reactivation point—essentially a pointer that tells a process routine where to begin execution after some time-delay command has been carried out. The executive program does more in these languages than in either the activity-oriented or event-oriented languages. Included in this category are SIMULA, OPS, and SOL.

Although flow charts are often a vital tool in describing the logic and interrelation of the elements of systems modeled in the three previously described types of discrete-change languages, they are not basic to their programming or theory. In them, the user constructs a simulation model by writing statements which define conditions that must hold for certain actions to transpire, describe results of these actions, and specify the time relationships between system elements and activities in which they participate. In the last type of language described, the concepts of the process-oriented languages apply, but the systems are modeled by tracing flows of transactions through stylized activity blocks. Simulated time advances as the transactions pass through the blocks, and decisions are made as the logic of the simulated system is reproduced. The languages fitting this category are termed transaction flow languages. A person already familiar with flow chart concepts finds it easy to learn to model in these languages. Specialized blocks are assembled into structures representing the logic and flow of the system being modeled. The system is represented in terms of the blocks; the program creates transactions, moving them to the specified blocks, and executes the actions associated with each block. Since these specialized modeling blocks are also the basic programming statements, construction of a flow chart model is equivalent to writing a program. The price paid for the ease of learning to use these languages is a loss in flexibility. Languages in this category are GPSS and BOSS.

3.7 Decision flow diagram

The two-phased screening process proposed in the section entitled Factors Pertinent to Language Selection and the classification scheme discussed in the preceding section provide us with the background for the presentation of another aid in selecting a simulation language. In this section, we present a decision tree used in conjunction with a scramble book approach for language selection. Prior to using the decision tree, the analyst should have accomplished the earlier steps of analyzing the system, breaking it down into subsystems, and modeling the subsystems. He is at that point sufficiently knowledgeable about his particular problem to be able to select a language based upon what he wants the simulation to do.

The information discussed in previous sections can be arranged in the flow diagram form of Fig. 3.4. Hopefully, this decision tree and the accompanying discussion and information in Section 3.8 will allow the analyst to diagnose easily which language is most applicable to solve his problem. Termination in a block suggests or recommends that the languages contained in that block be seriously considered. At each step of the decision tree is a number in the left-hand side of the element. It is a numerical key to the appropriate section of a complementary scramble book approach, presented in Section 3.8.

The scramble book discusses each decision point and tries to help guide the user to a termination block. Upon reaching a termination point, the reader will find certain languages recommended, hardware with which they are compatible, and references

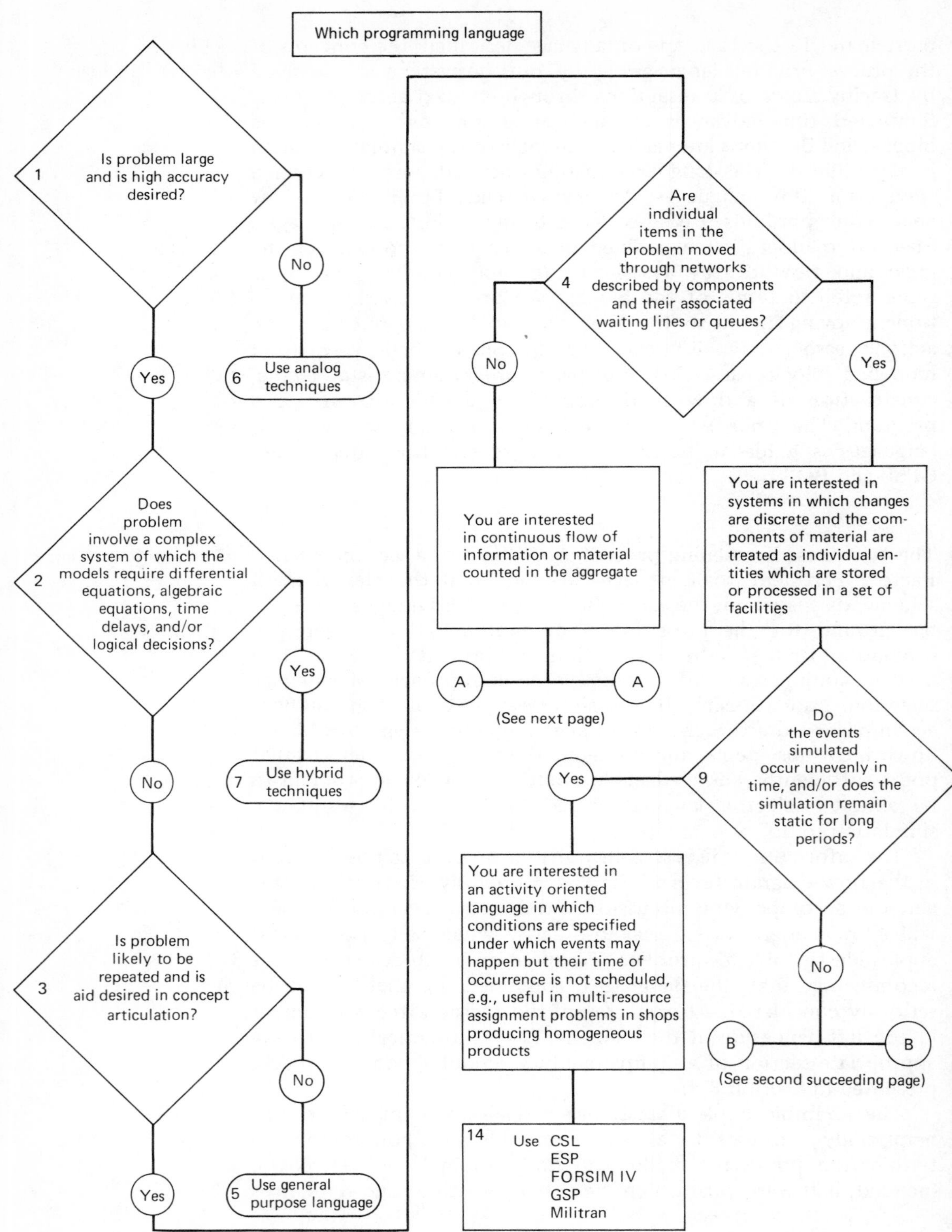

Figure 3.4 Decision flow diagram.

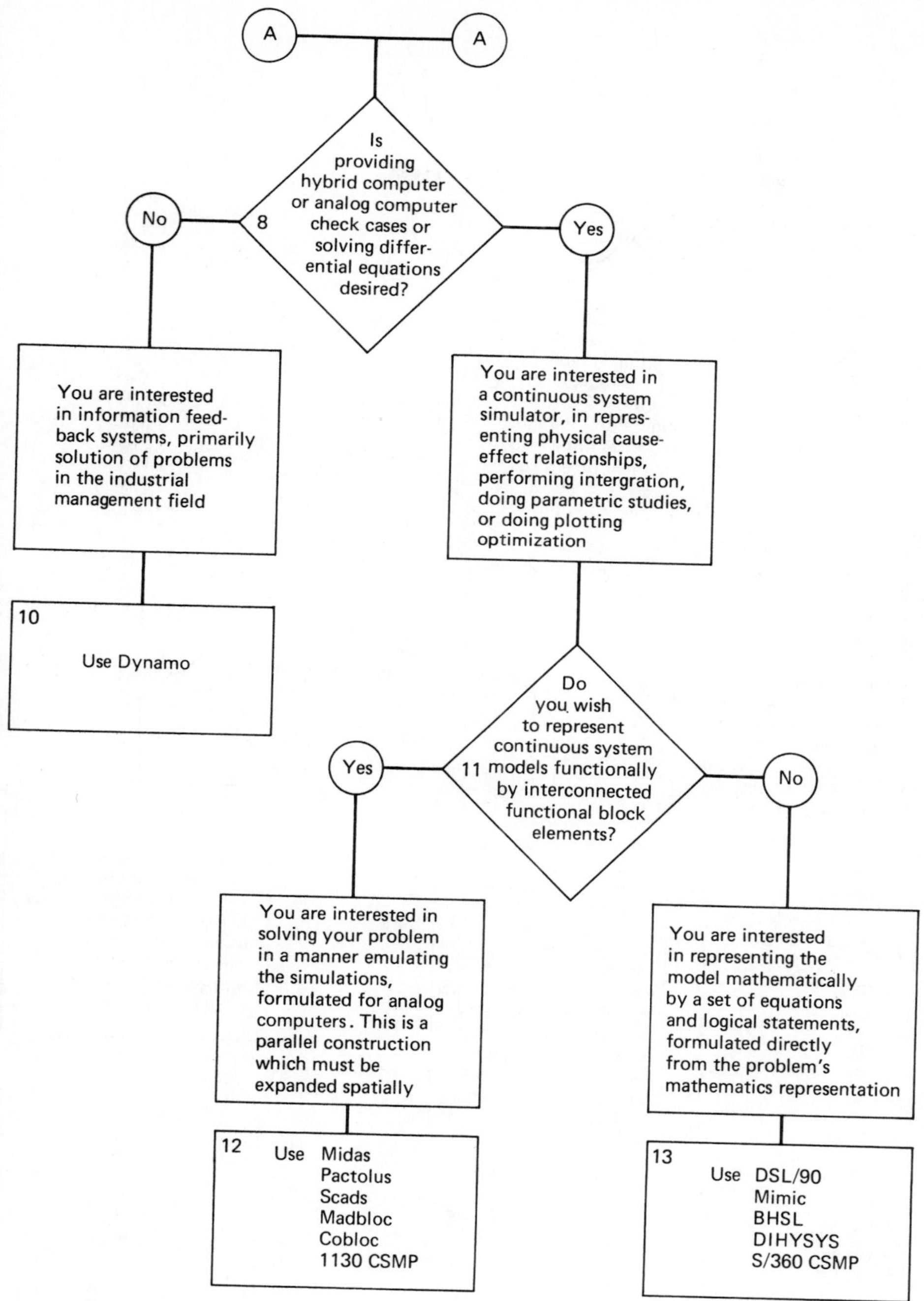

Figure 3.4 *cont.*

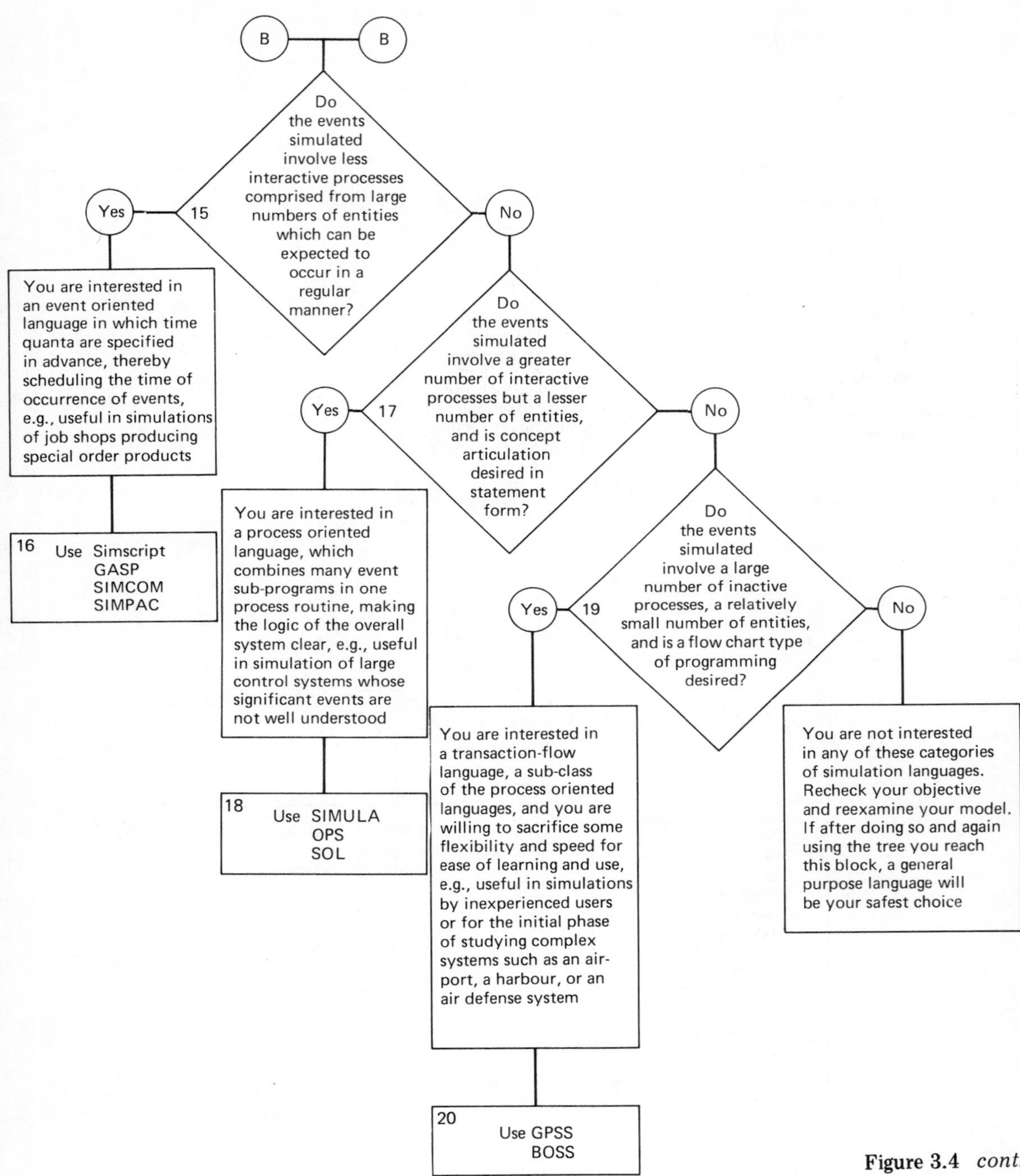

Figure 3.4 *cont.*

to assist him in finding more information about the languages he has selected. Obviously, not all possible languages are listed for each termination point; we have attempted to list only the more widely used and well-documented languages. When several languages accomplish essentially the same purpose or utilize similar features, the listings reflect the more popular of the languages and/or those which are compatible with the widest range of hardware to which the reader may have access. In other words, a conscious effort is made to list sufficient languages in each category (where they exist) to permit the analyst to find a language available on his hardware. We recognize that developers and users of languages that have been omitted will be disappointed that they were not included. For these omissions we apologize, and offer the excuse that space limitations forced the exclusion of many other excellent and useful languages.

3.8 Scramble book

The following descriptions and discussions are numerically keyed to the decision tree shown in Figure 3.4. To commence the selection of a language *go to Block 1.*

BLOCK 1

Underlying the desire to select a programming language might be the assumption that the problem is large, complex, and that there is an overriding requirement for high precision. Is this the case?

1 Yes. *Go to Block 2.*

2 No. *Go to Block 6.*

BLOCK 2

If the simulation is one in which the information flow in the model is both continuous and discrete; if the overall system requires high precision, but there are numerous inner loops not readily lending themselves to digital approaches; or if both differential equations and algebraic equations or logical decision are required; then hybrid techniques are recommended. Is this true?

1 Yes. *Go to Block 7.*

2 No. *Go to Block 3.*

BLOCK 3

Table 3.1 compared the advantages and disadvantages of general-purpose languages and special-purpose simulation languages. Essentially, if the problem is to be repeated and if aid is desired in concept articulation, then a special purpose simulation language is a better choice. Are these requirements met?

1 Yes. *Go to Block 4.*

2 No. *Go to Block 5.*

BLOCK 4

If it is possible to view the model as containing individual items that are moved through networks, and if the items being processed or stored, as well as the elements doing the processing, should be represented as individual entities as opposed to comprising an aggregate, then one of the discrete-change languages is indicated. Is this the case?

1 Yes. *Go to Block 9.*

2 No. *Go to Block 8.*

BLOCK 5

You are interested in using a general-purpose or procedural-type language. You have very few restrictions placed on what you can do with such languages. The major shortcomings you must keep in mind are that you will not receive hèlp in concept articulation and debugging aids to check your simulation logic will not be available. Versions of these languages are compatible with most hardware systems. Pertinent references to assist you in writing simulation programs in these languages are given in references at the end of this chapter [1, 6, 7, 8, 15, 18, 19, 20, 23, 24, 26].

BLOCK 6

You appear to be interested in using analog techniques. These techniques are useful in some of the following applications:

1 Classroom demonstrations. Dramatic illustrations of the effect of changing parameters and variables.

2 System predesign. Effects of various tradeoffs investigated and demonstrated for review boards or customers.

3 Solution of large set of simultaneous differential equations. High speed is more important than great precision.

4 Empirical simulation development. Insufficient a priori information to support design of a mathematical model.

5 Man-machine interaction. Systems where intimate rapport between the user and the machine is essential.

No language is used in the simulation itself; however, some digital languages are helpful in calculating the proper scale factors, potentiometer settings, and patchboard connections, as well as in running static checks. *Go to Block 11* and/or *Go to the next section of this chapter and see APACHE and APSE.*

BLOCK 7

You are interested in using hybrid techniques. If you have available either the medium scale EAI 690 or the larger EAI 8900 Hybrid Computing Systems, then you can use language HOI. If you have available a digital and analog hardware combination, then you will use a digital language, such as FORTRAN, with a library of subroutines for data conversion and control.

Should you desire checks on your system or aid in calculating the setup of the analog computer portion, you may wish to see Block 11 or APACHE and APSE in the next section of this book. References are [2, 9, 11, 16]. *Go to these references and/or Go to the next section of this chapter and see APACHE and APSE.*

BLOCK 8

Having decided that you are interested in systems considered in the aggregate, and in the continuous flow of information or material, you now must decide whether providing check cases for hybrid computers or analog computers or solving differential equations fits what you want to accomplish. Does it?

1 Yes. *Go to Block 11.*

2 No. *Go to Block 10.*

BLOCK 9

You are attempting to decide which method of timing or synchronization will best suit your simulation. If the mean length of events is relatively long, if the events occur unevenly in time, and/or if the simulation remains static for long periods, then you should use a next event type of mode. Activity-oriented languages will be most efficient with next event type models. Does this fit your problem?

1 Yes. *Go to Block 14.*

2 No. *Go to Block 15.*

BLOCK 10

You are interested in information feedback systems. They may be econometric models, models of a firm, models of all the firms of a particular industry, models of medical facilities, models of environmental problems, models of military engagements, or a variety of other applications. Whatever the model's field, the model is of an aggregate nature. The best language available is DYNAMO, although MIMIC and CSMP (see Block 13) can also be used. DYNAMO is compatible with IBM 709, 7090; Burroughs B5500; Nippon Electric Company NEAC 2200/200; UNIVAC 1107, 1108; and IBM 360. A multitude of references are available: T. H. Naylor, *et al.*, *Computer Simulation Techniques*, John Wiley & Sons, Inc., New York, 1968, pp. 32, 38, 124, 188, 220, 222, 233; M. Klerer and G. Korn, *Digital Computer User's Handbook*, McGraw-Hill Book Company, 1967, pp. 270–73; J. E. Sammet, *Programming Languages*, Prentice-Hall, Inc., Englewood Cliffs, N.J., 1964, pp. 651–53; R. B. Vail, *Military Applications of DYNAMO*, unpublished master's thesis, Georgia Institute of Technology, 1968; A. L. Pugh, *DYNAMO II User's Manual*, Massachusetts Institute of Technology Press, 1970; J. W. Forrester, *Industrial Dynamics*, Massachusetts Institute of Technology Press, 1961, pp. 369–81; and W. G. Browne, "Industrial Dynamics—Looking at the Total Firm," *Fourth Annual Simulation Symposium*, Mar. 1971, pp. 135–65. *Go to References.*

BLOCK 11

You have decided that your interest lies in a continuous system simulator, in representing physical cause-effect relationships, performing integration, doing parametric studies, solving systems of differential equations, or plotting optimization. Now you are interested in deciding between two basic language types, both of which solve these problems. If you wish to represent your continuous system model functionally by interconnected functional block elements, then you wish to use a block-oriented language. Is this true of your problem?

1 Yes. *Go to Block 12.*

2 No. *Go to Block 13.*

BLOCK 12

You are interested in solving your problem in a manner emulating the simulations formulated for analog computers. The components of an analog computer are replaced component-by-component by operational codes. The interconnections accomplished by the electrical linking of the patchboard of the analog computer are accomplished by the block-oriented language. Languages of this type are hereby listed, with hardware compatible with them.

MIDAS	IBM 7090/94, 7040/44, 1620
	UNIVAC 1107, 1108
PACTOLUS	IBM 1620
SCADS	CDC G-20
MADBLOC	IBM 7090
COBLOC	CDC 1604
1130 CSMP	IBM 1130

A few references will be listed for each language:

MIDAS: R. J. Harnett, F. J. Tansom, and L.M. Warshwasky, "MIDAS . . . An Analog Approach to Digital Computation," *Simulation*, Sept., 1964, pp. 16—43; G. E. Blechman, "An Enlarged Version of MIDAS," *Simulation*, Oct., 1964, pp. 41—54; and G. H. Burgin, "MIDAS III . . . A Compiler Version of MIDAS," *Simulation*, Mar., 1966, pp. 160-68.

PACTOLUS: R. D. Brennan and H. Sano, "PACTOLUS: A Digital Analog Simulation Program for IBM 1620," *Proceedings of the Fall Joint Computer Conference*, 1964, pp. 299-312; R. D. Brennan, "PACTOLUS—A Simulator Language Which Makes a Digital Computer Feel Like an Analog Computer," *Simulation*, Aug., 1964, pp. 12-19; and G. C. Willems, "A Note on PACTOLUS," *Simulation*, Vol. 3, No. 5, 1964, pp. 3-4.

SCADS: D. A. Gall, "Introduction to Continuous System Simulation," *Proceedings of the Summer Computer Simulation Conference*, 1970, pp. 3-9; and J. C. Strauss and W. L. Gilbert, *SCADS, A Programming System for the Simulation of Combined Analog Digital Systems*, 2nd ed., Carnegie Institute of Technology, Pittsburgh, Pa., Mar., 1964.

MADBLOC: V. C. Rideout and L. Tavernini, "MADBLOC, A Program for Digital Simulation of a Hybrid Computer," *Simulation*, Jan., 1965, pp. 20-24.

COBLOC: J. J. Skiles, R. M. Janoski, and R. L. Schaefer, "COBLOC," paper presented at Joint Meeting of Midwestern and Central States Simulation Councils, May, 1965; and R. M. Janoski, R. L. Schaefer, and J. J. Skiles, "COBLOC: A Program for All Digital Simulation of a Hybrid Computer," *IEEE Transactions on Electronic Computers*, Feb., 1966, pp. 78-82.

1130 CSMP: R. D. Brennan, "Continuous System Modeling Programs: State of the Art and Prospectus for Development," *Proceedings of the IFIP Working Conference on Simulation Programming Languages*, 1968, pp. 371-96; and *1130 CSMP Program Reference Manual*, IBM 1130-CX-13X.

Go to References.

BLOCK 13

You probably already have a mathematical representation formulated; therefore, to reformulate the model in functional terms would be an extra step requiring considerable time and effort. This extra step is eliminated when an equation-oriented language is used. Languages of this type are listed, with hardware compatible with them.

DSL/90	IBM 7090/94
MIMIC	IBM 7090/94
	UNIVAC 1107, 1108
	CDC 6000 series
BHSL	EAI 8400/8900
DIHYSYS	Burroughs B5500
S/360 CSMP	IBM 360

A few references are listed for each language:

DSL/90: Y. Chu, *Digital Simulation of Continuous Systems*, McGraw-Hill, Book Co., N.Y., 1969, pp. 345-417; and W. M. Syn and D. G. Wyman, *DSL/90—Digital Simulation Language, User's Guide*, IBM System Development Division, Technical Report TR 02. 355, Jul., 1965.

MIMIC: H. E. Peterson, F. J. Sansom, and L. M. Warshawsky, *MIMIC—A Digital Simulation Program*, SESCA Internal Memo 65-12, Wright-Patterson Air Force Base, Ohio, May, 1965; and Y. Chu, *Digital Simulation of Continuous Systems*, McGraw-Hill Book Co., N.Y., 1969, pp. 24-64.

BHSL: J. C. Strauss, "Basic Hytran Simulation Language—BHSL," *AFIPS Conference Proceedings*, Vol. 29, 1966, pp. 603-11; and *HYTRAN Simulation Language (HSL) Programming Manual*, Electronic Associates, Inc., Publ. No. 008270006-0, 1967.

DIHYSYS: J. Leon, C. O. Alford, and J. L. Hammond, Jr., "DIHYSYS . . . A Hybrid Systems Simulator," *Proceedings of the 1970 Summer Computer Simulation Conference*, Jun., 1970, pp. 42-48.

S/360 CSMP: R. D. Brennan, "Continuous System Modeling Programs: State-of-the-Art and Prospectus for Development," *Proceedings of the IFIP Working Conference on Simulation Programming Languages*, 1968, pp. 371-96; G. A. Pall, *Introduction to Scientific Computing*, Appleton-Century-Crofts, N.Y., pp. 588-600; and *Continuous System Modeling Program*, IBM Data Services Center, Report GE19-0036-0, 1970.

Go to References.

BLOCK 14

Your model uses a next-event or variable time increment type of timing mechanism. Your problem contains highly interactive processes involving a fixed number of entities with events happening irregularly. You desire an activity-oriented language in which activity scan specifies event occurrence. Languages of this type are listed with hardware compatible with them.

CSL	IBM 7090/94, 1620, 1410
	Honeywell 400/1400, 200/2200
ESP	Elliot 503, 803
FORSIM IV	IBM 7030
GSP	Elliot 503
	Ferranti Pegasus
MILITRAN	IBM 7090/94

A few references are listed for each language:

CSL: J. N. Buxton and J. G. Laski, "Control and Simulation Language," *Computer Journal*, Vol. 5, 1964, pp. 194-99; *Control and Simulation Language—Introductory Manual*, IBM United Kingdom, Ltd., and Esso Petroleum Co., Ltd., Mar., 1963; and *Control and Simulation Language—Reference Manual*, IBM United Kingdom, Ltd., and Esso Petroleum Co., Ltd., Mar., 1963.

ESP: J. W. J. William, "E.S.P.—The Elliot Simulator Package," *Computer Journal*, Vol. 6, 1964, pp. 328-31.

FORSIM IV: E. Famolari, *FORSIM IV Simulation Language User's Guide*, The MITRE Corp., ESD-TDR-64-108, May, 1964.

GSP: K. D. Tocher, *Handbook of the General Simulation Program*, Vol. I (rev.) and Vol. II, Dept. Operations Research and Cybernetics, Report 77/ORC 3/Tech. and Report 88/ORC 3/Tech, The United States Steel Companies, Ltd., Sheffield, England.

MILITRAN: *MILITRAN Reference Manual*, Systems Research Group, Inc., 1964; *MILITRAN Programming Manual*, Systems Research Group, Inc., 1964; and *MILITRAN Operation Manual*, Systems Research Group, Inc., 1964.

Go to References.

BLOCK 15

You have decided that your system can best be modeled using a fixed time increment timing mechanism. Now you are interested in deciding between an event-oriented language and a process-oriented language or a transaction-flow language. If there are a relatively large number of entities and a relatively small number of interactive processes, then an event scheduling language is indicated. Is this the case?

1 Yes. *Go to Block 16.*

2 No. *Go to Block 17.*

BLOCK 16

You are interested in a language in which time quanta are specified and events are thus scheduled as to time of occurrence. The major differences between models that can most efficiently use such languages and other models, also using the fixed time increment concept, are characterized by the degree of interactive processes involved and the number of entities. Your model contains a lesser degree of interactive processes and a larger number of entities. Event-oriented languages are listed, with compatible hardware.

SIMSCRIPT	IBM 7090/94, 360
	UNIVAC 1107, 1108
	CDC 3600, 6000 series
	GE 626, 635
	Philco-Ford 210, 211, 212
	RCA 70/45/55
	NCR 314 RMC
GASP	Any system with a FORTRAN IV compiler
SIMCOM	IBM 709
SIMPAC	IBM 7090

A few references are listed for each language:

SIMSCRIPT: F. P. Wyman, *Simulation Modeling: A Guide to Using SIMSCRIPT*, John Wiley & Sons, Inc., New York, 1970; M. A. Geisler and H. M. Markowitz, *A Brief Review of SIMSCRIPT as a Simulating Technique*, the Rand Corporation, Santa Monica, Cal., Aug., 1963; H. M. Markowitz, *Simulating With SIMSCRIPT*, the Rand Corporation, Santa Monica, Cal., Nov., 1963; H. W. Karr, *A Quick Look at SIMSCRIPT*, paper presented at Fourth Phi Delta Kappa Symposium in Educational Research, Nov., 1962; H. M. Markowitz, B. Hauser, and H. W. Karr, *SIMSCRIPT—A Simulation Programming Language*, Prentice-Hall, Inc., Englewood Cliffs, N.J., 1963; and P. J. Kiviat, R. Villauveva, and H. Markowitz, *The SIMSCRIPT II Programming Language*, The Rand Corporation, Santa Monica, Cal., Oct., 1968.

GASP: A. A. B. Pritsker and P. J. Kiviat, *Simulation With GASP II: A Fortran Based Simulation Language*, Prentice-Hall, Inc., Englewood Cliffs, N.J., 1969; and P. J. Kiviat and A. Colker, *GASP—A General Activity Simulation Program*, The Rand Corporation, Santa Monica, Cal., 1964.

SIMCOM: T. G. Sanborn, "SIMCOM—The Simulation Compiler," *Proceedings of the Eastern Joint Computer Conference*, 1959, pp. 139-42; and *SIMCOM User's Guide*, Information Systems Operations, TR-65-2-149010, General Electric Company, 1964.

SIMPAC: R. P. Bennett, *et al.*, *SIMPAC User's Manual*, System Development Corporation, TM-602/000/00, Apr., 1962; and M. R. Lackner, "Toward a General Simulation Capability," *1962 Western Joint Computer Conference*, pp. 1-14.

Go to References.

BLOCK 17

You have decided that your model can most efficiently be programmed in a language favoring a greater number of interactive processes. The decision facing you now is whether you prefer a statement type language or one in which programming is accomplished by the flow charting function itself. If your system is quite large or if you cannot afford the inflexibility of the fixed nature of the blocks in a flow chart language, you will desire a process-oriented language. Is this the case?

1 Yes. *Go to Block 18.*

2 No. *Go to Block 19.*

BLOCK 18

You are interested in a process-oriented language. More particularly, you do not want to be tied down by the inflexibility of using only certain types of blocks to describe your system as with the transaction flow languages. Nor do you want to preallocate much of the available memory space of the computer as is required for a block-structured language. Languages fitting your requirement are listed, with compatible hardware.

SIMULA	UNIVAC 1107, 1108
	Burroughs B5500
	CDC 6000 series
OPS	IBM 7090/94
SOL	UNIVAC 1107, 1108
	Burroughs B5500

A few references are listed for each language:

SIMULA: O. J. Dahl and K. Nygaard, "SIMULA—An ALGOL Based Simulation Language," *Communications of the ACM*, Sept., 1966, pp. 671-78; O. J. Dahl and K. Nygaard, "Class and Subclass Declarations," *Proceedings of the IFIP Working Conference on Simulation Programming Languages*, 1968, pp. 158-74; *SIMULA Programmer's Reference*, UNIVAC UP-7556, Sperry Rand Corporation, 1967; and O. J. Dahl, B. Myhrhaus, and K. Nygaard, *SIMULA 67 Common Base Language*, Norwegian Computing Center, Report S2, May, 1968.

OPS: M. Greenberger, "A New Methodology for Computer Simulation, *Project MAC*, MAC-TR-13, MIT, 1964; and M. Greenberger and M. Jones, "On Line, Incremental Simulation," *Proceedings of the IFIP Working Conference on Simulation Programming Languages*, 1968, pp. 13-32.

SOL: D. E. Knuth and J. L. McNeley, "SOL—A Symbolic Language for General Purpose Systems Simulation," *IEEE Transactions on Electronic Computers*, 1964, pp. 401-14.

Go to References.

BLOCK 19

Are you still interested in a system containing a relatively large number of interaction processes and a relatively small number of entities? Is a flow chart type of language desirable to you to cut down on the amount of time required in programming and debugging, and/or is such a language desirable because it is relatively quick and easy to learn? If you answer yes, your requirements can be satisfied by a transaction flow language. Is this the case?

1 Yes. *Go to Block 20.*

2 No. *Go to Block 21.*

BLOCK 20

You are either relatively inexperienced in programming and desire the ease of using this type of language, or you wish to use this language in the initial phase of studying a complex system. The attractiveness of the block-structured language for the purpose of your problem is not overshadowed by its relative inflexibility, its greater memory space requirements, and its slower running time. Languages of the transaction flow type are listed with compatible hardware.

GPSS	IBM 7090/94, 7040/44, 360
	UNIVAC 1107, 1108
	CDC 3600, 6000
BOSS	Burroughs B5500

A few references for each language are:

GPSS: G. A. Pall, *Introduction to Scientific Computing*, Appleton-Century-Crofts, Inc., New York, 1971, pp. 574-87; *General Purpose Simulation System/360 User's Manual*, IBM H20-0326, 1967; and J. R. Emshoff and R. L. Sisson, *Design and Use of Computer Simulation Models*, The MacMillan Company, New York, 1970, pp. 117-19, 144-45, 234-38, 266-70.

BOSS: P. F. Roth, "The BOSS Simulator—An Introduction," *Fourth Conference on Applications of Simulation*, Dec. 1970, pp. 244-50; and P. F. Roth and A. J. Meyerhoff, "BOSS Simulation of a Time Sharing Message Processing System for Bank Applications," *Proceedings of the Third Annual Simulation Symposium*, Jan., 1970.

Go to References.

BLOCK 21

You have not settled on a categorization that fits your problem. You should now recheck the objective of your simulation, reexamine your model, and again attempt to use the decision tree. If after so doing you again arrive at this block, recognize that you are dealing with a problem which can more safely be programmed using a general-purpose language. Is this the first time you have reached this block concerning this simulation problem?

1 Yes. *Go to Block 1.*

2 No. *Go to Block 5.*

3.9 Useful extensions and dialects

Since the requirements of programming languages in general and simulation languages in particular are not rigorously defined, there is little agreement on what are or are not truly languages. In addition to languages in their own right, there are dialects, language or *L*-like languages, and language extensions. A dialect is a minor variation of a particular language. An *L*-like language is a language similar in spirit and notation to language *L*, but differing in its features sufficiently that it cannot be considered a dialect. The prime distinction between the two definitions is one of degree. If it is a minor variation, it is a dialect. A language *E* is an extension of language *L* if *L* is a subset of *E*.

This presents us with problems when we try to classify languages. Consider, for example, the existence of SIMSCRIPT I, I.5, II, II.5, and II Plus. These languages are clearly related and yet different. The problem is, are they five separate languages or only one? Is each a new language or merely a new dialect? In the preceding discussions, we evaded this difficulty by using only generic family names.

However they are labeled, some useful programs not included in the previous sections deserve mention. Included are GERTS (*G*raphical *E*valuation and *R*eview *T*echnique *S*imulation Program), APACHE, APSE, QUIKSIM (no exact meaning given; a quicker SIMSCRIPT), ECSS (*E*xtendable *C*omputer *S*ystem *S*imulator), and SOLPASS (*S*imulation-*O*riented *L*anguage *P*rogramming *A*nd *S*imulation *S*ystem).

GERTS is useful in the modeling and analysis of any stochastic network problem, such as assembly line operations, project management networks, conveyer systems, general test and checkout processes, problem solutions involving stochastic processes, and job shop loading studies. The elements of GERT networks are directed branches, which stand for activities, and logical nodes, which represent events. Each branch commences in one node and terminates in another. The branches are described by two parameters. One parameter indicates the conditional probability that a branch is taken, given that the node from which it emanated was realized. Another parameter represents the time required to accomplish the activity that the branch represents. The time parameter can be a random variable. GERTS can accommodate networks having OR and AND logical operations on the input side of the nodes and deterministic or probabilistic operations on the output side of the nodes. The branches utilize one parameter to show the conditional probability that a branch is taken, given the node from which it emanates was reached. Another parameter utilizes a probability density function to represent the time required to accomplish the activity. Added in later GERTS are two more characteristics of the activities simulated: a counter-type number and an activity number. Each time an activity occurs (a branch is realized), the counter is increased by 1. Activity numbers permit network modifications, which involve replacing a node by another node. The node to be replaced is deleted from the network when and if the network modification activity is realized. The activities caused to occur thereafter are from the inserted nodes. GERTS also provides a queue node, which allows modeling and collecting statistics about situations involving waiting lines at particular nodes. The queue nodes have parameters specifying maximum queue lengths before shunting occurs to another node and queue discipline. GERTS is not a language but a method of analysis and a technique for modeling, which is, in turn, programmed in a general-purpose language such as FORTRAN or in a simulation language such as GASP.*

APACHE and APSE both carry out the tedious arithmetic involved in programming analog computers and in accomplishing the static check calculations for analog computation. The input language is equation-oriented in both cases and is written as an extension to a general-purpose digital language. By using either of these languages, we can reduce the programming time on an analog computer, but neither language makes full use of the capability of the digital computer in simulating a model, as compared with a continuous-change language. APACHE and APSE provide a service, but they do not perform the entire job. Some prefer them, however, because they do provide the convenience of setup

* References for GERT are, A. A. B. Pritsker and W. W. Happ, "GERT: Graphical Evaluation and Review Technique, Part I. Fundamentals," *Journal of Industrial Engineering,* May, 1966, pp. 267-74; A. A. B. Pritsker and G. E. Whitehouse, "GERT: Graphical Evaluation and Review Technique, Part II. Probabilistic and Industrial Engineering Applications," *Journal of Industrial Engineering,* Jun., 1966, pp. 293-301; G. E. Whitehouse and A. A. B. Pritsker, "GERT: Part III—Further Statistical Results; Counters, Renewal Times, and Correlations," *AIIE Transactions,* Mar., 1969, pp. 45-50.

calculation by digital programming yet retain the high speed/low cost problem solution capability of the parallel analog system.†

QUIKSIM and ECSS are both extensions to SIMSCRIPT. QUIKSIM attempts to combine the advantages of both a block-oriented discrete language and a statement-oriented discrete language. It is modular in design, and a block routine can be added, deleted, or altered without changing any of the other block routines. QUIKSIM contains 27 already prepared blocks, and the user can write any additional blocks he may desire in SIMSCRIPT or FORTRAN. The QUIKSIM blocks are sufficient for most simulation models, but the user can write his own block routines. The number of debugging runs is usually small, and since learning and programming is quick and relatively easy, obtaining a usable run is quite rapid. We achieve such a run more rapidly than we would have by using SIMSCRIPT; however, run times have been lengthy.*

ECSS is an extension of SIMSCRIPT designed to aid in constructing simulation models of computer systems. Since programming and debugging a simulation model of a complex computing system eat up so much time and money, they are not attempted often. Some tries at designing packages or simulators preceded the design of ECSS, but they were inflexible. ECSS is designed to enable simple models and parts of complex models to be built merely by requesting necessary facilities and providing appropriate input data. If the user cannot find the capability he needs, he can go outside the framework of ECSS and utilize the full power of the simulation language SIMSCRIPT for these unique features or special requirements. ECSS is built around a translator that converts the special features and commands into SIMSCRIPT (specifically SIMSCRIPT II) for compilation into object code. Detail work, except in certain complex areas, is removed from the user's tasks, allowing him to concentrate on the essential logic of the simulation. By reducing the tremendous time

† References for APACHE and APSE are, *APACHE: Analog Programming and Checking Programmer's Manual*, EURATOM Report EUR 2436 e, 1965, CETIS, Ispra Establishment, Italy; G. W. T. White, "Digital Simulation Languages for the Solution of Process Control Problems," *Proceedings of the IBM Scientific Computing Symposium on Digital Simulation of Continuous Systems*, 1964, pp. 72-76; G. H. Burgin, "APACHE: Some Encouraging Experiences," *Simulation*, Jan., 1966, pp. 16-19; C. Green, H. D'Hoop, and A. Debroux, "APACHE—A Breakthrough in Analog Computing," *IRE Transactions on Electronic Computers*, Oct., 1962, pp. 699-706; and *Automatic Programming and Scaling of Equations: Simulation Language Programming for Analog/Hybrid Computers*, Electronic Associates, Inc., West Long Branch, N.J.

*Reference for QUICKSIM is D. G. Weamer, "QUICKSIM—A Block Structured Simulation Language Written in SIMSCRIPT," *Third Conference on Applications of Simulation*, Dec., 1969, pp. 1-11.

requirement involved in developing and debugging simulation models of complex computer systems, we make analysis of these systems feasible. ECSS is still under development and will likely incorporate changes to improve on these and other areas of omissions and inflexibilities as they are pointed out in further implementations.*

SOLPASS is a simulation system developed by using the simulation language SOL to enable efficient simulations of large networks. Although its objective when designed was to permit the Army to gather communication traffic statistics better and to simulate the traffic through the various system configurations to determine component requirements of data links and nodes, it has a wide range of other applications. Examples of applications can be divided into four classes, as follows:

1 Discrete simulations
  - Traffic simulations
    - Road traffic studies
    - Air traffic studies
    - Communications traffic studies

2 Systems simulations
  - Control systems
  - Computer systems
  - Communication systems
  - Transmission systems

3 Hardware simulations
  - Error detection/correction
  - Encoding/decoding devices
  - Logical circuits

4 Continuous simulations
  - Flood control systems
  - Analog processes
  - Electronic circuits
  - Transient analysis

*References are N. R. Nielsen, *ECSS: An Extendable Computer System Simulator*, The Rand Corporation, RM-6132-NASA, 1970; N. R. Nielsen, "ECSS: An Extendable Computer System Simulator," *Third Conference on Applications of Simulation*, Dec., 1969, pp. 114-29; and D. W. Kosy, "Experience With the Extendable Computer System Simulator," *Fourth Conference on Applications of Simulation*, Dec., 1970, pp. 235-43.

Because of the anticipated simultaneous processing of up to 10,000 messages through the network, SOLPASS was designed to utilize on-line disk storage. Instead of the number of transactions it can process being a function of the core storage capability, it is a function of disk memory availability. It was also designed to permit operation from on-line terminals without sacrificing significantly large-scale simulation speed. Its design permits the number of queueing variables and the number of simultaneous queues to be virtually unlimited.*

3.10 User opinions

In 1971, Henry Kleine surveyed users' views of discrete simulation languages. These data were later reanalyzed utilizing Shannon and Wyatt's weighted majority rule method [14, 27]. In the survey, Kleine very wisely recognized the fact that even when he asked a group of "experts" to evaluate a group of entities (in this case, discrete-simulation languages) he found wide discrepancy in the degree of the judges' familiarity regarding each entity judged. He therefore asked each respondent to assess his own level of familiarity or expertise on a scale from 0 to 4, defined in the following discussion. The respondents (103 returns) were asked to rank order the languages on the basis of personal preference for writing simulation languages. They were also asked to evaluate the degree of difficulty in learning and using the special simulation features provided by the language. This was to be done on a scale of 1 through 10, where 1 meant "impossible" for practical purposes and 10 meant very, very easy. Finally, the respondents were asked to evaluate the capability of each language again on a scale from 1 to 10, where 1 meant it had no advantage and 10 meant it had every desirable feature.

In compiling this survey, Kleine left the list of discrete simulation languages to be compared open-ended. Languages not specifically designed for simulation programming (i.e., FORTRAN, PL/1) were to be evaluated with respect to their use in writing simulation programs. Languages that received very light attention were not included in the data reduction and analysis.

Reference [14] presents the results when the data were analyzed, using the ratings as additive cardinal utility scales. As many readers know, the validity of interpersonal comparisons of cardinal utilities has been debated seriously by economists, philosophers, psychologists, and mathematicians since the eighteenth century, when the Marquis de Dondorcet discussed the problems of amalgamating group preferences in his study of election procedures. The original survey data were therefore

* Reference is J. Armstrong, *et al.*, "SOLPASS: A Simulation Oriented Language Programming and Simulation System," *Third Conference on Applications of Simulation*, Dec., 1969, pp. 24-37.

reanalyzed by two methods using ordinal utilities. By "using ordinal utilities" we mean that if a respondent gave language A a rating of 9 on some scale (e.g., ease of use) and language B a rating of 6, we entered the data by merely saying the respondent preferred A to B for that scale.

The first method used was that of simple majority rule, as proposed by Arrow and others. By "majority rule" we mean that if we let $N_{(x,y)}$ be the number of individuals such that $x \geqslant y$, and $N_{(y,x)}$ be the number of individuals for whom $y \geqslant x$ where $x > y$ is read "x is preferred to y" and $x = y$ is read "there is no difference," then

$$x \geqslant y \quad \text{if and only if} \quad N_{(x,y)} \geqslant N_{(y,x)}$$

The method of majority rule as proposed by Arrow gives equal weight to each judge (i.e., one respondent = one vote). The second method used, which we shall call weighted majority rules, was developed by Shannon and Wyatt. Their philosophy was to accept a nonexpert's opinion, but to weigh it according to his degree of expertise. The rationale is that the respondent's ability to prefer accurately one language over another is directly related to the least amount of knowledge he has of either language. One might prefer a language in which he has a high degree of expertise over a language he knows less because of his lack of knowledge and appreciation for the strengths of the lesser known language. We therefore make two assumptions:

1 If two languages are compared for preference in the method of majority with different degrees of expertise, the weighted vote to be used will be that of the lowest level of expertise or the smallest weight of the two factors.

2 The decisionmaker (analyst) can assign a subjective weight to each level of expertise that will be recognized in the amalgamated order.

In the case of the present analysis, we have subjectively made the following value judgments as to the relative weight that should be given to the different levels of expertise.

Level	Familiarity with the Language	Weight
0	Do not know it well enough to make an evaluation.	0.00
1	Have never used it but understand the concepts well enough to make an evaluation.	0.25
2	Have had a little experience with it.	0.50
3	Know the language very well.	0.75
4	I am considered to be an expert.	1.00

The results are presented as three rank orderings for each of the scales of measurement: (1) method of majority rule; (2) weighted majority rules; (3) Kleine's cardinal utility analysis. The third rank order was derived on the basis of the overall means or total points. As before, the notation $A > B$ is to be read "A is preferred to B," and $A = B$ is to be read "there is no difference between A and B." The code identifying the 10 languages is A = GPSS, B = SIMULA, C = Simscript I.5, D = Simscript II, E = GASP, F = SOL, G = FORTRAN, H = PL/1, I = APL.

I. Capability of Language

Majority rule:	$D > B > C > A > E > H > F > G > I$
Weighted majority:	$D > B > C > A > E > H > F > G > I$
Kleine:	$D > C > B > A > F > E > H > G > I$

II. Ease of Use of Language

Majority rule:	$A > C = E > D > B > F > H = I > G$
Weighted majority:	$A > D > E > B > F > C > H = I > G$
Kleine:	$A > E > C > D > H > B > I > J > F$

III. Preferred Language

Majority rule:	$C > A = B = D = E = G > H > F > I$
Weighted majority:	$D > C > B > A > E > G > H > F > I$
Kleine:	$C > A > D > G > B > I > E > I > F$

3.11 Concluding remarks

In this chapter, we have tried to discuss the problems associated with translating our simulation model into a language that the computer can understand and execute. It is not an easy problem to solve owing to the large number of special- and general-purpose languages available. In our experience, most analysts and professional computer programmers know from one to three languages. The question then naturally arises as to whether the benefits to be gained outweigh the efforts required to learn a new language. Each person must answer that question for himself. Hopefully, the preceding discussions will help make that decision easier and provide some guidance if it is made in the affirmative. Other survey papers or discussions that the reader might find helpful are Chapter 6 of reference [4] and the papers by Kiviat [13], Linebarger and Brennan [19], Sammet [25], Teichroew and Lubin [28], and Tocher [29]. It is often useful to see some *simple* models programmed in the particular language one is considering. Table 3.2 indicates where the reader can find such models.

Table 3.2 *Simple example models in various languages.*

Source	Examples
J. R. Emshoff and R. L. Sisson *Design and Use of Computer Simulation Models* The MacMillan Company, New York, 1970	DYNAMO GPSS
G. S. Fishman *Concepts and Methods in Discrete Event Digital Simulation* Interscience Publishers, New York, 1973	GPSS SIMSCRIPT SIMULA
G. Gordan *Systems Simulation* Prentice-Hall, Inc., Englewood Cliffs, N.J., 1969	DYNAMO GPSS SIMSCRIPT 360/CSMP 1130/CSMP
H. Maisel and G. Gnugnoli *Simulation of Discrete Stochastic Systems* Science Research Associates, Inc., Chicago, 1972	GPSS SIMSCRIPT PL/1
R. C. Meir, W. T. Newell, and H. L. Pazer *Simulation in Business and Economics* Prentice-Hall, Inc., Englewood Cliffs, N.J., 1969	DYNAMO GPSS SIMSCRIPT
T. H. Naylor, J. L. Balintfy, D. S. Burdick, and K. Chu *Computer Simulation Techniques* John Wiley & Sons, Inc., New York, 1966	DYNAMO GPSS SIMSCRIPT GASP
J. W. Schmidt and R. E. Taylor *Simulation and Analysis of Industrial Systems* Richard D. Irwin, Inc., Homewood, Ill., 1970	GPSS SIMSCRIPT

REFERENCES

1. BAUER, CHARLES R., ANTHONY P. PELUSO, and DAVID A. GOMBERG, *Basic PL/I Programming Self-Instructional Manual and Text*, Addison-Wesley Publishing Co., Inc., Reading, Mass., 1968.

2. BEKEY, GEORGE A., and WALTER J. KARPLUS, *Hybrid Computation*, John Wiley & Sons, Inc., New York, 1968.

3. CONWAY, R. W., B. M. JOHNSON, and W. L. MAXWELL, "Some Problems of Digital Systems Simulation," *Management Science*, Vol. VI, Oct., 1959.

4. EMSHOFF, JAMES P., and ROGER L. SISSON, *Design and Use of Computer Simulation Models*, MacMillan & Co., Ltd., London, 1970.

5. EVANS, G. W., G. F. WALLACE, and C. L. SUTHERLAND, *Simulation Using Digital Computers*, Prentice-Hall, Inc., Englewood Cliffs, N.J., 1967.

6. FARINA, MARIO V., *COBOL Simplified*, Prentice-Hall, Inc., Englewood Cliffs, N.J., 1968.

7. FORMAC (*Operating and User's Preliminary Reference Manual*), IBM Program Information Dept., No. 7090 R2IBM 0016, IBM Corporation, Hawthorne, N.Y., Aug. 1965.

8. HIGMAN, BRYAN, *A Comparative Study of Programming Languages*, American Elsevier Publishing Co., Inc., New York, 1967.

9. *HYTRAN Simulation Language (HSL) Programming Manual*, Electronic Associates, Inc., Publishing No. 008270006-0, 1967.

10. JOHNSON, C. L., *Analog Computer Techniques*, McGraw-Hill Book Company, New York, 1956.

11. KARPLUS, WALTER J., and WALTER W. SOROKA, *Analog Methods Computation and Simulation*, 2nd ed., McGraw-Hill Book Company, New York, 1959.

12. KIVIAT, PHILIP J., "Development of Discrete Digital Simulation Languages," *Simulation*, Vol. VIII, No. 2, Feb., 1967.

13. KIVIAT, P. J., "Digital Computer Simulation: Computer Programming Languages," *RAND Report RM-5993-PR*, Jan., 1969.

14. KLEINE, HENRY, "A Second Survey of Users' Views of Discrete Simulation Languages," *Simulation*, Vol. 17, No. 2, Aug., 1971.

15. KLERER, MELVIN, and GRANINO A. KORN, *Digital Computer User's Handbook*, Mc Graw-Hill Book Company, New York, 1967.

16. LANDAUER, J. PAUL, *EAI-640 Based Hybrid Systems Software*, Scientific Computation Dept. Report No. 11-71, Electronic Associates, Inc., Apr., 1971.

17. LAVE, ROY E., Jr., "Timekeeping for Simulation," *The Journal of Industrial Engineering*, Vol. XVII, No. 7, Jul., 1967.

18. LECHT, CHARLES PHILIP, *The Programmers PL/I*, McGraw-Hill Book Company, New York, 1968.

19. LINEBARGER, R. N., and R. D. BRENNAN, "A Survey of Digital Simulation—Digital Analog Simulator Programs," *Simulation*, Vol. 3, No. 6, Dec., 1964.

20. McCRACKEN, DANIEL D., *A Guide to Fortran Programming*, John Wiley & Sons, Inc., New York, 1961.

21. McCRACKEN, DANIEL D., *A Guide to ALGOL Programming*, John Wiley & Sons, Inc., New York, 1962.

22. NAYLOR, THOMAS H., *et al.*, *Computer Simulation Techniques*, John Wiley & Sons, Inc., New York, 1968.

23. PERSTEIN, M. H., *Grammar and Lexicon for Basic JOVIAL*, System Development Corporation, TM-555/005/00, Santa Monica, Calif., May, 1966.

24. SAMMET, JEAN E., *Programming Languages: History and Fundamentals*, Prentice-Hall, Inc., Englewood Cliffs, N.J., 1969.

25. SAMMET, J. E., "Programming Languages: History and Future," *Communications of the ACM*, Vol. 15, No. 7, Jul., 1972.

26. SAMMET, J. E., and E. BOND, "Introduction to FORMAC," *IEEE Transactions on Electronic Computers*, Vol. EC-13, No. 4, Aug., 1964.

27. SHANNON, R. E., and M. W. WYATT, "Discrete Simulation Languages User's Survey Revisited," *Simulation*, Vol. 19, No. 4, May, 1973.

28. TEICHROEW, D., and J. F. LUBIN, "Computer Simulation—Discussion of the Technique and Comparison of Languages," *Communications of the ACM*, Vol. 9, No. 10, Oct., 1966.

29. TOCHER, K. D., "Review of Simulation Languages," *Operational Research Quarterly*, Vol. XVI, Jun., 1965.

EXERCISES

1. Read case study III, Appendix A.
   a) This model could be programmed on an analog computer, but if you were going to do it on a digital computer, what language would be best?
   b) How could the author's complaints about excessive amounts of data be partially overcome?
2. The author of case study I in Appendix A programmed his model in GPSS. Do you think his choice was wise?
3. What language would you choose to program the model discussed in case study II?

chapter four

DESIGN OF COMPUTER SIMULATION EXPERIMENTS

An experiment is worthwhile only if it increases the profitability of the decision more than it costs.

Ronald A. Howard

4.1 Importance of design

The design of a computer simulation experiment is essentially a plan for purchasing a quantity of information that may be acquired at varying prices depending upon the manner in which the data are obtained. The effective use of experimental resources is profoundly affected by the choice of design because:

1 The design of the experiment determines in great measure the form of statistical analysis that can be used appropriately to analyze the results.

2 The success of the experiment in answering the questions of the experimenter (without excessive expenditure of time and resources) depends largely upon the right choice of design.

Computer simulation experiments are expensive in terms of time and labor of the experimenter as well as cost of machine time. Since the more effort he expends on one investigation the less he can spare for another, it is important that the researcher plan for obtaining as much information as possible from each experiment. The primary purpose of conducting simulation studies is to learn the most about the behavior of the system being simulated for the lowest possible cost. To do so, we must plan and design carefully not only the model but also how it is to be run or used. It follows, then, that we must concern ourselves with the strategic planning of how to design our experiments so as to yield the desired information at the least cost. Since the first publication in 1935 of R. A. Fisher's book *The Design of Experiments*, a great number of books and papers on experimental design have appeared, and designed experiments have found widespread application in the biological, physical, and systems sciences. The purpose of using these designs is twofold: (1) they are economical in terms of reducing the number of experimental trials required; and (2) they provide a structure for the investigator's learning process.

The running of a simulation experiment is the process of exercising or running the model so as to observe and analyze the resulting information to obtain the desired answers. The experimental design selects a particular approach to gathering the information needed to allow us to draw valid inferences. A complete discussion of the design of experiments is beyond the scope of this book. This chapter provides the reader with the necessary terminology and background to consult the extensive literature on the design of experiments to obtain help on any particular problem.

Several excellent books treat in detail the problems of constructing and analyzing experimental designs in connection with research programs. Outstanding among these are Bartee [3], Cochran and Cox [9], Davis [12], Dixon and Massey [13], and Hicks [16]. In addition, a volume edited by Naylor [20], containing papers presented at a symposium held at Duke University in October, 1968, on the design of experiments for computer simulation, is very useful in pointing out some of the difficulties involved.

Some types of experimental design are now so standardized that we may select one by referring to a handbook. Like ready-made clothes, ready-made designs meet many tasks very well and adequately. But just as the tailor must be prepared to fit a garment to an unusually shaped person, so must the researcher be prepared to modify existing designs to suit problems of special complexity.

4.2 Differences between physical and computer experiments

Although the underlying objectives of designing computer simulation experiments are essentially the same as those for conducting physical experiments, some differences must be considered. Among the more important of these are the following:

1 difficulties in defining a single datum point or sample;

2 ease with which experimental conditions may be repeated or reproduced;

3 ease of stopping and resuming experimentation;

4 presence or absence of correlation between subsequent data points;

5 control of variability—in physical experiments stochastic variability is beyond the control of the experimenter; in simulation, variability is deliberately built into the model by the experimenter.

In determining how to run our model and analyze the results, one of the first issues we must decide upon is what we shall consider a single datum point or sample to be. There are several possibilities, including:

1 A complete run of the model. This may entail considering the mean or average value of the response variable for the entire run as being the datum point.

2 A fixed time period during the run in terms of simulated time. For example, the model might be run for n time periods, where n is measured in hours, days, weeks, etc., and the mean or average value of the response variable for each time period considered a datum point.

3 Each transaction considered a separate sample. For example, turn-around time for each job or the total time in the system of each customer is considered a separate datum point.

4 Transactions aggregated into groups of fixed size. For example, we might take the turn-around time of each 25 jobs flowing through the system, and then use the mean time of the group as a single datum point.

Each of these has advantages and disadvantages. Depending upon the model, the first definition of a single datum point or response may result in excessive use of computer time owing to run lengths, or in having to discard the early part of each run owing to startup transient conditions. Chapter 5 discusses startup conditions and their influence on stochastic convergence in detail. The second method of fixed time increments introduces the possibility of correlation of the data samples from one time increment to the next (as do the last two).

The ease with which experimental conditions can be repeated

or reproduced in a computer model is often a distinct advantage of computer simulation over physical experimentation. If we are interested in comparing two alternatives in a relative manner only, we can run the model in such a way that each alternative is compared under identical conditions (same sequence of events). This is accomplished by repeating or using the same series of random numbers for each alternative, which reduces the residual variation in mean performance of the alternatives and requires considerably smaller sizes to establish statistically significant differences in the response. On the other hand, if we are interested in evaluation of absolute system performance, we can use a new stream of random numbers for each run.

Another difference between experimentation with computer models vs. that with physical systems is the ease with which we can stop and resume the experimentation. This facility allows us to use sequential or heuristic experimental methods that might not be feasible in a real world system. With the computer model we can always stop the experiment while we analyze the results and decide whether to change the parameters or continue as we were. The ability to put the model back on the shelf (so to speak) while we think about what is occurring can be a distinct advantage not readily available in the real world. Again, the question of starting conditions (see Chapter 5) may turn this advantage into a disadvantage.

Analysis of computer simulation experiments often presents some difficult problems, because outputs are sometimes auto- or serially correlated. Autocorrelation arises when the observations in the output series are not independent of each other (one of the assumptions of many experimental designs). In many simulation models, the value of one output observation depends upon the value of the previous observation or upon some other past observation. Thus, not as much information is contained in that observation as there would be if the two were completely independent. Since most experimental designs found in the literature assume independence of observations, many common statistical techniques are not directly applicable to autocorrelated simulation results.

4.3 Role of experimental design

In a well-conducted study, there are two areas of interface between the experimental planning design and the total computer simulation process. Figure 4.1 (an elaboration of Fig. 1.3, Chapter 1) illustrates these two interface areas with a block diagram representing the overall simulation study process from the genesis of the problem to the final documentation and implementation of the study results. The two blocks representing the experimental planning/design functions are shown in heavy outline. Once the

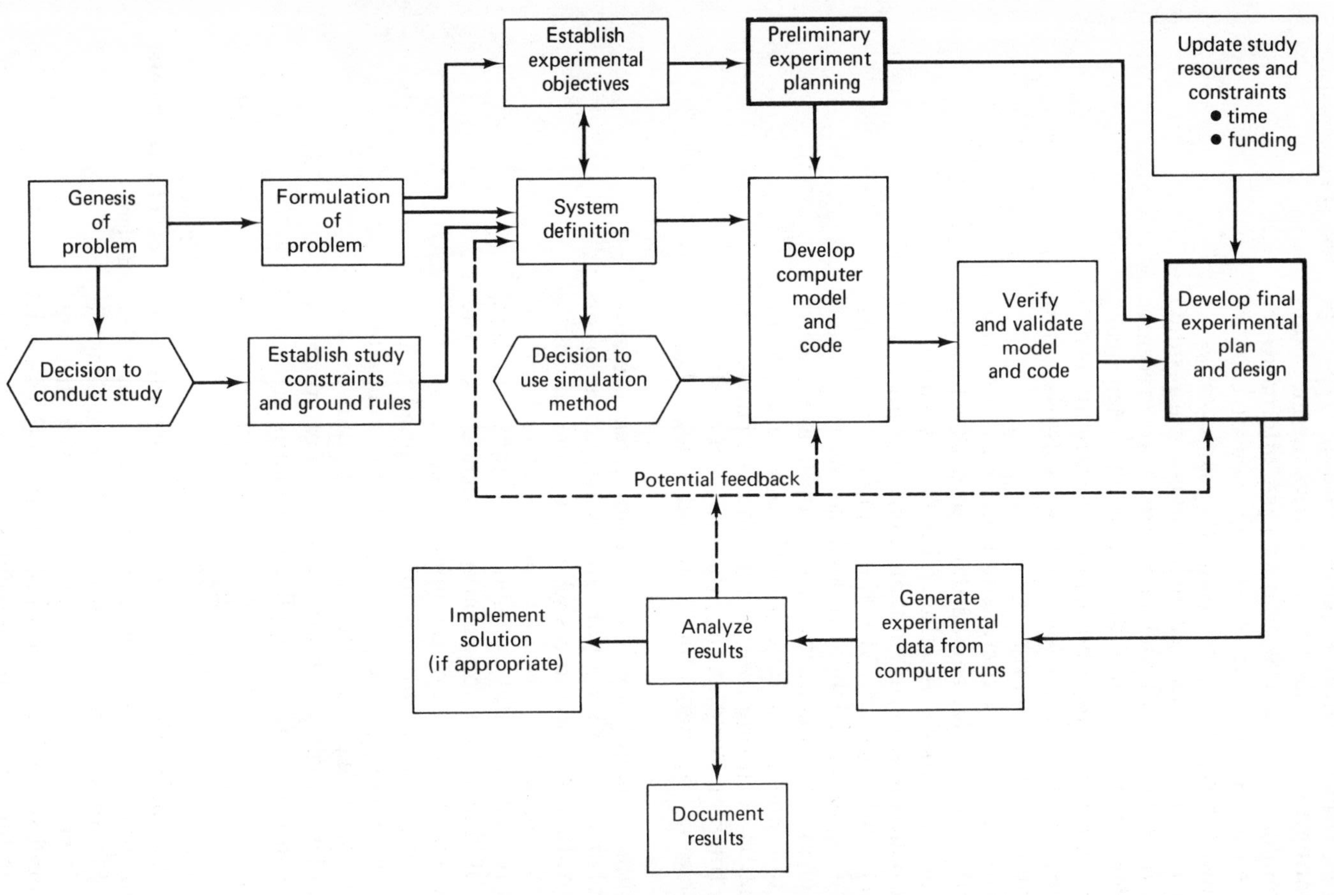

Figure 4.1 Simulation process [21].

experimental objectives and a system definition have been established and the decision is made to employ computer simulation, careful preliminary experimental planning at an early stage in the development of the computer model can be very helpful. It is good to have a fairly detailed idea of the experimental plan early so that the model itself can be better planned to provide efficient generation (and possibly partial analysis) of the desired experimental data. Since computer time is expensive, knowledge of the magnitude and special requirements of the desired data output may have a significant impact on the concept and details of the model. After the model is designed and coded is a poor time to find out that the operating philosophy and output formats need to be changed to meet the experimental data generation needs.

As indicated in Fig. 4.1, the second and principal function of experimental design is that of providing the final strategic and tactical plans for the execution of the experiment. Here the project constraints on time (schedule) and costs must be updated to the current conditions and imposed upon the design. Even though careful planning and budget control may have been exercised from the beginning of the project, now is the time to take a good hard, realistic look at the resources remaining and how best to utilize them. Whether the objectives of a given study are effectively and efficiently accomplished depends to a significant degree upon the care taken and skill exercised in the experimental design. The larger and more complex is the simulation, the more critical this phase becomes.

Very few studies do not have resource limitations imposed upon them in terms of time (schedule), dollars, and computer availability. In most situations, these constraints place very severe restrictions on the experimental design and often override academic statistical considerations. An excellent example is given by Bidwell in his discussion of a simulation study of a highway paving operation using a mobile central-mix plant with different truck speeds and fleet size combinations for hauling. Regarding the application of the computer program, Bidwell states;

> After the computer runs were completed, it was apparent the cost of computer time was going to be a limiting factor in contractor acceptance and use of the study. That is, if a contractor was interested in finding his least cost configuration and procedure, data from his equipment spread would have to be used in the simulation. Say he simulated nineteen different haul fleet sizes (1-18 and 21 trucks), four different rates of paving progress (½, 1, 1½, and 2 miles per day) and three average truck speeds (15, 30, 45 mph), he would have more than 200 hours of simulation time involved. [4]

The example cited involves $19 \times 4 \times 3 = 228$ combinations, requiring 200 hours of computer time if all combinations are to be run. If we assume the computer time is available and costs \$300 per hour, we find that the experiment will cost \$60,000, and even then have very low statistical validity owing to the lack of replication. If by careful analysis and design it were possible to achieve significant results using three fleet sizes (1, 11, and 21 trucks), three paving rates (½, 1¼, and 2 miles per day), and three average truck speeds (15, 30 and 45 mph), then the number of computer runs could be reduced to $3 \times 3 \times 3 = 27$. The computer time (and associated costs) would decrease proportionally. This, of course, is not the only design tradeoff possible, since we could also consider using a faster or cheaper computer, using only one average truck speed, etc.

In most complex simulation studies, the number of possible combinations of factors and factor levels of interest is almost infinite; hence, a large number of design tradeoffs are made to stay within the resource constraints. The type of design the experimenter should choose is again very much dictated by the purpose or goal of the study and the type of statistical analysis required to fulfill those goals. Depending upon the specific purpose of the experimenter, several different types of analysis may be required, among the more common of which are,

1 comparing means and variances of alternatives;

2 determining the importance or effect of different variables and their limitations;

3 searching for the optimal values of a set of variables.

Designs to accomplish the first type of analysis are generally so-called single-factor experiments and are fairly straightforward, with the major concerns of the experimenter being such matters as sample size, starting conditions, and the presence or absence of autocorrelation. Chapter 5 discusses these matters in more depth. The second type of analysis is one toward which most textbooks on design and analysis of experiments are directed. These designs primarily utilize analysis of variance and regression techniques for the interpretation of the results. The third type of analysis usually requires sequential or search techniques of experimentation. In the following sections, we briefly discuss some of the techniques appropriate to these different types of experimentation.

4.4 Design approach

In the design and execution of simulation experiments, we are concerned with two types of variables, which in this chapter we call *factors* and *responses*. We can distinguish them clearly if we consider a simple experiment entailing only two variables, x and y, in which the purpose of the experiment is to answer the question, "How does a change in x affect y?". In this case, x is a factor and y is a response. The literature also refers to factors as treatments or independent variables and responses as yields or dependent variables. Earlier, Chapter 1, Section 1.7, we also used the terms exogenous (input) and endogenous (output or status) in the same sense as factors and responses. Thus, we see that the terms *factor, treatment, independent variable, input variable*, and *exogenous variable* all refer to the same thing, as do *response, yield, dependent variable, output variable, status variable*, and *endogenous variable*. This terminology derives from the fact that much of the early interest in statistical experimentation came from agricultural research followed by the biological sciences. Authors in each of these fields have tried to use terms that were most meaningful to their readers.

To select an appropriate experimental design, we need to follow a three-step process, summarized as [3]:

1 determine the experimental design criteria;

2 synthesize the experimental model;

3 compare model to standard experimental designs and choose the optimal design.

The design of a simulation experiment should be approached in a systematic manner, just as in any other design problem. We therefore discuss the synthesis of the experimental design in three steps, borrowing heavily from the approach proposed by Bartee [3]:

1 design of the *structural* model;

2 design of the *functional* model;

3 design of the *experimental* model.

As in any design problem the ultimate character of the final design is dictated by the design criteria, which are determined to be pertinent. Among the criteria to be considered are the following:

1 The number of factors to be varied.

2 The number of levels (values) to be used for each factor.

a) Are the levels of the various factors quantitative or qualitative?

b) Are the levels of the various factors to be fixed (controlled) or random (uncontrolled)?

c) Are nonlinear effects to be measured?

d) Are all factors to be set at an equal number of levels?

3 The number of measurements of the response variable to be taken.

a) Are interactions between the factors to be measured?

b) Do resource limitations exist owing to lack of time, money, or computer time?

c) What precision is required?

To make sense of the available literature, we again define some of the terms we have just introduced. A *quantitative* factor or variable is one which occurs in various degrees that can be measured on a specified (either interval or ratio) scale. Examples would be temperature, length, velocity, cost, time, etc. A *qualitative* variable, on the other hand, is one whose occurrences cannot be placed in an order of magnitude, i.e., it is measured on an ordinal or nominal scale. Examples of a qualitative variable would be machines, policies, geographic areas, organizations, decision rules, etc. The terms random or fixed levels are fairly self-evident. If we decide to control or set the levels or values of the variable for each run of the model, the levels are fixed. If we let the levels or values vary randomly (perhaps using a Monte-Carlo sampling technique), the levels are random. If the mathematical model used for the experimental design contains factors that are all fixed, it is called a *fixed model.* If all the factors are allowed to vary randomly, it is called a *random model.* If the experiment contains both fixed and random factors, the model is called a *mixed model.* An *interaction effect* may be defined as the combined influence of two or more factors on the response, which is in addition to the individual influence or effect of these factors separately.

4.5 The structural model

In the approach we follow in our discussion, the first two criteria are of prime importance in determining the structural model. The structural model is described by and is a function of:

1 the number of factors;

2 the number of levels of each factor.

These should be determined on the basis of the objectives of the experiment, the measurability of the factors, the interest in nonlinear effects, etc. They should not be dictated by limitations upon the total number of measurements possible owing to resource constraints. Such limitations are the concern of the functional model. The structural model should be developed on the basis of what should be done—the functional model on the basis of what can be done.

The structural model of an experiment is, therefore,

$$N_s = (q_1)(q_2)(q_3)\ldots(q_k) \tag{4.1}$$

where

N_s = number of cells in the experiment
k = number of factors in the experiment
q_i = number of levels of ith factor, $i = 1, 2, 3, \ldots k$

We call the basic structural unit or building block of an experiment a *cell*, defined as being the simplest form of an experiment or the case of one factor at one level—i.e., $k = 1, q_1 = 1, N_s = 1$.

Several considerations are necessary in determining the number and identity of the factors to be used. The first and foremost is the objective or goal of the study. This, of course, means that we must decide which response variable or variables are our primary interest—i.e., what do we wish to measure to obtain the answers we seek? For example, if we are modeling an information storage and retrieval system, we might decide that the important response is turn-around time or how long it takes the system to answer an inquiry. On the other hand, we might wish to measure the maximum number of inquiries that could be handled in a given time period, or some other measure of the system's effectiveness.

After deciding upon the response variable(s), we next determine which factors we believe influence the response. We generally make a comprehensive list and then try to reduce it to the significant few. As the reader is undoubtedly aware, for any given system and response variable the number of factors that *can* have an influence is very large. Unfortunately, the less we know about a system the more possibly influential factors we can see. We agree with Ackoff and Sasieni [1] that, "In general, the extent to which a phenomenon is understood is inversely proportional to the number of variables required to explain it." Most systems operate according to the *Pareto principle*, that in terms of performance and effectiveness there are a few significant factors and many insignificant ones. In fact, the rule of thumb is that in most systems 20% of the factors will account for 80% of the performance, whereas the other 80% of the factors contribute the remaining 20% of the performance. Our problem is to decide which are the significant few.

Having determined the response variable(s) and drawn up a list of what we think are the significant contributing factors, we must now classify these factors as to how we will integrate them into the design. For each factor there are three possible ways it can contribute to the experiment [3]: (1) the variable can be held constant and therefore become a part of the boundary conditions of the experiment; (2) the factor can be allowed to vary without being controlled and become a part of the experimental error; or (3) the factor can be measured and controlled in the experiment. Members of the third classification are the ones of interest in our structural model. At this point, the reader should recognize the significance of including experimental design considerations early in the simulation process as well as just prior to the actual execution of the experiment(s). The analyst *has* to know what variables he possibly wants to measure and control during the design phases; otherwise he may find that he has to redesign his model.

The next step in designing the structural model is to decide the levels at which each selected factor should be measured and set. The minimum number of levels for any selected factor is two, since if only one level is used the factor becomes a constant. The number of levels of each factor should obviously be kept at the minimum number that will accomplish the goals of the experiment. Every additional level adds cost and should be carefully evaluated. If it is possible to set an equal number of levels for each factor (and particularly if that number is either two or three), certain analytical advantages can be gained. When the factor is qualitative, it is by its nature usually set at a certain number of levels; thus, for example, if one of the factors is the priority rules used for processing arrivals to the system, then the levels are each of the policies under consideration. Even though we may for convenience designate the levels of a qualitative variable as 1, 2, 3 or A, B, C, we should remember that these orderings are arbitrary. In other words, these alternatives cannot logically be placed on a quantitative scale since qualitative factors provide no information regarding their proper sequence or order on a scale.

When a factor is quantitative, we must consider the range of interest and whether we are interested in nonlinear effects. If the analyst is interested in only linear effects, then two levels of the quantitative variable at the extremes of the range of interest are sufficient. On the other hand, if the analyst suspects or is interested in quadratic effects, he must use three levels; if he is interested in cubic effects, he must use four levels; etc. This obviously results from the number of independent constrasts that can be made and the number of points necessary to define the function. The analysis of the data is greatly simplified if the levels of a quantitative variable are set at equal intervals, owing to the fact that equal spacing allows us to provide *orthogonal* contrasts, thereby simplifying the determination of the constants in a polynomial function. Thus, two levels of a quantitative variable are usually set at the extreme points of the range of interest, and any intermediate levels (depending upon the interest in nonlinear effects) are set so as to provide equal spacing or intervals between the levels.

As mentioned earlier, considerable analytical advantage is to be gained in an experiment if it is feasible to set all the factors at the same number of levels, particularly if the number of levels is either two or three. Such structural models are symmetrical and are described as

$$N_s = q^k \tag{4.2}$$

4.6 The functional model

The functional model determines how many of the cells in the structural model will actually contain a response measurement—i.e., how many data points (without replication) we actually get. Such functional models are defined by Bartee as being either complete or incomplete [3]. A functional model is considered to be complete when all cells contain a response—i.e., $N_f = N_s$. A functional model is incomplete when the number of responses actually taken is less than the number of cells—i.e., $N_f < N_s$. Obviously, the ideal situation is for the functional model to be equal to the structural model, since in essence the structural model is what we have said we would like to have. Unfortunately, most simulation studies have resource limitations imposed upon them in terms of time (schedule), dollars, and computer availability. In many situations, these constraints severely restrict the experimental design and often override statistical considerations. The functional model should help us make the necessary design tradeoffs between our desires (the structural model) and our resources.

In most complex simulation studies, the number of possible combinations of factors and factor levels of interest is almost infinite. Hence, we must make a large number of design tradeoffs to stay within the resource constraints. Figure 4.2 provides a four-quadrant nomograph that we can use in preliminary experimental planning to perform approximate tradeoffs of the number of factors, levels, replications, computer speed, and cost. The two left-hand quadrants of the chart are based on the following expression for total number of computer runs required for a replicated, symmetrical experiment:

$$N = pq^k \tag{4.3}$$

where

k = number of factors (input parameters or variables)
q = number of factor levels
p = number of replications
N = total number of computer runs required

An example is shown in Fig. 4.2. We enter the upper left quadrant of the figure by choosing the number of factors. We then

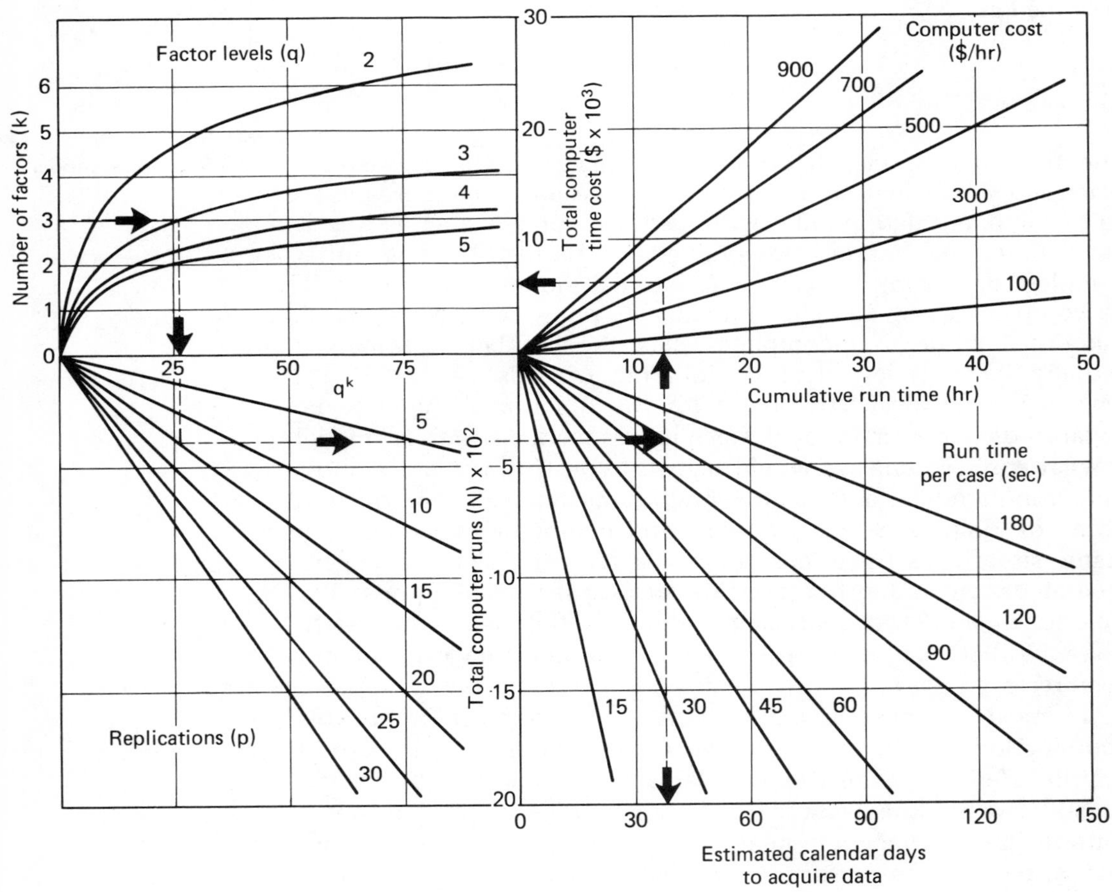

Figure 4.2 Nomograph for cost estimates [21].

project horizontally to a curve for the desired number of factor levels and read vertically downward to the left lower quadrant to a curve of desired number of replications. This point is projected horizontally into the lower right quadrant, where the total number of computer runs required is read on the vertical axis. The estimated or known average run time per case of the simulation program is used to read the total computer run time requirement and estimated calendar days to acquire data on day scale is based upon the assumption that no more than 20 minutes per day of computer time can be dedicated to the project. A continuation of the vertical projection into the upper right quadrant plot provides an estimate of the total computer time costs by projecting horizontally to the vertical scale from one of the curves (or interpolation between) of computer cost in dollars per hour.

The example in Fig. 4.2 is based on a single experiment of 3 factors, 3 levels, 15 replications, 120 seconds time per simulation run, and computer costs of $500 per hour. This design requires 400 computer runs, approximately 13½ hours of computer time, about 40 days for data generation, and $6000 for computer time. We demonstrate the importance of careful experiment design in choosing the number of factor levels by reducing the preceding experiment to 3 factors at 2 levels. The new design requires 135 computer runs, 4½ hours computer time, 13½ days for data generation, and only $2250 for computer time (a 625% reduction).

The nomograph allows the designer to evaluate experimental design tradeoffs he might be considering quickly. For example, if the dollars available are fixed, then, given the computer cost per hour and the computer run time per case, the designer can quickly find the maximum number of runs available to him. He can then look at the possible tradeoffs among factors, levels, and replications. Or if he thinks he sees a way to cut the run time of each case from, say, 120 seconds to 60 seconds, by reprogramming, he can quickly tell if the effort is justified (a savings of $2667 on computer time and 20 days on the schedule for our example).

Assume that because of constraints on costs or computer time, the designer finds it necessary to reduce the total number of cases run. It is then desirable to analyze the initial structural design to determine whether factors, levels, or replications are dominant in reducing the number of cases required. One way to do this is to find out which variable (factors, levels, or replications) gives the greatest savings in terms of total number of samples required per unit change. We accomplish this by the partial differentiation of Equation 4.3 with respect to p, k, and q, and then by consideration of the ratios of the resulting equations [21].

If we take the two equations obtained from the partial derivatives of N with respect to k and q, we obtain, after simplification,

$$\frac{\partial N/\partial k}{\partial N/\partial q} = \frac{q \ln q}{k} \tag{4.4}$$

In a given case, if

$$\frac{\partial N}{\partial k} > \frac{\partial N}{\partial q}$$

then a unit change in the number of factors is dominant over a unit change in the number of levels in reducing or decreasing the total number of computer runs. Conversely, if

$$\frac{\partial N}{\partial k} < \frac{\partial N}{\partial q}$$

then a unit change in number of levels is dominant over a unit change in the number of factors.

If we now consider the ratio obtained from the partials of N with respect to p and q, we obtain, after reduction,

$$\frac{\partial N/\partial p}{\partial N/\partial q} = \frac{q}{\text{kp}} \tag{4.5}$$

Thus, in a given case, if

$$\frac{\partial N}{\partial p} > \frac{\partial N}{\partial q}$$

then a change in the number of replications is dominant over a unit change in the number of levels in reducing or decreasing the total number of computer runs. Conversely, if

$$\frac{\partial N}{\partial p} < \frac{\partial N}{\partial q}$$

then a unit change in the number of levels is dominant over a change in the number of replications.

Finally, if we take the ratio of the partials of N with respect to p and k, after simplification we get

$$\frac{\partial N/\partial p}{\partial N/\partial k} = \frac{1}{p \ln q} \tag{4.6}$$

Thus, if

$$\frac{\partial N}{\partial p} > \frac{\partial N}{\partial k}$$

then a unit change in the number of replications is dominant over a unit change in the number of factors in reducing the total number of computer runs required. Conversely, if

$$\frac{\partial N}{\partial p} < \frac{\partial N}{\partial k}$$

then a unit change in number of factors is dominant over a unit change in number of replications.

On the basis of Equations 4.4, 4.5, and 4.6, the conditions determining which rate of savings or cost is dominant in a given situation may be stated as follows:

1 If $kp > q$ and $k > q \ln q$, then the *levels* are dominant; i.e., a change in the number of levels will result in the greatest savings in total computer runs per unit decrease.

2 If $kp > q$ and $k < q \ln q$, then the *factors* are dominant.

3 If $p < q$ and $p \ln q < 1$, then the *replications* are dominant.

The analysis presented provides a basis for developing a rapid graphical technique to determine the dominant variable in a given experimental design. The technique is based on the use of the graphical plots of Equations 4.4, 4.5, and 4.6, which are shown in Figs. 4.3, 4.4, and 4.5, respectively. Figure 4.3 is a plot of the ratio $(q \ln q)/k$ as a function of number of levels q, for a number of factors from 1 to 5. If, for a given experimental design, the ratio $(q \ln q)/k$ exceeds 1 (or unity) for the given values of q and k, the factors are the dominant variable. A value of the ratio less than unity indicates that the levels are dominant.

Figure 4.4 is a plot of the ratio q/kp vs. the number of levels q, for values of the product kp from 1 to 5. If in a given case $q/kp > 1$, the replications are the dominant variable. On the other hand, if $q/kp < 1$, the levels are dominant. Note that if $kp \geqslant 5$, the levels are always the dominant variable.

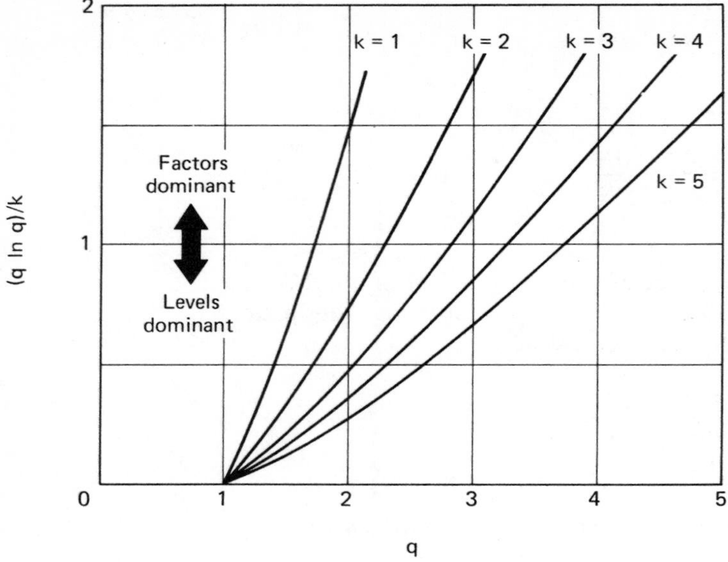

Figure 4.3 Factors vs. levels [21].

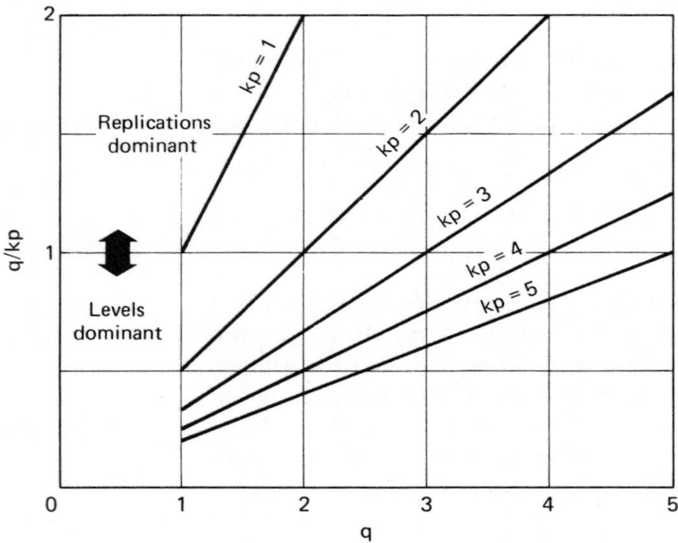

Figure 4.4 Replications vs. levels [21].

Figure 4.5 is a plot of the ratio $1/(p \ln q)$ as a function of the number of levels q, for replications p of 1, 2, and 10. If in a given problem $1/(p \ln q) > 1$, the replications are the dominant variable. If $1/(p \ln q) < 1$, the factors will always be dominant over replications if the number of replications $p \geqslant 2$ and the number of

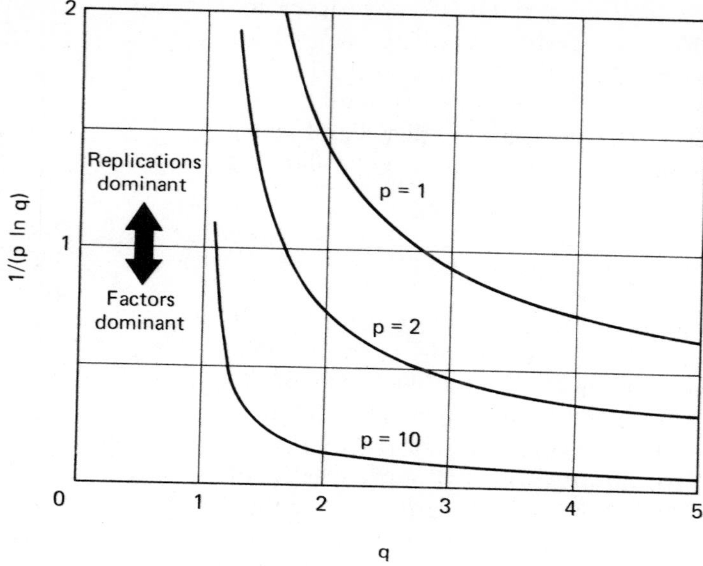

Figure 4.5 Replications vs. factors [21].

levels $q \geqslant 2$. These figures, with the previously stated conditions, provide all that is necessary to establish rapidly whether a unit decrease in number of factors, levels, or replications will result in the maximum reduction in total required computer runs.

EXAMPLE 4.6.1

What is the dominant variable in reducing the total number of computer runs required in a replicated factorial experiment where the current structural design is based on 4 factors ($k = 4$), 3 levels ($q = 3$), and 8 replications ($p = 8$)?

Enter Fig. 4.3 with $q = 3$ and $k = 4$. The ratio $(q \ln q)/k$ is seen to be less than 1; therefore, the levels dominate factors—i.e., a unit reduction in the number of levels is dominant over a unit reduction in number of factors in reducing the total number of computer runs required in the experiment.

Now examine Fig. 4.4 with $q = 3$ and $kp = 4 \times 8 = 32$. Obviously, the ratio q/kp is less than 1 and the levels are the dominant variable also over replications. Thus, the answer to the problem is that a reduction in the number of levels from 3 to 2 will result in the maximum savings in total number of computer runs required in the experiment. Of course, this can only be done if we are willing to settle for an analysis of linear effects only. Figures 4.4, 4.5, and 4.6 should be used for guidance only and not to make the decisions.

4.7 Single-factor experiments

The simplest design is where only one factor is varied, and it is referred to as a single-factor experiment. The levels of the factor to be investigated can be quantitative or qualitative, fixed or random. The factor being varied may be different operating policies, different configurations of a system, or different levels of an input or exogenous variable. The number of observations or runs for each level of a treatment or factor is determined by cost considerations and the power of the test or statistical significance desired. We can represent this situation with the following mathematical model:

$$X_{ij} = \mu + T_j + \epsilon_{ij}$$

where X_{ij} represents the ith observation ($i = 1, 2 \cdots n$)on the jth treatment or level ($j = 1, 2 \cdots k$ levels). For example, X_{42} represents the fourth observation or run using level two of the factor. A common effect for the whole experiment is μ; T_j represents the effect of the jth treatment; and ϵ_{ij} represents the random error present in the ith observation of the jth treatment. In most experimental models found in the literature, ϵ_{ij} is assumed to be normally distributed with zero mean and variance σ^2, the variance being the same for all j. For more complex experimental designs, we elaborate on this basic model by including additional variables on the right-hand side of the equation to represent the effects of other factors or imposed constraints.

If there are only two levels of the treatment or factor (e.g., the current configuration and a proposed change), then only straightforward hypothesis testing utilizing standard tests (t, F, χ^2 or tests for proportions) is required. Table 4.1 shows a typical layout for a single-factor experiment with k levels of the factor.

Table 4.1 *Population layout for one-way ANOVA.*

Factor level	1	2 . . . j . . . k
	X_{11}	$X_{12} \ldots X_{1j} \ldots X_{1k}$
	X_{21}	$X_{22} \ldots X_{2j} \ldots X_{2k}$
	X_{31}	$X_{32} \ldots X_{3j} \ldots X_{3k}$
	.	. . .
	.	. . .
	.	. . .
Means	X_{i1}	$X_{i2} \ldots X_{ij} \ldots X_{ik}$
	$X_{.1}$	$X_{.2} \ldots X_{.j} \ldots X_{.k}$

If more than two levels of the factor or treatment are used, the usual analysis of a single-factor experiment consists in a one-way analysis of variance test (ANOVA), where the null hypothesis is that $T_j = 0$ for all j. The number of observations or runs under each of the factor levels need not be the same. If the hypothesis is true, then varying the levels of the factor has had no effect and each observation X_{ij} is made up of its population mean μ and a random error ϵ_{ij}. If the hypothesis is rejected, then we can run several other tests to determine which levels of the factor have had a significantly different effect. We can handle this in two ways, depending upon when the selection is made (before or after the data are gathered), deciding which comparisons or contrasts between means are of interest. If we make the selection ahead of time, such comparisons can usually be drawn without disturbing the α risk of the original ANOVA. The number of such comparisons cannot exceed the number of degrees of freedom between the treatment means (i.e., $k - 1$). The method usually used is called *the method of orthogonal contrasts.* If we postpone the decision regarding what comparisons are to be made until after we examine the data, it is more appropriate to use the *Duncan New Multiple Range Test.* [3]

Most classical experimental designs in the literature are based upon using either analysis of variance or regression analysis upon the data after they are gathered. In general, analysis of variance is used if any qualitative factors are present, and regression analysis if all the factors are quantitative. A very good discussion of the relationship between regression analysis and analysis of variance is found in F. A. Graybill's book [15]. Other excellent discussions of analysis of variance can be found in any of the books on experimental design previously recommended, and a very good treatment of regression analysis as well as the required computer programs is contained in a book by Bartee [2].

4.8 Factorial designs

In the preceding section, we were primarily concerned with the effect of a single factor upon the response. We now consider the situation where two or more factors are of interest to the experimenter. One traditional method is to vary the levels of one factor at a time while keeping all others constant. Under such a scheme, we would make a series of computer runs varying the levels of only one factor while keeping the others unchanged, then another series of runs varying another factor, and so forth until each of the factors had been explored while the others remained unchanged. An experiment on one factor would seldom be considered as adequately replicated unless it had about 8 samples at each level. If we looked at two factors A and B and two levels of each we would then have a situation similar to Table 4.2 (where each x is a datum point) with a total sample of 32.

Table 4.2 *Varying one factor at a time.*

Experiment I		Experiment II		ANOVA for Each Experiment	
A_1	A_2	B_1	B_2	*Source*	*d.f.*
x	x	x	x	Between levels	1
x	x	x	x	Within levels	14
x	x	x	x	Total	15
x	x	x	x		
x	x	x	x		
x	x	x	x		
x	x	x	x		
x	x	x	x		

However, we could run this experiment as a symmetrical full factorial experiment, such as is shown in Table 4.3. A *factorial experiment* is one in which all levels of a given factor are combined with all levels of every other factor in the experiment. By "symmetrical" we mean that each factor has an equal number of levels. Examining Table 4.3, we see that even though we have cut the sample size from 32 to 16, we still have 8 measurements of factor A at each level and 8 measurements of factor B at each level. In addition, we now have a measure of any possible interaction between A and B, which we did not have before, while retaining almost the same number of degrees of freedom for the error term.

Table 4.3 *Factorial design of 2 factors, 2 levels.*

	A_1	A_2
B_1	xxxx	xxxx
B_2	xxxx	xxxx

ANOVA

Source	*d.f.*
A effect	1
B effect	1
AB interaction	1
Error	12
Total	15

It would be an unimaginative experimenter who could not think of additional factors highly relevant to his investigation. So we might consider what would happen if we add a third factor, C, also at two levels. Such a design in matrix form might look like Table 4.4. As a general rule of thumb, we would like to keep the degrees of freedom for the error term at ten (10) or above; therefore, we have increased the sample size to 24. Even so, we find that we now have 12 measurements for each level of the three factors (instead of 8 measurements for each level of two factors), as well as a measurement of all second-order interactions and the third-order interaction, while running eight less samples than originally. We can now begin to see the power and advantage of good experimental design.

Table 4.4 *Factorial design of 3 factors, 2 levels.*

	A_1		A_2	
	C_1	C_2	C_1	C_2
B_1	xxx	xxx	xxx	xxx
B_2	xxx	xxx	xxx	xxx

ANOVA

Source	*d.f.*
A effect	1
B effect	1
C effect	1
AB interaction	1
AC interaction	1
BC interaction	1
ABC interaction	1
Error	16
Total	23

If we wish to study the result upon the model response of changing two or more factors, then, in general, the most efficient method is to use a factorial design. By this we mean that factorial designs allow us to obtain the required information having the required degree of precision with the minimum expenditure of effort. Thus, the factorial design plays an important role in most simulation studies. We can summarize the advantages of factorial design over the classical "one factor at a time" approach as follows:

1 Maximum efficiency in the estimation of the effects of the variables.

2 Correct identification and interpretation of factor interactions if they exist.

3 The effect of a factor is estimated at several levels of the other factors, and thus the conclusions reached hold over a wide range of conditions.

4 Ease of use and interpretation.

4.9 Fractional factorial designs

At the beginning of most simulation studies, we emphasize the speedy determination of which variables are important and which are not. One of the difficulties is that with a full factorial, the number of combinations can grow unmanageable if more than a very few variables are to be investigated. For example, if we look at seven factors each at 2 levels, then $2^7 = 128$ data points are required with no replications. Obviously each replication doubles this. Along with this increase in total number of combinations comes in increase in the number of higher order interactions. Some of these high-order interactions may be used as error since those above second order (three way) would be very hard to explain, even if found significant. Table 4.5 shows the number of main effects and higher order interactions that can be recovered if a complete 2^n factorial is run.

Table 4.5 2^n *Factorial effects [16].*

			Interactions				
n	2^n	*Main Effects*	*1st*	*2nd*	*3rd*	*4th*	*5th*
5	32	5	10	10	5	1	
6	64	6	15	20	15	6	1
7	128	7	21	35	35	21	7
8	256	8	28	56	70	56	28

Since the use of a full factorial design can easily lead to excessive computer time requirements, we need a design that can be used for screening—i.e., for identifying the most important subset of variables influencing the response. It turns out that if we are not interested in some of the higher order interactions, we can glean a great deal of information from running only a portion or fraction (½, ¼, ⅛, etc.) of the total combinations. When only a fraction of a full factorial replicate is run, the design is called a *fractional replication* or a *fractional factorial.* This method allows the analyst to design a series of short experiments that permit him to screen rapidly with relatively few runs a large number of variables for their effect upon the response variable; he then concentrates upon the most significant.

Anytime that we run fewer samples than required by the full factorial design, we pay the penalty of confounding of effects. By *confounding* we mean that the statistic which measures one effect also measures another effect if it is present. For example, if a main effect is confounded with a higher order interaction, the two effects are so mixed that we cannot separate them or distinguish between them. Thus, if our analysis shows that some effect is present, we cannot be sure whether it is the main effect, the interaction effect, or some additive combination. To design an experiment with fewer treatments than are required by the full factorial, the experimenter must first decide on what effects he is willing to confound. In general, it is better to confound a higher order interaction than a main effect. The hope, of course, is that the higher order interaction is not present and that meaningful information on main effects and lower order interactions can still be found. When two or more effects are confounded, they are said to be *aliases* of one another. Success in the use of fractional factorials depends upon choosing the fraction to be run in such a way that no important effect has another one as its alias.

When the number of factors is small (less than 5, generally), a fractional design is usually not desirable because the aliases are bound to prevent discrimination between main effects and important interactions. The basic principles and methods of constructing fractional factorial designs are presented in almost any good text on experimental design [9, 12, 16]. Tables of some of the available designs can be found in Cochran and Cox [9] and publications in the applied Mathematics Series of the National Bureau of Standards [10, 11, 14]. Table 4.6 shows what effects it is possible to retrieve in 2^n designs with various fractional factorial designs. Davis [12, p. 476] presents a similar table for 3^n designs. Remember that Table 4.6 does not show an allowance of any degrees of freedom for the error estimate needed to perform an analysis of variance (ANOVA). It would therefore be necessary to use some of the higher order interactions for this purpose.

For demonstration purposes, let us consider a design for a ¼ replicate of a 2^5 factorial design. One possibility would be as shown in Table 4.7. For each observation, a −1 under a particular factor indicates it should be set at its low value and a +1 indicates it should be set at the higher level.

This particular design was derived in the following way. We want to confound the higher interactions, and since we are going to take a ¼ replicate we need to divide the 32 combinations called for in the full factorial into four blocks of equal size (8 each) and then choose one of the blocks for our experiment. This means we must confound three effects in the process of putting the observations into blocks. We do this by choosing two, three-factor interactions with one factor in common. In this case we chose BCE and ADE, which have the factor E in common. When we confound these two interactions, we also automatically confound $ABCD$. We therefore have the so-called *identity* or *defining contrast*, $I = BCE = ADE = ABCD$, which was used to divide the 32 observations into four blocks of 8 observations each. This means we have lost the possibility of recovering these effects from the data. Having used this identity as the basis for confounding in our design, we would now end up with the following aliases:

$$A = DE = BCD = ABCE$$

$$B = CE = ACD = ABDE$$

$$C = BE = ABD = ACDE$$

$$D = AE = ABC = BCDE$$

$$E = BC = AD = ABCDE$$

Table 4.6 *Fractional replicates in 2^n designs.*

Number of Factors	Number of Observations: 8	16	32	64
3	All main and interaction effects—full factorial	Two replicates of full factorial	Four replicates of full factorial	Eight replicates of full factorial
4	All main effects and 3 of the 2-factor interactions	All main and interaction effects—full factorial	Two replicates of full factorial	Four replicates of full factorial
5	All main effects and the interaction of 1-factor with each of 2 others	All main effects and 2-factor interactions	All main and interaction effects—full factorial	Two replicates of full factorial
6	All main effects and 1 2-factor interaction	All main effects and not more than nine 2-factor interactions*	All main and 2-factor interaction effects plus not more than ten 3-factor interactions*	All main and interaction effects—full factorial
7	All main effects	All main effects and not more than eight 2-factor interactions*	All main effects and not more than 18 of the 2-factor interactions and 6 of the 3-factor interactions*	All main effects, all 2-factor interactions (21) and all 3-factor interactions (35)

* No two of which are from the same alias group.

Table 4.7 *Design for five factors in eight observations*

	Factor				
Observation	*A*	*B*	*C*	*D*	*E*
1	1	1	1	1	1
2	1	-1	-1	1	-1
3	-1	1	1	-1	-1
4	1	1	1	1	-1
5	1	1	-1	-1	1
6	-1	1	-1	1	1
7	1	-1	1	-1	1
8	-1	-1	1	1	1

$$AB = CD = ACE = BDE$$

$$AC = BD = ABE = CDE$$

This means that if there are *no* interactions, we could recover from our eight data points the information on each of the five main effects (A, B, C, D, and E) and use the data from the AB and AC interaction effects as an estimate of the error to perform our ANOVA analysis. A complete discussion of how to select the identity, develop the blocks, determine the aliases, and analyze the data after the experiment, can be found in any number of references [3, 9, 12, 16]. Bonini [5] has an example of the use of a fractional factorial design in simulation studies, as do Hunter and Naylor [18].

It is also sometimes desirable to use fractional mixed-level factorial designs. Mixed-level designs are for experiments where some factors are at two levels and other factors are at three levels. Connor and Young [10] present a whole family of such designs constructed so that the grand mean, all main effects, and all two-factor interaction effects can be estimated without aliasing among them. It is assumed that all higher order interaction effects are negligible.

In summary, we use fractional factorial designs primarily during the early stages of a study to explore the effect of variables when we are faced with a large number of factors (five or more) of possible interest. If, on the basis of this screening, we can reduce the number to four variables or less, we then use full factorial designs for more detailed analysis.

4.10 Determination of optimum conditions

In many simulation projects, the goal is to find those values or levels of the independent variables that will yield the optimum (maximum or minimum) values of the response or dependent variable. If the dependent and independent variables are both quantitative and continuous, *response surface methodology* (RSM) is usually the most appropriate approach to use. Response surface methodology was initially proposed by Box and Wilson [8] in 1951. The underlying philosophy and use of RSM is very well discussed in a number of textbooks, including Davies [12], Cochran and Cox [9], and Hicks [16]. The discussion in Davies is particularly good for someone beginning the study of RSM.

The concept of a response surface entails a dependent or response variable (y) and several independent or controlled variables ($x_1, x_2, \cdots, x_k$). If all these variables are assumed to be quantitative, continuous, and measurable, then, if the value of the response depends upon the levels of the k factors, the response surface can be expressed as

$$y = F(x_1, x_2, \cdots x_k)$$

It is useful to visualize the relation between the response and the factor levels geometrically, such as in Fig. 4.6. In this figure we have two factors or independent variables x_1 and x_2 plotted on the abscissa and ordinate, respectively, and y, the response variable, plotted perpendicular to the x_1 x_2 plane. An alternative and very useful two-dimensional representation is obtained if we connect the values of x_1 and x_2 that yield the same value of y, such as in Fig. 4.7, yielding a series of equal response lines or contours. These are similar to the contours of equal height on topographical maps or the isobars showing the distribution of atmospheric pressure on weather charts. The methods developed for RSM are in no way limited by the fact that geometrical

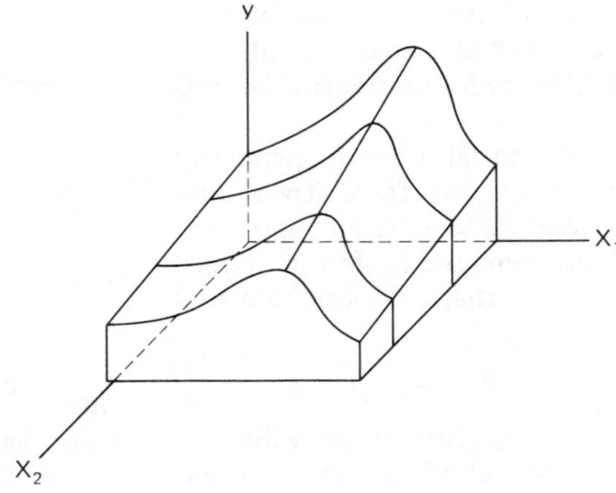

Figure 4.6 Response surface.

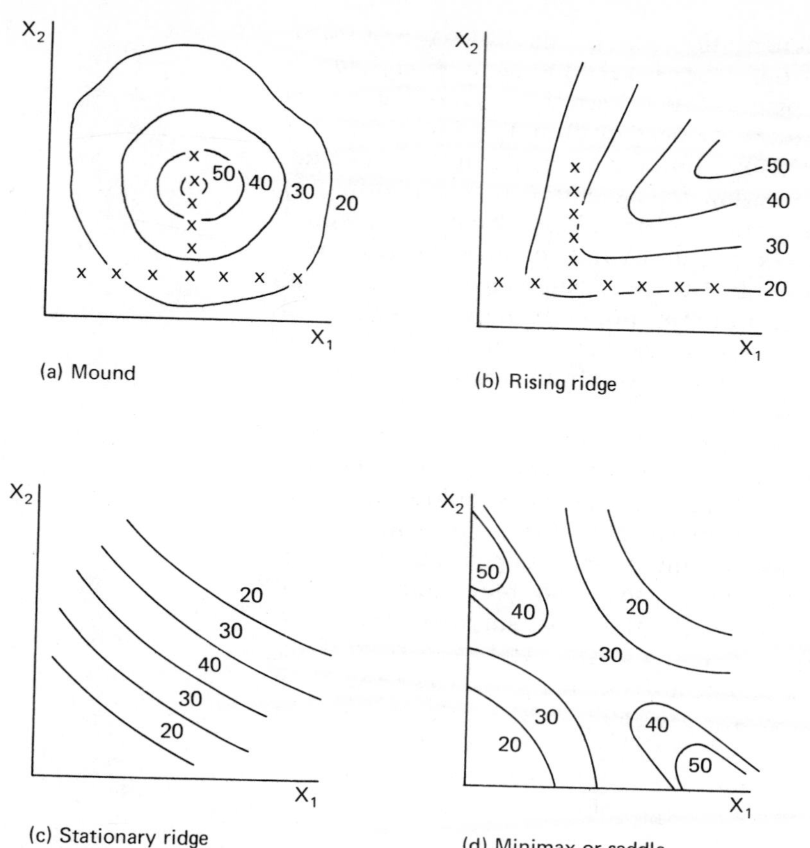

Figure 4.7 Typical response contours.

representation is only possible up to three dimensions. However, in the following discussions we limit ourselves to surfaces of the form $y = f(x_1, x_2)$ because familiarity with the geometrical representation for two and three factors makes the general situation for k factors (which cannot be geometrically visualized) more readily understood.

Response surface methodology usually involves a series of small experiments utilizing full or fractional factorial designs to explore the response surface. We start by running a small experiment, and are then faced with two problems. The first is to determine from this experiment where to move for the next experiment, so as to come closer to the optimal point on the underlying response surface. The second problem is that having located the optimum or near optimum of the surface, we would like to determine the equation of the response surface in the area near this optimum point.

One method of experimentation that is sometimes used to seek the optimum point is the traditional one-factor-at-a-time or sectioning method. As shown in Fig. 4.7(a), if x_2 is fixed and x_1 is varied, we find the optimum (or near optimum) value for x_1 for the specified value of x_2. Now, holding x_1 at its optimum value, we could vary x_2 until we find its optimum (or near optimum) setting.In the case of the mound in Fig. 4.7(a), this method would lead eventually to the peak of the mound (or near it). However, this same method, when applied to the rising ridge shown in Fig. 4.7(b), fails to lead to the optimum point. In most experimental work, the type or form of the response surface is usually unknown; thus, a better method is needed if the optimum set of conditions is to be found for an unknown surface.

If the exact mathematical form of the response function is known, finding the optimum would be relatively easy through analytical means. Of course, if this was true, we shouldn't be doing a simulation study in the first place. Therefore, our problem is to move to the area that is near optimal as soon as possible, and then, as the optimum is approached, use some technique of local exploration to find more precisely the most suitable conditions. Since we do not generally know the form of the response surface, some form of flexible, graduating function is usually used as an approximation. The two most commonly used forms are the first order polynomial

$$y = a_0 + a_1 x_1 + a_2 x_2 + \cdots + a_k x_k$$

and the second-order polynomial

$$\begin{aligned} y = a_0 &+ a_1 x_1 + a_2 x_2 + \cdots + a_k x_k + a_{12} x_1 x_2 + a_{13} x_1 x_3 \\ &+ \cdots + a_{k-1,k} x_{k-1} x_k + a_{21} x_1^2 + a_{22} x_2^2 + \cdots + a_{kk} x_k^2 \end{aligned}$$

where the parameters a_0, $a_1 a_2$, etc., are to be estimated from the data. It has been shown that the flexibility of such first- and second-order models can sometimes be increased substantially by transformations of the x's and y's [6, 7].

The method most commonly used in RSM is called the *path of steepest ascent method.* The idea is to run a simple experiment (usually some form of factorial) over a small area of the response surface, where for all practical purposes we can consider the surface to be a plane. We then determine the equation of this plane (by fitting the data to a first-order polynomial) and from it decide the direction we should go for the next experiment so as to move toward the optimum of the surface. Since the next experiment should be in the direction in which we hope to scale the height the fastest, we refer to it as the path of steepest ascent.

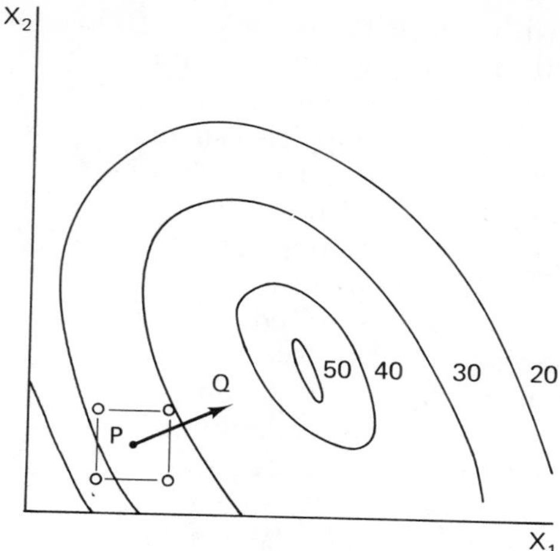

Figure 4.8 Responses contour with 2^2 factorial design.

This technique does not determine how far away from the original experiment succeeding experiments should be run, but it does indicate to the analyst the direction along which the next experiment should be conducted.

We can perhaps see this concept if we consider the response contours of Fig. 4.8. Suppose the analyst has run a 2^2 factorial experiment plus two data points at the center, as shown at point P. From these data points he estimates a_0, a_1, and a_2, which along with their signs give him the slopes of the plane and allow him to calculate the direction of steepest ascent or greatest slope up the plane. This is the direction at right angles to the contour lines and shows the relative amounts by which the factors must be varied in order to give an increase in the response. He then proceeds to point Q in this direction, where he repeats the process. By this step-by-step procedure, he attains points of higher and higher response.

As we approach the optimum area, this procedure becomes self-defeating, because slopes a_1 and a_2 become more gradual and thus more difficult to estimate. In addition, as we begin to approach the peak, the plane approximation (first-order polynomial) becomes less satisfactory and we need a higher order polynomial. In the example shown in Fig. 4.8, a simple 2^2 factorial would be sufficient to estimate the coefficients a_0, a_1, and a_2. However, by running two or more data points at the geometric center of the factorial, we obtain not only a better regression line but also some degrees of freedom for testing the significance of the estimated parameters. This could also be accomplished by replicating the whole experiment.

The successive use of 2^n factorials with center points will not accurately locate a maximum but it will bring the analyst rapidly into what might be called a *near stationary region.* A near stationary region is one where the surface slopes are small compared with the errors of estimation. Once we reach such a region, we need to explore its nature and the possible location of a maximum point nearby by other techniques, including second-order or higher polynomials to account better for curvature in the response surface. This transition is usually prompted by a significant lack of fit of the first-order model.

The fact that a region is near stationary does not, of course, necessarily mean that we are in the neighborhood of a maximum, such as that in Fig. 4.7(a). We may, in fact, have arrived at a ridge that is slowly rising, as in Fig. 4.7(b), or stationary, as in Fig. 4.7(c). Another possibility is that the point reached is near a minimax (sometimes called a saddle point), which is a maximum in some directions but a minimum in others [such as Fig. 4.7(d)]. This latter occurrence (a saddle point) might provide a useful warning that we are dealing with a response surface having two or more peaks and that we may have climbed up one of the minor peaks and not be in the region of the true maximum at all.

The method of steepest ascent has the potential shortcoming of not guaranteeing that we will reach the maximum if the surface has more than one peak. In such cases, the particular scaled peak would depend upon the starting point for the experiment. If one suspects the possibility of multiple peaks, or if one appears to have reached a saddle point, it would be a good idea to explore a larger portion of the surface by trying some other starting points. The presence of multiple peaks in the response surface would imply that some fundamental changes in the nature of the response of the system occur under certain combinations of factors. Although this is certainly always a possibility. We know of no actual real world cases where multiple peaks have been encountered.

As mentioned, when exploring a near stationary region for an optimum point, we wish to try and fit at least a second-degree polynomial. Experimental arrangements for this purpose are called *second-order designs.* The general form of a quadratic (second-degree) polynomial for two factors is given by

$$y = a_0 + a_1 x_1 + a_2 x_2 + a_{11} x_1^2 + a_{22} x_2^2 + a_{12} x_1 x_2$$

To estimate the regression coefficents in this model, we must measure each factor or variable for at least three different levels. This would suggest the use of 3^n full or fractional factorial designs, which indeed are often used for this purpose. Unfortunately, if more than three variables are being explored, the number of runs required can become very large, even when we use fractional designs. In addition, the coefficients of the squared terms are estimated with relatively low precision from 3^n factorial designs. Therefore, Box and Wilson [8] developed new designs specifically for fitting second-order response surfaces. Their most useful designs are called *central composite* or *rotatable* designs, and are constructed by adding further data points to those obtained from a 2^n factorial. In a rotatable design, the standard error is the same for all points that are at the same distance from the center of the region. Such designs exist for any number of factors and consist in regular or semiregular geometric figures plus center points.

A fairly simple and very useful design is the *cube plus star plus center points.* Such a design for two variables is shown in Fig. 4.9.

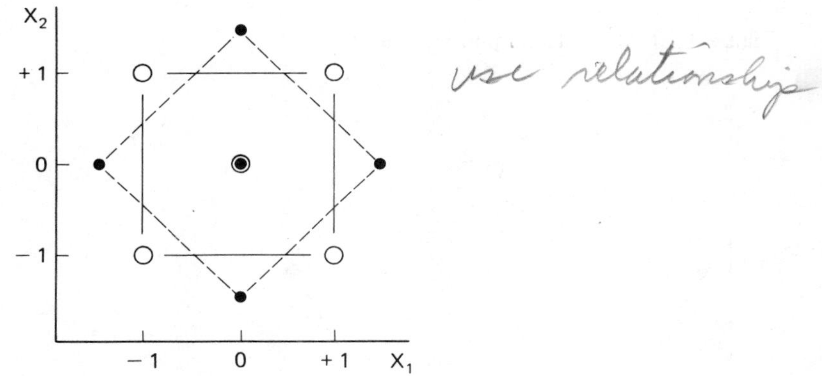

Figure 4.9 2^2 **factorial plus star and center points.**

Notice that this is a 2^2 factorial (the cube) plus four extra points (the star), which are equally spaced around the circumference of a circle in the x_1x_2 plane with center (0, 0) plus center points. The same concept for a three-factor design is shown in Fig. 4.10. Other rotatable geometric designs are sometimes used, such as the *hexagonal plus center points* design shown in Fig. 4.11.

Finally, we must mention the use of *canonical analysis* as a method for interpreting the approximating quadratic equation. This method of analysis, which is discussed in Davies [12] as well as Cochran and Cox [9], helps us gain insight into the nature of the response surface. It is particularly useful in exploring whether one has arrived at the peak of a mound or, on the contrary, a minimax or a rising ridge. It is particularly handy when we deal with complex systems containing three or more factors.

The literature on response surface methodology is very extensive and has been given very cursory treatment in this section. The best treatment of the one-at-a-time or sectioning

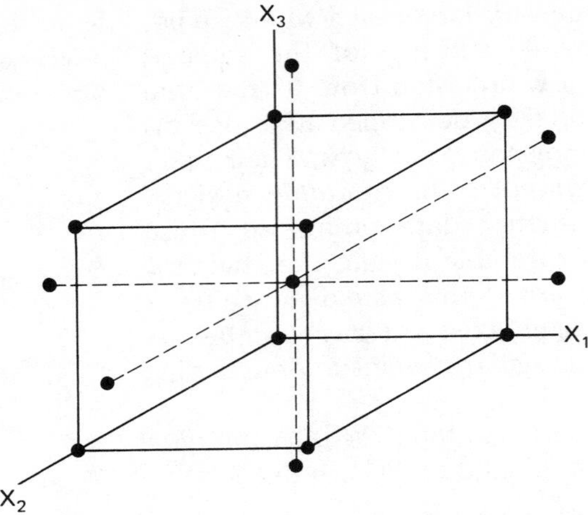

Figure 4.10 3^3 factorial plus star and center points.

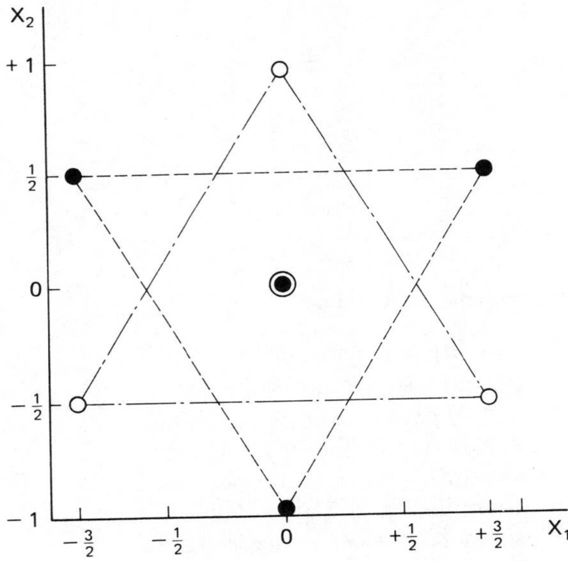

Figure 4.11 Hexagonal design.

approach is given by Schmidt and Taylor [22]. For the method of steepest ascent, we recommend Davies [12] and/or Cochran and Cox [9]. A discussion more directly tied to simulation is given by Mirham [19]. The best guide to the journal literature through 1965 is contained in a review by Hill and Hunter [17].

REFERENCES

1. ACKOFF, R. L., and M. W. SASIENI, *Fundamentals of Operations Research*, John Wiley & Sons, Inc., New York, 1968.
2. BARTEE, E. M. *Statistical Methods in Engineering Experiments*, Charles E. Merrill Books, Inc., Columbus, Ohio, 1966.
3. BARTEE, E. M., *Engineering Experimental Design Fundamentals*, Prentice-Hall, Inc., Englewood Cliffs, N. J., 1968.
4. BIDWELL, E. L., "Simulation in GPSS/360 of a Highway Paving Operation Using a Mobile Control-mix Plant with Different Haul Truck Speeds and Fleet Size Combinations," *Proceedings of Winter Simulation Conference*, 1971.
5. BONINI, C. P., *Simulation of Information and Decision Systems in the Firm*, Prentice-Hall, Inc., Englewood Cliffs, N. J., 1963.
6. BOX, G. E. P., and D. R. COX, "An Analysis of Transformations," *Journal of the Royal Statistical Society*, Ser. B, Vol. 26, 1964.
7. BOX, G. E. P., and P. W. Tidwell, "Transformations of the Independent Variables," *Technometrics*, Vol. 4, 1962.
8. BOX, G. E. P., and K. B. WILSON, "On the Experimental Attainment of Optimum Conditions," *Journal of the Royal Statistical Society*, Ser. B, Vol. 13, No. 1, 1951.
9. COCHRAN, W. G., and G. M. COX, *Experimental Designs*, 2nd ed., John Wiley & Sons, Inc., New York, 1957.
10. CONNOR, W. S., and S. YOUNG, "Fractional Factorial Designs for Experiments with Factors at Two and Three Levels," *National Bureau of Standards Applied Mathematics Series*, No. 58, U.S. Government Printing Office, Washington, D. C., 1961.
11. CONNOR, W. S., and M. ZELEN, "Fractional Factorial Experiment Designs for Factors at Three Levels," *National Bureau of Standards Applied Mathematics Series*, No. 54, U.S. Government Printing Office, Washington, D. C., 1959.
12. DAVIES, O. L. (ed.), *The Design and Analysis of Industrial Experiments*, Hafner Publishing Co., New York, 1963.
13. DIXON, W. J., and E. J. MASSEY, Jr., *Introduction to Statistical Analysis*, McGraw-Hill Book Company, New York, 1957.
14. "Fractional Factorial Experiment Designs for Factors at Two Levels," *National Bureau of Standards Applied Mathematics Series*, No. 48, U.S. Government Printing Office, Washington, D. C., 1957.
15. GRAYBILL, F. A., *An Introduction to Linear Statistical Models*, Vol. I, McGraw-Hill Book Company, New York, 1961.

16. HICKS, C. R., *Fundamental Concepts in the Design of Experiments*, 2nd ed., Holt, Rinehart & Winston, Inc., New York, 1973.

17. HILL, W. J., and W. G. HUNTER, "A Review of Response Surface Methodology: A Literature Survey," *Technometrics*, Vol. 8, No. 4, Nov., 1966.

18. HUNTER, J. S., and T. H. NAYLOR, "Experimental Designs for Computer Simulation Experiments," *Management Science*, Vol. 16, No. 7, Mar., 1970.

19. MIRHAM, G. A., *Simulation: Statistical Foundations and Methodology*, Academic Press, New York, 1972.

20. NAYLOR, T. H. (ed.), "*The Design of Computer Simulation Experiments*, Duke University Press, Durham, N. C., 1969.

21. ODOM, P. R., and R. E. SHANNON, "Nomographs for Computer Simulation," *Industrial Engineering*, Vol. 5, No. 11, Nov., 1973.

22. SCHMIDT, J. W., and R. E. TAYLOR, *Simulation and Analysis of Industrial Systems*, Richard D. Irwin, Inc., Homewood, Ill., 1970.

EXERCISES

1. Read case study IV, Appendix A.
   a) The authors ran a series of small experiments in this study. Do you think this was a good approach or should they have run one big factorial design?
   b) Show what a full factorial design would look like for the conditions shown in Fig. 4.4.
   c) How many cells would you have for your design in (b)?
2. A researcher has developed a simulation model that requires 2 minutes per case to run. He has $10,000 left in his budget for computer time, which costs $300 per hour. His structural model for running the experiment consists of 4 factors each at 3 levels in a factorial design. How many replicates of the basic design can be run?
3. The researcher in Problem 2 believes he can improve the model by reprogramming in such a way that it will not affect the validity of the model but will halve the run time. He calculates that the cost of reprogramming will be about $2000 (cost of labor, etc.). Should he redo the model? Assume he wishes to retain the same number of replicates calculated in Problem 2.
4. Assume that the results of the study must be available in about 100 working days and that no more than 20 minutes per day of computer time can be devoted to the project. The researcher can only devote one programmer to the reprogramming effort at a cost of $5 per hour. Assuming the programmer can work only 8 hours per day and

disregarding computer time used during the reprogramming, would you change your decision reached in Problem 3?

5. The researcher in Problem 2 discovers an accounting error has been made and he only has $5000 available for computer time. After reconsidering his experimental design, he decides that although undesirable, he could probably reduce the number of factors, levels, or replications. Disregarding the information given in Problem 3 and 4, which should he consider first?

6. Put all the information given in Problems 2 through 5 together and recommend what the researcher should do.

7. In case study V, Appendix A, the authors ran an experiment with various combinations of operators (1 to 3), mechanics (1 or 2) and helpers (1 to 3). Design a full factorial experiment for this situation. Fill in the results (total cost) from Table III, case A-V, and determine whether their lack of running a full factorial experiment may have hurt their results.

chapter five

TACTICAL PLANNING

Reality is neither in the world nor in our models, but in the process of working back and forth between world and model.

E. A. Singer

5.1 Problems of variability

In addition to the problems of basic experimental design discussed in Chapter 4, another group of closely related topics might be called tactical problems [3]. Since variablility is a characteristic of all stochastic simulation models, replication (the repetition of the set of factor treatments in the experiment) is required for accuracy and precision in our results. Computer time is not free, and the running time to execute a complex simulation experiment can be great; therefore, we must get as much information as possible from the smallest number of runs. The investigator has to conduct his experiment in such a way that he not only obtains results but also can estimate the precision of his results and the confidence attributable to the conclusions he draws from the data.

The degree of precision is defined as the variability associated with the factor or treatment mean (the variance of a treatment mean). Given that a specified treatment difference is important, the experimenter must decide the risk he is willing to take for asserting that (1) a true difference of the stated size does not exist when it is actually present (type II error or β) and (2) a larger than stated value exists when no true difference exists (type I error or α). The degree of precision desired may be expressed in several ways, such as (a) fraction of the standard deviation, (b) percent of the mean, or (c) in absolute terms. The degree of precision desired depends upon the nature of the treatment and the magnitude of the treatment differences expected for a specified factor. The amount of simulation effort and the accuracy of the results are very much interrelated. Thus, if one can reduce the amount of simulation effort required while still maintaining a certain level of accuracy, or if one can increase the accuracy without increasing the effort, the cost of obtaining a usable solution can be reduced.

Two alternative methods reduce the error of the estimate of a treatment effect. One is to refine the technique (i.e., use variance reduction techniques) associated with how we draw our samples and the manner in which we generate the random variates. Alternately, we can replicate (repeat) the experiment a number of times and take the average of the results. Replication increases the accuracy obtainable by averaging the results of several observations. Unfortunately, this process is inefficient, for the error of the average (standard error) is inversely proportional to the square root of the number of observations. Thus, averaging 4 observations brings the error of the average down to half that of a single observation, whereas the average of 16 observations decreases this to only one-fourth of the error of a single observation.

In determining the number of replicates for an experiment, the analyst should list the response variables of interest with their estimated standard deviations. The number of replicates required should then be determined for the response variable of greatest interest. If several outputs are important we should determine the number of replicates based upon the response with the greatest variability. This number will then be larger than the required number of replicates for all the other output variables. If the expected size of treatment differences varies considerably, and if different levels of confidence are desired for the various response variables, we may have to compute the required number of replicates for all the responses individually and then choose a number large enough to cover all the conditions of interest.

We can only assess the right number of replicates to use from considering the expected standard errors in relation to the expected magnitudes of the effects of interest. An estimate of the experimental variance must be available, and the size of the minimum difference of interest must be specified. For computing, the number of replicates required to detect a specified minimum difference between parameters at a specified significance level, we have several methods available, the choice of which is determined by the specific situation. In the remainder of this chapter, we address some of the important considerations that affect the possible cost-accuracy-precision tradeoffs. Among the questions to be considered are those of convergence, starting conditions, sample size, and variance reduction techniques.

5.2 Starting conditions and equilibrium

As we pointed out in Section 4.2, one of the primary ways that experimentation by simulation differs from real world physical experimentation is the ease with which experimental conditions may be repeated or reproduced and the ease of starting, stopping, and resuming experimentation. This is both an advantage and a disadvantage—an advantage because it gives the experimenter greater control over his experiment, and a disadvantage because he must think and consider carefully how to start up his model and when to start taking data. We run most simulation models with the idea of studying steady-state or equilibrium operating conditions, i.e., we want to study the system under typical day-to-day operating conditions. In most stochastic models there is unfortunately an initial bias or transient condition that is not typical of steady-state or equilibrium conditions, owing to the fact that it takes some time for the model to overcome the atypical or artificial situation created by the abrupt beginning of operation. Probably the most common way to start a simulation model is with the system empty and idle, i.e., no transactions are going on and the system is completely clear of activity. For some real world systems, this condition is true only when the system is activated for the first time. For example, a hospital is never empty, a busy airport almost always has some planes on the ground and in the air, the manufacturing plant always has work in progress (even when shut down for the night), etc.

There are at least three ways that we can reduce the effect of this initial transient period so as to minimize the biasing effect it might have on our data:

1 Use long enough computer runs so that the data from the transient period are insignificant relative to the data from the steady-state condition.

2 Throw out or exclude some appropriate part of the initial period of the run from our considerations.

3 Choose initial starting conditions that are more typical of the steady-state condition and thus reduce the transient period.

Each of these presents its own problems in terms of implementation. The first approach of using long runs may be satisfactory if the model runs very fast, i.e., a large sample size can be easily run with minimal computer time. But if we are dealing with a complex model where each datum point is costly or the initial period of instability is long, the cost of computer time can easily become prohibitive. More often than not, it will be more satisfactory to use one or both of the latter two approaches.

The second approach i.e., throwing away or discarding the initial part of each run (the nonequilibrium or non-steady-state portion), also presents some problems. First, we are discarding or negating a part of each computer run with the attendant cost in machine time. Second, there is evidence that in some cases, even though we decrease the bias, we can, in fact, increase rather than decrease the variance. Fishman [11] has shown that in some cases the standard error increases when truncation is used. Third, we have the basic question of deciding when equilibrium has been attained so as to determine how much to truncate or discard.

No foolproof method determines when steady-state or equilibrium conditions have been obtained. As Conway [3] states, "It is also important to recognize that equilibrium is a limiting condition which may be approached but actually never attained." Also, steady-state conditions do not mean that the response variables have settled out to a steady or constant value. The response variables will continue to show their inherent variability despite the number of samples taken. By *equilibrium* or *steady-state*, we mean a condition of regularity or stability in which opposing forces or influences are balanced. We assume that for each model there is a limiting probability distribution of the response that is characteristic of the system. The model is in equilibrium when the output agrees with that limiting or characteristic probability distribution.

We said that in most studies we were interested in the performance of the system under steady-state conditions. Obviously this is not always true; there are cases in which we are mainly interested in the startup period and its characteristics. Our problem remains, however, since now instead of trying to decide how much of the early part of a run to throw away, we need to decide at what point the transient period has ended so as to stop the run.

We know of no completely satisfactory way to decide when equilibrium has been achieved. In general, all the rules of thumb that have been used and proposed are based upon examination of a pilot or preliminary run or runs. If possible, decisions should be made on the basis of more than one pilot run. Gordon [13], Fetter and Thompson [9], and others have demonstrated that the period of instability is highly sensitive to the initial choice of a seed for the random number generator. In deciding when equilibrium has been achieved (i.e., how much of the initial part of a run to discard), we ordinarily should decide upon a period of stabilization based upon examination of several runs and then discard or delete the same amount from all succeeding runs.

How do we decide when equilibrium has been achieved? We can only offer the reader several suggestions made by others.

1 Conway [3]. "Truncate a series of measurements until the first of the series is neither the maximum nor minimum of the remaining set."

2 Emshoff and Sisson [7]. "A simple method we have found useful involves examining a sequence of observations of a run. If the number of observations in which the output is greater than the average is about the same as the number in which it is less, then steady-state conditions are likely to exist."

3 Fishman [10]. For an autocorrelated time series, Fishman suggests calculating a time period k, the longest time lag for which the data are significantly correlated, and then deleting this time period (k) from the beginning of each succeeding run.

4 Tocher [29]. "A general requirement is that the longest cycle in the plant should have been executed at least 3 or 4 times before transient abnormal behavior induced by non-sensible starting conditions can be expected to have died away."

5 Gordon [13]. Gordon suggests plotting the variance of the sample mean vs. sample size on log-log paper. If the initial bias is not present, the data should lie on a straight line, sloping downward at the rate of about 1 in 2, indicating that in the absence of initial bias, the standard deviation can be expected to be inversely proportional to $n^{1/4}$.

6 Emshoff and Sisson [7]. "Another method is to compute a moving average of the output and to assume steady state when the average no longer changes significantly over time."

Inasmuch as suggestions 5 and 6 are based upon cumulative statistics, we know that the output will ultimately settle to a steady-state value (see Fig. 5.1). This, unfortunately, can cause unnecessarily great quantities of data to be discarded, since cumulative statistics lag behind the current state. The greater is the magnitude of the fluctuations during the startup period, the greater is the lag between current state conditions and the status indicated by the cumulative statistic.

We can shorten (but not eliminate) the stabilization period if we choose initial starting conditions that are more typical of the steady-state condition. This approach which was proposed by Conway [3], has a great deal to recommend it. The danger, of course, is that we may bias the result to preconceived conclusions if the data are collected before the transients have been removed. There is also a further complication when the purpose of the experiment is to compare alternatives. Suppose we want to

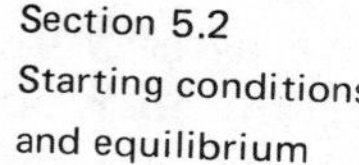

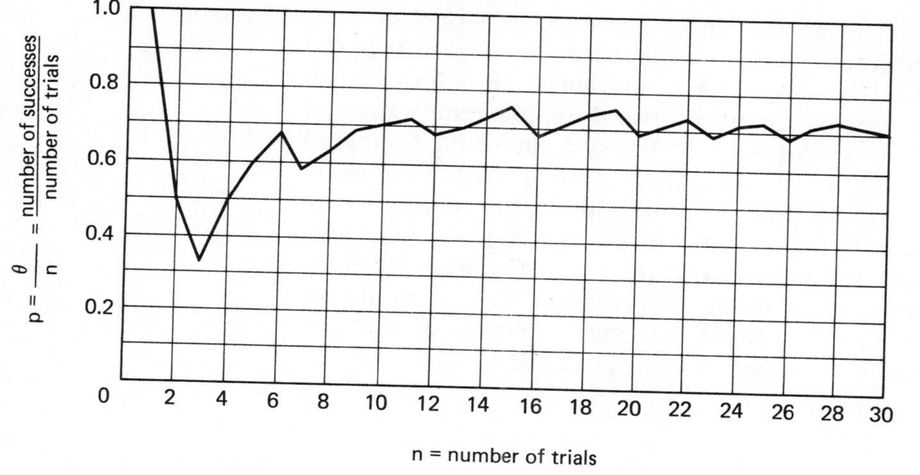

Figure 5.1 Cumulative estimate of p.

compare alternatives *A* and *B*. Assume we estimate that the steady-state conditions for each alternative are significantly different. What should we use for initial starting conditions? Should we use what we expect for alternative *A* as initial conditions when we run *A*, and for alternative *B* when we run *B*? As Conway [3] points out, the analyst has at least three different choices respecting starting conditions:

1 Test each system starting empty and idle.

2 Test each system using a common set of starting conditions that are a compromise among those that appear most reasonable for each system.

3 Test each system with its own most reasonable starting conditions.

The second choice is clearly more efficient than the first, and the third is more efficient than the second if we can come anywhere close in our estimate of reasonable starting conditions. Unfortunately, the use of the third strategy (although most efficient) puts us in the position of having to report our results as follows. "I wished to compare two systems; A and B. I anticipated that System A would yield a greater mean value of attribute M than would system B. I performed an experiment in which the initial value of attribute M for System A was set greater than that for System B. The experimental results demonstrate that the mean value of attribute M for System A is significantly greater than that for System B." [3]

We agree with Conway that most investigators would be very reluctant to report their results in such a manner, even if it did result in greater efficiency. Our preference, therefore, is to construct a set of starting conditions based upon a compromise of "reasonable" conditions for each of the systems to be compared and then use these compromise conditions for all runs. Even when we have a great deal of a priori knowledge about the system we are simulating, it is no trivial task to establish a set of reasonable starting conditions. We agree, however, with Conway when he says, "But at least there is the consolation that it is almost impossible to do worse than empty and idle conditions so that any effort will allow some reduction in computing time." [3]

5.3 Determination of sample size

When we use simulation to study a stochastic system, we represent one or more of the variables in the model by probability distributions from which we draw samples. The analyst makes little progress in planning an experiment before being faced with the problem of selecting the sample size. Probably the question most frequently asked a statistician is, "How many samples must I take to get statistical significance?" Unfortunately, we cannot answer this question unless we know how much information the decisionmaker is willing to buy or pay for. Since these samples are randomly drawn, we have some degree of imprecision in the result, which is highly influenced by the choice of sample size. The choice of an appropriate sample size to achieve the desired level of precision and yet to minimize the cost of operating the model is an extremely difficult yet very important one. Unfortunately, as Duncan suggests [5], the sample size taken is usually a function of the money allocated for the testing. However, since we shall use the information furnished by the simulation experiment as the basis for decisions regarding the operation of the real system, we want this information to be as accurate and precise as possible, or we at least want to know the degree of imprecision present. It is therefore essential that a statistical analysis be conducted to determine the required sample sizes.

It is easier said than done, since in most cases if we know precisely the information we need for the analysis, there would be no need for the experiment. If we are interested in estimating output parameter values or shifts of those variables, we need to be able to specify the following:

1 How large a shift in the parameter or distribution do you wish to detect? Or, how close an estimate of the population parameter do you desire?

2 How much variability is present in the population?

3 What size risks are you willing to take?

4 Are the outputs autocorrelated?

5 What is the form or distribution of the outputs?

It is impossible to cover all the ramifications and approaches to sample size determination in this chapter. Instead, we examine some of the more common situations and suggest possible approaches. The sample size may be determined in either of two ways: (1) prior to and independently of the operation of the model; (2) during the operation of the model and based upon the results generated by the model. We address the problem of prior determination first.

It is frequently possible to justify the use of certain forms of prior analysis based upon knowledge of the model. Many forms of analysis are based upon the assumption that the responses of the model are independent and normally distributed. These assumptions can frequently be justified because of the *central limit theorem* from probability theory. Historically, a series of central limit theorems has developed, and we refer the reader to the two volumes by Feller for an excellent discussion [8]. The essence of each of these is that a random response variable Y, which is the result of the sum of a large number of independent random variables from a common probability distribution function, is going to be normally distributed. Both Diananda [4] and Mihram [22] have shown that the usual condition of independence of many contributing random variables is not necessary, and they need not arise from a common probability density function. It is usually sufficient that the response is the additive sum of a large number of contributing effects. This allows us to assume that the response variable of most complex simulation models, which is the additive result of a number of different random variates, will be normally distributed. Another important consideration in deciding whether we are justified in invoking the central limit theorem is the manner in which we have defined a single sample. If each sample is itself a mean (see Section 4.2), then the central limit theorem holds and we can assume normality of the response.

In the most straightforward case, where we can invoke the central limit theorem and assume no autocorrelation, we can take a confidence limit approach to determining the sample size required for estimating the parameters to a specified level of precision. These parameters are population mean, proportion, and population standard deviation. The sample size estimation methods are now presented, with examples under the headings of the parameter being estimated.

5.4 Estimating the population mean

Suppose we wish to determine an estimate $\bar{X}$ of the true population mean μ, such that

$$P\left\{ \mu - d \leqslant \bar{X} \leqslant \mu + d \right\} = 1 - \alpha \qquad (5.1)$$

where $\bar{X}$ is the sample mean, μ is the true population mean, and $1 - \alpha$ is the probability that the interval $\mu \pm d$ contains $\bar{X}$. The problem is to determine the sample size such that Equation 5.1 holds. If our assumptions of normality are valid, we can show that

$$n = \frac{(\sigma Z_{\alpha/2})^2}{d^2} \qquad (5.2)$$

where $Z_{\alpha/2}$ is the two-tailed standardized normal statistic for the probability we seek. The derivation of Equation 5.2 is found in Mize and Cox [23]. To solve Equation 5.2, we must know σ, $Z_{\alpha/2}$, and d—i.e., (a) How much variability is there in the system? (b) What risks are we willing to take? (c) How big a difference do we want to allow between our estimate and the true parameter? In many cases, we must either guess at the value of σ or run a short pilot experiment. If we have some idea of what the highest and lowest possible responses of the system might be (i.e., the feasible range), then we can take advantage of the fact that the range is approximately equal to 4σ to estimate σ.

EXAMPLE 5.4.1

Suppose we want to estimate the average daily yield of a chemical processing plant such that the error of the estimate is less than $\pm$ 4 tons with a probability of 0.95. That is, we wish to have a 95% probability that our estimate $\bar{X}$ will be within the interval $\mu \pm 4$ tons. We estimate that the feasible range of outputs covers 80 tons. Thus, $4\sigma = 80$ or $\sigma = 20$, $d = 4$, $Z_{\alpha/2} = 1.96$.

$$\begin{aligned} n &= \frac{(\sigma Z_{\alpha/2})^2}{d^2} \\ &= \frac{[20\,(1.96)]^2}{4^2} \\ &= 96 \end{aligned}$$

EXAMPLE 5.4.2

Assume now that we do not know the feasible range of outputs for the problem in Example 5.4.1 and have no idea of the value of the true σ. We decide that rather than set an arbitrary value on d, we want our estimate to lie within the interval $\mu \pm \sigma/4$ with a probability of 0.95. Thus, $d = \sigma/4$ and $Z_{\alpha/2} = 1.96$.

$$n = \frac{(1.96\sigma)^2}{(\sigma/4)^2}$$

$$= \frac{(4)^2\ (1.96)^2\ \sigma^2}{\sigma^2}$$

$$= 61$$

Thus, we can estimate the required sample size without specifying σ if we are willing to state our d in terms of fractions of the true σ.

If feasible, a more desirable approach if the variance is unknown is to run a pilot study or take a preliminary sample and obtain an estimate of the variance (s^2), and then compute the total number of observations necessary. Thus, when σ^2 is unknown, we take a sample and estimate n by

$$n = \frac{t^2 s^2}{d^2} \tag{5.3}$$

where

t = tabulated t value for the desired confidence level and the degrees of freedom of the initial sample

d = the half-width of the desired confidence interval

s^2 = the estimate of the variance obtained in the sample or pilot run

5.5 Use of Tchebycheff's theorem

Tchebycheff's theorem of Inequality says that given a number k greater than or equal to 1 and a set of n measurements $x_1, x_2, \ldots, x_n$, at least $(1 - 1/k^2)$ of the measurements will be within k standard deviations of their mean. This theorem holds for all samples and thus makes no assumptions about normality of the underlying population. If we are unwilling to assume normality of our response output, we can use this theorem to estimate our required sample size in the following way. Tchebycheff's theorem states

$$P \left\{ |x-\mu| > k\sigma \right\} \leqslant \frac{1}{k^2} \tag{5.4}$$

Let us assume we wish our estimate again to be within $\sigma/4$ of the true μ with a probability of 0.95—i.e., we want

$$P \left\{ |\bar{X} - \mu| > \frac{\sigma}{4} \right\} \leqslant 0.05 \tag{5.5}$$

Relating this to Equation 5.1 of Section 5.4, we have

$$P \left\{ |\bar{X} - \mu| > \frac{\sqrt{n}}{4} \frac{\sigma}{n} \right\} \leqslant 0.05 = \frac{4^2}{n} \tag{5.6}$$

since

$$k = \frac{\sqrt{n}}{4} \text{ and } \frac{1}{k^2} = \frac{4^2}{n}$$

We then solve the right-hand side for n

$$n = \frac{4^2}{0.05} = 320$$

This gives us a considerably larger required sample size—the price we must pay for being unable to assume a normally distributed output. Table 5.1 shows the difference in sample sizes for various values of d using the central limit theorem and Tchebycheff's theorem for $\alpha = 0.05$:

Table 5.1 *Sample size required to determine mean within* $\mu \pm d$ at $\alpha = 0.05$.

Difference d	n Central Limit	n Tchebycheff's Theorem
$\sigma/2$	15	80
$\sigma/4$	61	320
$\sigma/6$	138	720
$\sigma/8$	246	1280
$\sigma/10$	384	2000
$\sigma/12$	553	2880
$\sigma/20$	1537	8000

5.6 Estimating proportions

In many simulation models, the output is in the form of a proportion, fraction, or percentage. For example, in simulations of military weapons we are often interested in the percentage of attackers or targets destroyed under certain situations. This is usually called the "kill probability." In industrial situations, we are sometimes interested in estimating the percent of defective units produced or the percent of jobs finished on time. Responses of this type are characteristically classified into one of two states—i.e., success-failure; go-no go; present-not present; destroyed-not destroyed; cured-not cured; etc. Such responses are called *Bernoulli variables* and are characterized by the binomial distribution.

If we let p equal the true probability that an event will happen in a single trial and $q = 1 - p$ the probability that it will fail to happen in any single trial, then, if n equals the number of trials and θ equals the number of successes observed, Bernoulli's theorem tells us that the difference, d, between the proportion of successes in n trials and the true probability of a success in a single trial tends to zero as n approaches ∞. Another way of saying this is the relation

$$\left|\frac{\theta}{n} - p\right| \leqslant d \tag{5.7}$$

as $n \to \infty$, $d \to 0$.

Our problem is that we wish to determine an estimate of the true proportion of success such that

$$P\left\{\left|\frac{\theta}{n} - p\right| \leqslant d\right\} = 1 - \alpha \tag{5.8}$$

where θ/n is our estimate of p and $1 - \alpha$ is the probability that our estimate does not deviate from p by more than d.

A number of people have demonstrated that if n is large and if neither p nor q is too close to zero, the binomial distribution can be closely approximated by a normal distribution. In practice, the approximation is very good if both np and nq are greater than 5 or if $npq \geqslant 25$. If these assumptions hold, we can show that

$$n = \frac{Z_{\alpha/2}^2}{4d^2} \tag{5.9}$$

where $Z_{\alpha/2}$ is the two-tailed standardized normal statistic for the probability we seek. Again, the derivation of Equation 5.9 is found in Mize and Cox [23]. To solve this equation, we must be given $1-\alpha$ and d, i.e., what risks are we willing to take, and how big a difference do we want to allow between our estimate θ/n and the true parameter p?

EXAMPLE 5.6.1

What sample size is needed for a probability of 95% or better that the observed relative frequency of successes will differ from the true probability, p, by no more than 0.02? Thus, $d = 0.02$ and $Z_{\alpha/2} = 1.96$, and

$$n = \frac{(1.96)^2}{4(0.02)^2} \geqslant 2400$$

EXAMPLE 5.6.2

How many engagements must we use to estimate the kill probability of an antitank weapon such that we have a 99% probability that our observed relative frequency of target kills will differ from the true probability of a target kill by no more than 0.05? Here, $d = 0.05$ and $Z_{\alpha/2} = 2.58$, and

$$n = \frac{(2.58)^2}{4(0.05)^2} \geqslant 666$$

Table 5.2 shows some additional minimum n values for selected α and d levels.

Table 5.2 *Minimum sample size for Bernoulli estimates.*

$1-\alpha$	$Z_{\alpha/2}$	d	n
0.90	1.65	0.10	68
0.95	1.96	0.10	96
0.99	2.58	0.10	167
0.90	1.65	0.08	107
0.95	1.96	0.08	150
0.99	2.58	0.08	261
0.90	1.65	0.05	272
0.95	1.96	0.05	384
0.99	2.58	0.05	666
0.90	1.65	0.02	1700
0.95	1.96	0.02	2400
0.99	2.58	0.02	4163

5.7 Estimating population variance

Since the standard deviation of a population is important as a measure of dispersion and in setting confidence bands on parameter estimates, we are often concerned with the precision of the estimated variance. In such circumstances, we wish to determine an estimate of the population variance such that the following probability statement is true:

$$P\left\{(1 - d)\sigma^2 \leqslant s^2 \leqslant (1 + d)\sigma^2\right\} = 1 - \alpha \quad (5.10)$$

where d is a number between 0 and 1 that expresses how close we require the estimate (s^2) to be to the true variance (σ^2).

Mize and Cox [23] show that we can manipulate Equation 5.10 to obtain the χ^2 statistic, $(n-1)s^2/\sigma^2$ with $n-1$ degrees of freedom, which does not involve σ^2 in the terms at each end of the braces. Furthermore, if n is large so that the χ^2 distribution is approximated by the normal distribution, we can derive the following quadratic equation, which can be solved for n:

$$Z_{\alpha/2} = \frac{d(n - 1)}{\sqrt{2(n - 1)}} \quad (5.11)$$

or

$$n = 1 + \frac{2(Z_{\alpha/2})^2}{d^2} \quad (5.12)$$

EXAMPLE 5.7.1

Assume we wish to estimate σ^2 within 5% with a probability of 0.95. Equation 5.12 yields

$$n = 1 + \frac{2(1.96)^2}{(0.05)^2}$$
$$= 3074$$

EXAMPLE 5.7.2

Suppose we are willing to decrease our precision to 10% with a probability of 0.95. Equation 5.12 then yields

$$n = 1 + \frac{2(1.96)^2}{(0.10)^2}$$
$$= 769$$

5.8 Comparison of two distributions

Occasionally we are interested in determining to some desired accuracy whether there is a significant difference in the distribution of the response of our simulation model from some other distribution of interest. The scenario in which we wish to compare two distributions could derive from validation runs where we wish to compare the response of our model of a system with historical data from the real world system. We could also want to compare the distributions of the responses under two different policies.

As in Sections 2.12 and 2.13, if we wish to compare the distribution of a set of data with some other distribution, we can use the Kolmogorov-Smirnov goodness of fit test. In this section, our interest is similar, except we are using reverse reasoning in that we want to determine the sample size at which a specified maximum difference (desired accuracy or allowable error) would become significant. By desired accuracy or allowable error we mean the maximum allowable decimal difference between the two cumulative relative frequency distributions being compared at any point. For example, if we choose $d = 0.10$ and use the calculated sample size, and if the absolute difference between the two cumulative distributions being compared exceeds 0.10 at any point, the Kolmogorov-Smirnoff test will indicate a significant difference at the chosen α level.

Following a suggestion by Spalding [26], we can calculate the desired sample size to be used for each sample as follows:

$$\alpha = 0.01 \qquad n = \left(\frac{1.63}{d}\right)^2 \tag{5.13}$$

$$\alpha = 0.05 \qquad n = \left(\frac{1.36}{d}\right)^2 \tag{5.14}$$

$$\alpha = 0.10 \qquad n = \left(\frac{1.22}{d}\right)^2 \tag{5.15}$$

Tabulated results appear in Table 5.3. Our one restriction is that if we compare two different samples, we must use the same sample size for each.

5.9 Autocorrelated data

In the preceding sections on estimating required sample sizes, we assumed the samples to be independent and uncorrelated. For many simulation models, this is a poor assumption. The term *autocorrelated* indicates those situations where a future value of the output will be directly influenced by the present value. When autocorrelation is present, not as much information is contained in a single observation as there would be if it was completely independent. Fishman [10] and others have shown that required sample sizes are very sensitive to the amount of autocorrelation present within each run. Estimates of the variance in particular are too low if the correlation is positive and not taken into account.

Table 5.3 *Sample size necessary to simulate a system to a desired accuracy.*

Desired Accuracy	Level of Significance .10	.05	.01
.10	149	185	266
.05	596	740	1,063
.04	913	1,156	1,655
.03	1,650	2,053	2,950
.02	3,721	6,084	6,643
.01	14,884	18,496	26,569
.001	1,488,400	1,849,600	2,656,900

Basically, there are two methods of dealing with autocorrelated data:

1 Dividing the simulation run into equal subgroups and treat each subgroup as a single independent observation.

2 Estimate the autocorrelation function and include its effects in the estimation of the parameters.

Here we concern ourselves only with the latter approach, where the n observations or measurements are related to time. If a production run yields n correlated observations, the estimate of the parameters is given by

$$\bar{X} = \sum_{i=1}^{n} \frac{X_i}{n} \tag{5.16}$$

$$\sigma_{\bar{x}}^2 = \frac{\sigma_x^2}{n}\left\{1 + 2\sum_{p=1}^{m}\left(1 - \frac{p}{m+1}\right)\rho_{p,x}\right\} \tag{5.17}$$

see class notes

where

σ_x^2 = the variance of the population
$\rho_{p,x}$ = the pth coefficient of autocorrelation
m = the maximum time lag for which autocorrelations are computed, such that $p = 1, 2, 3, \cdots, m$

To calculate the correlation coefficients, we must make a trial run and obtain output data under steady-state conditions. We can then use these data to estimate the lag coefficents by

$$\rho_{p,x} = \frac{\sum_{i=1}^{n-p} \left\{ (X_i - \bar{X})(X_{i+p} - \bar{X}) \right\}}{s_x^2 \ (n-1)} \tag{5.18}$$

where s_x^2 is the variance of the n measurements in our trial run.

As a good rule of thumb, the maximum lag for which autocorrelations are computed should be approximately 10% of the number of n measurements, although each $\rho_{p,x}$ could be tested to determine if it is significantly different from zero [12].

Geisler [12] then shows that we can calculate the minimum sample size required by

$$n = \frac{(t_{\alpha/2})^2 \ s^2 \left\{ 1 + 2 \sum_{p=1}^{m} \left(1 - \frac{p}{m+1}\right) \rho_{p,x} \right\}}{(d\bar{X})^2} \tag{5.19}$$

EXAMPLE 5.9.1

A test run was made of the model in case study I, Appendix A. Observations numbered 500. The mean was 205.74 minutes and the variance $(s^2) = 101,921.54$. Estimates of the lag coefficients were computed, using Equation 5.18, as: $\rho_{1,x} = 0.3301$, $\rho_{2,x} = 0.2993$, and $\rho_{3,x} = 0.1987$. Calculate the minimum sample size to assure the estimate lies within ±10% of the true mean with $\alpha = 0.05$.

$$\begin{aligned} n = \ & \left[(1.96)^2 \ (101{,}921.54) \left\{ 1 + 2 \left[\left(1 - \tfrac{1}{4}\right) 0.3301 \right.\right.\right. \\ & \left.\left.\left. + \left(1 - \tfrac{2}{4}\right) 0.2993 + \left(1 - \tfrac{3}{4}\right) 0.1987 \right] \right\} \right] \Big/ (0.1)^2 \ (205.74)^2 \\ \approx \ & 1757 \end{aligned}$$

5.10 Use of automatic stopping rules

In the immediately preceding sections, we discussed possible methods of estimating the sample size required to achieve desired levels of precision. A second approach that is sometimes used is to determine the confidence intervals for output values as they are generated during a simulation run and then terminate the execution when a predetermined confidence interval has been attained. In this way, we can avoid the inefficiencies of runs that are either too long or too short. On the other hand, it is obvious that by including stopping rules and confidence interval calculation routines in the model we increase the required computation time per datum point. Thus, another design tradeoff is required.

Another problem with incorporating automatic stopping rules into the model itself is premature stoppage or estimates based upon very small sample sizes. Unless we are careful, under certain circumstances the random characteristics of the model output can cause the stopping rules to be satisfied when a very small number of data points have been generated. In addition, we must consider the problem of the transient period.

At least two basic approaches incorporate automatic stopping rules into the model:

1 Run the simulation in two stages. First, run a sample of size n. Use the results to estimate n^* by one of the methods previously described. If $n^* < n$, the run is over. Otherwise, extend the run by $n^* - n$.

2 Use sequential sampling by specifying a minimum n, after which time a subroutine calculates the sample standard deviation, s. Then compare the quantity

$$\frac{(s)t_{1-\alpha,\, n-1}}{\sqrt{n}}$$

with d, and the simulation stops when the quantity is $\leqslant d$ for the first time. It is usually advantageous to do the calculations after every y output rather than after each datum point.

5.11 Variance reduction techniques

During the early days of simulation (1940's and 1950's), when computer speeds were much slower, investigators found themselves in a position where it was very expensive to decrease variability by increasing the sample size. Consequently, interest grew in developing sample-estimating procedures that could either increase the precision of estimates for a fixed sample size or, conversely, decrease the sample size required to obtain a fixed degree of precision. These methods were often referred to as Monte-Carlo techniques, but since the use of this term led to great confusion, the term *variance reduction techniques* has recently gained wide acceptance. Its underlying principle is the utilization of knowledge about the structure of the model and properties of the inputs. This knowledge is used to change or distort the original problem so that special techniques can be used to obtain the desired estimates at a lower cost.

The purpose of the next several sections is to introduce briefly the basic techniques developed in variance reduction methods. Since these techniques tend to become rather sophisticated we describe them in their most general terms with minimal use of mathematical notation, referring to other sources for detailed explanations. Most of the techniques to be discussed were proposed and developed in the period from the early 1940's to the early 1960's. The underlying statistical approaches had, in most cases, been in use much earlier for different purposes (particularly the social sciences), but during this 20-year period they were refined for specific use as variance reduction techniques in computer simulations. Of course, it is not accidental that interest in these techniques was highest when computers were relatively slow and declined as computer speed increased; however, the increasing complexity of computer simulations has tended to offset the speed factor, and there is a renewed interest in the use of these techniques. With some of the models of complex systems requiring 1½ hours of computer time just to obtain a *single* sample [2], variance reduction techniques can make the critical difference between success and failure.

The books by Hammersley and Handscomb [14] or Spanier and Gebbard [27] present probably the most comprehensive overview of the use of variance reduction techniques. Less rigorous summaries are presented by Hillier and Lieberman [17], Ackoff and Sasieni [1], a series of reports [18, 21], and journal articles [6, 9, 20]. Handscomb [16] has suggested that since we can reduce the variance by increased work or accept a greater amount of variance with less work, we could define the efficiency with which a simulation estimates a parameter as

$$\text{Efficiency} = \frac{1}{(\text{variance} \times \text{work})}$$

This implies, of course, that a reduction in variance is not worthwhile if the work required to achieve it is excessive. Therefore, we should consider not only the potential reduction in variance of a particular method, but also the cost or work required to achieve this reduction. In real world problems, enough information is usually not available to estimate correctly the work involved or the potential reduction in variance for a given method. Consequently, the experience and intuition of the analyst are again important. We must also remember that these techniques are not mutually exclusive and can combine with one another in a single problem.

5.12 Stratified sampling

It has long been common practice among social scientists to increase the efficiency per unit cost in statistical surveys by using stratified sampling. In stratified sampling, the distribution function to be sampled is broken up into several pieces, each piece is then sampled separately, and the results are combined into a single estimate. The specification of the strata to be used is based upon prior knowledge about the characteristics of the population to be sampled. We want the elements within the resulting strata to be more homogeneous (having less variation) than the elements in the population as a whole. That is to say, we want to set up classifications or strata such that the sampling units in each group tend to be uniform and such that the groups or strata tend to be different from one another insofar as the characteristic of interest is concerned.

For example, in a queueing simulation, the numerical results obtained from sets of random numbers that produced large numbers of short times between arrivals might be relatively homogeneous and sharply different from the numbers obtained from sets of random numbers that led to long-time intervals between arrivals.

We can choose the size of each piece of the distribution in various ways. The simplest (but not necessarily best) way is to divide the total interval into a number of equally sized pieces. A better way (if sufficient a priori knowledge exists) is to choose the pieces or strata such that the variance is the same for each piece. The latter unfortunately requires a great deal of knowledge about the population to be sampled.

The next question is to determine the number of samples to be drawn from each stratum. The criterion should be the variability within each stratum. Obviously, if there is no variability within a particular stratum (i.e., the units are identical with respect to the characteristic of interest), then a single observation from that stratum tells us all we need to know. On the other hand, if a stratum is characterized by great variability, then we improve the precision of the estimate by taking a large number of samples from it. Tocher [29] suggests that in deciding how many observations to take from each stratum one should make N (the number of observations made in the jth stratum) proportional to the product of the probability of a random observation falling inside the jth stratum times the standard deviation within this stratum. These probabilities and standard deviations have to be estimated by pilot study, expert opinion, prior data, or some other technique. If there is no great difference in the variance from stratum to stratum, *proportionate stratified sampling* (sampling fraction is the same for all strata) is a satisfactory technique for sample allocation. That is, the number of observations in the sample is allocated among the strata in proportion to the relative number in each stratum in the population.

Stratified sampling has demonstrated variance reduction ratios of approximately 13:1, as compared with standard Monte-Carlo random sampling [14]. The reader might wish to read Clark's example [2] of the use of this technique in a queueing situation, or that of Hillier and Lieberman in sampling exponential distributions [17].

5.13 Importance sampling

The concept of importance sampling involves a distortion of the physical probabilities of the actual system so that events of particular interest occur more frequently in the model than in the actual physical state. In importance sampling, we use a sampling procedure in which the population elements have probabilities (or likelihoods) of inclusion within the sample which are different from the probabilities under simple random sampling. We draw samples from a distribution other than the one suggested by the problem and then use an appropriate weighting factor, which, when multiplied by the final results, corrects for having used the *wrong* distribution. A wrong distribution selected is one in which a larger proportion of the samples drawn will be from the interval of importance.

Suppose, for example, that we want to estimate the probability $P(x)$ of an event occurring after a time greater than x. Assume the actual value of $P(x)$ is 0.01. If we sample from the actual distribution of x, we get approximately one occurrence of the event in each 100 samples drawn. If we could sample from another distribution that had $P^*(x) = 0.05$, we would get the same number of occurrences with only one-fifth as many samples. This basic approach is used in importance sampling. Samples drawn from the distorted distribution will cause any statistical parameters estimated from the sample to be biased, and a correction will be necessary. We correct by multiplying each sample drawn by a weighting factor, which is a ratio of the likelihood (i.e., probability) of getting a single occurrence from the true distribution to the likelihood of getting a single occurrence from the distorted distribution—i.e., $W_{(x)} = P(x)/P^*_{(x)}$.

As an example, suppose a condominium apartment house wishes to estimate the mean annual expenditure of its residents in the barber shop and beauty parlor. It is known that the expenditures by women differ more widely than those by men. Most men get a haircut every 4 weeks at an annual cost of approximately \$50; expenditures of as much as \$100 or as little as \$25 per year are found occasionally. Annual expenditures by women can vary from nothing to over \$500. The variance for expenditures by women is therefore much greater and the mean expenditure more difficult to estimate than those amounts.

Assume that 80% of the residents are men and that a sample of size 15 is to be taken. If simple random sampling were employed, we would expect the sample to consist of approximately 12 men

(80% of 15) and 3 women; however, we could obtain a sample that included 5 men and 10 women instead. Suppose the annual expenditures of the members of the sample turned out to be

Men:	45, 50, 55, 40, 90
Women:	80, 50, 120, 80, 200, 180, 90, 500, 320, 75

It is intuitively clear that such data will lead to a more accurate estimate of the overall average than would the expenditures of 12 men and 3 women. To analyze these data, we calcuate the means of the sample data as

$$\bar{M} = \$56, \qquad \bar{W} = \$169.50$$

We must apply appropriate weights to the original sample data if we wish to obtain an appropriate overall mean. Thus, if M_i and W_i are used to designate the ith sample of men and women, respectively,

$$\bar{X} = \tfrac{1}{15}\left[\sum_{i=1}^{5} \tfrac{0.80}{0.33} M_i + \sum_{i=1}^{10} \tfrac{0.20}{0.66} W_i\right] \approx \tfrac{1}{15}(678.78 + 513.62)$$

$$\approx \$79.49$$

The greatest difficulty in using importance sampling is in choosing the appropriate distorted distribution to sample. Hammersley and Handscomb [14] gives a good description of how to select a distorted distribution for this purpose. The variance reduction possible via importance sampling has been shown to be as great as 30 to 1 [14].

5.14 Russian roulette and splitting

Russian roulette and splitting were ideas of J. von Neumann and S. Ulan, who first used them in their work on particle diffusion problems. The combination of these two techniques can be extremely effective when there exists some knowledge about important and unimportant regions of a distribution. The basic ideas are similar to those used in sequential sampling schemes of quality control sampling. A simple illustration of Russian roulette is the case of tossing a pair of dice one at a time to estimate the probability of getting the number 3 as the sum of the two die. The number of tosses of an individual die can be reduced if we examine the number of spots appearing on the first die before tossing the second. If the number is not a 1 or a 2, then there is no point in tossing the second, since we cannot obtain the sum of 3.

We score a failure for that particular trial and proceed to the next. Thus, this technique makes it unnecessary to toss the second die two-thirds of the time. The overall results are the same, but we have reduced the sampling effort by approximately one-third [19].

In the more complicated problems, we can often examine the sample at each stage and classify it as being "interesting" or "uninteresting." We can delete the "uninteresting" sample from further consideration, allowing only the "interesting" ones to continue through the other stages. We can also classify the "interesting" samples by *splitting* them into subcategories, so as to pick the "most interesting" for further investigation. Thus, *splitting* (first used in connection with particle diffusion problems) is used in the good regions to separate each sample arriving in such a region into n independent branches for further study.

Spanier and Gebbard [27] state that splitting can be considered a special case of importance sampling, which is simpler to use since the selection of a distorted distribution is unnecessary. They further state that the technique can be extremely effective in achieving large reductions in variance when good definitions of the "interesting" and "uninteresting" regions can be found. They do not offer any indication, however, of what degree of reduction may be feasible. They do provide a formal derivation of the Russian roulette estimator and the effects of splitting.

5.15 Antithetic variates

One of the most important of the variance reduction methods to be created directly for simulation work is the antithetic variate technique, first proposed in a paper by Hammersley and Morton [15]. The term "antithetic variate" describes this technique because the estimates used mutually compensate for each other's variations. The underlying idea is to have two estimators, x_1 and x_2, of an unknown parameter y, such that x_1 has a negative correlation with x_2. The estimate of y is then equal to $(x_1 + x_2)/2$ and has a variance of

$$\tfrac{1}{4}(\sigma^2_{x_1} + \sigma^2_{x_2}) + \tfrac{1}{2} \text{ covariance } (x_1, x_2)$$

With negative correlation this value is smaller than that possible if x_1 and x_2 were independent estimates, and under certain conditions the variance reduction capability is truly astounding [14]. The technique is somewhat similar and might be considered a special case of correlated sampling, except that it requires a negative correlation, whereas correlated sampling needs a positive correlation.

An excellent example of the power and use of this technique is given by Page [25]. The model is of a single-channel queue, where

T_i represents the time between arrivals and S_i the service times. Page actually implemented two forms of the antithetic variate technique. In one form, the first run was made with randomly generated values of T_i and S_i; in the second run, these same values were used but the roles of these two variates were interchanged. In other words, if T_i^* and S_i^* denote the interarrival time and service time of the ith arrival during the first run, then, in the second run, $T_i^* = S_i$ and $S_i^* = T_i$. The computational efficiency was such that the same variance was achieved with half of the run size required without the technique. In the other form, Page let A_i and B_i be the uniformly distributed variates used to produce T_i and S_i, respectively, in the first run. Then, in the second run, he let $A_i^* = 1 - A_i$ and $B_i^* = 1 - B_i$. This scheme also reduced the required run size by 50%.

With this technique, variation reduction ratios of up to 3,980,000:1 have been reported for some special types of sampling distributions [14]. A major advantage of this technique is that it can be applied to any existing simulation program with only minor change. In practice, it is relatively easier to find negatively correlated unbiased estimators than it is to find an equally satisfactory distorted distribution for importance sampling or specialized functions required by some of the other variance reduction techniques. For this reason, the antithetic variate technique tends to be more efficient in practice and deserves serious consideration.

5.16 Correlated sampling

Correlated sampling is a technique of reducing variance that is especially applicable when the problem is to compare two or more alternatives. For example, the problem might be to determine if strategy A is better than strategy B, or if engineering design A is better than engineering design B, etc. In each case, we are interested in the *relative difference* between the two alternatives rather than in the absolute values of either. An estimate of the mean value of the random variable $\overline{Z} = \overline{A} - \overline{B}$ is desired so as to compare the two. If completely random sampling were used to obtain $\overline{A}$ and $\overline{B}$, then $\text{var}(Z) = \text{var}(A) + \text{var}(B)$. If correlated sampling were used, however, we could consider that the sample was drawn from a new probability distribution of a random variable Z^* with the same mean as Z but with a much smaller variance, given by $\text{var}(Z^*) = \text{var}(A) + \text{var}(B) - 2$ covariance (A, B). The gains from this technique are governed by the relative size of the variances of A and B and the problems encountered in constructing a sampling design that will make covariance (A, B) large and positive.

The most straightforward approach to obtaining this correlation is through the control of the random numbers during the two runs. In the preceding type of problem, we can establish the necessary correlation between samples or runs by using the same random number sequences for both alternatives. Thus, both alternatives will undergo the same histories and be compared under the same conditions. Since the random numbers generated in a computer are pseudorandom, (i.e., the program will generate the same random number stream each time the same seed is used), we can easily generate the same sequence of events in the model.

Thus, we can run two or more sets of experiments using correlated data and reduce the random error. But an even better method of conducting this experiment would be to run only one experiment and apply both sets of rules simultaneously. In this case, not only is the variance reduced, but also the amount of actual simulation is reduced significantly. This, of course, is not always possible and depends upon the particular circumstances. By using the same stream of random numbers and thus identical conditions, we can sometimes run and analyze the experiment as a paired comparison with a great gain in power of the statistical test [28].

5.17 Use of variance reduction

As stated in the preceding sections, variance reduction techniques seek either an increase in the precision (decreased variance) for a fixed sample size or a decrease in the sample size required to obtain a fixed degree of precision. These techniques are not new, but a review of the literature shows that they are not widely practiced. The basic ideas in some cases are almost trivial, but the results we can sometimes gain for the amount of additional effort required are almost unbelievable. Unfortunately, these techniques do not get the attention they deserve, probably owing to the fact that few textbooks discuss the techniques and that the analyst becomes bogged down in the elegance of his model and the problems of programming and debugging.

Hammersley and Handscomb [14] and Moy [24], as well as Ehrenfeld and Ben-Tuvia [6], have all run experiments and prepared tables comparing the performances of several of the techniques. These studies show that when applied to rather simple, artificial problems, these techniques have produced very impressive, even startling results. When we move to the arena of more complex, real world models, however, the gains achieved, although still worthwhile, are not quite as startling [14]. Several authors, including Kahn [19], have cautioned us in using these techniques. He states that if properly used, these techniques can provide a tremendous increase in the efficiency of the model; however, if the intuition is faulty and the analyst does not use a reasonable design, the techniques can also be very unpredictable and actually increase the variance in the cases of importance sampling, Russian roulette and splitting, and stratified sampling. On the other hand, techniques such as correlated sampling (when appropriate) are more stable, because it is almost impossible to worsen the variance even with faulty intuition.

REFERENCES

1. ACKOFF, R. L., and M. W. SASIENI, *Fundamentals of Operations Research*, John Wiley & Sons, Inc., New York, 1968.

2. CLARK, C., "The Utility of Statistics of Random Numbers," *Journal of Operations Research*, Vol. 8, No. 2, Mar.-Apr., 1962.

3. CONWAY, R. W., "Some Tactical Problems in Digital Simulation," *Management Science*, Vol. 10, No. 1, Oct. 1963.

4. DIANANDA, P. H., "Some Probability Limit Theorems with Statistical Applications," *Proceedings Cambridge Philosophical Society*, Vol. 49, 1953, pp. 239-46.

5. DUNCAN, A. J., *Quality Control and Industrial Statistics*, Richard D. Irwin, Inc., Homewood, Ill., 1959.

6. EHRENFELD, S., and S. BEN-TUVIA, "The Efficiency of Statistical Simulation Procedures," *Technometrics*, Vol. 4, No. 2, May, 1962.

7. EMSHOFF, J. R., and R. L. SISSON, *Design and Use of Computer Simulation Models*, The MacMillan Company, New York, 1970.

8. FELLER, W., *Introduction to Probability Theory and Its Applications*, Vols. I and II, John Wiley & Sons, Inc., New York, 1965.

9. FETTER, R. B., and J. D. THOMPSON, "The Simulation of Hospital Systems," *Operations Research*, Vol. 13, No. 5, Sept.-Oct., 1965.

10. FISHMAN, G. S., "Estimating Sample Size in Computor Simulation Experiments," *Management Science*, Vol. 18, No. 1, Sept., 1971.

11. FISHMAN, G. S., "Bias Considerations in Simulation Experiments," *Operations Research*, Vol. 20, No. 4, Jul.-Aug., 1972.

12. GEISLER, M. A., "The Sizės of Simulation Samples Required to Compute Certain Inventory Characteristics with Stated Precision and Confidence," *Memorandum RM-3242-RP*, The RAND Corporation, Santa Monica, Calif., Oct., 1962.

13. GORDON, G., *System Simulation*, Prentice-Hall, Inc., Englewood Cliffs, N. J., 1969.

14. HAMMERSLEY, J., and D. HANDSCOMB, *Monte Carlo Methods*, John Wiley & Sons, Inc., New York, 1964.

15. HAMMERSLEY, J. M., and K. W. MORTON, "A New Monte Carlo Technique: Antivariate Variates," *Proceedings Cambridge Philosophical Society*, Vol. 52, 1956.

16. HANDSCOMB, D. C., "Monte Carlo Techniques: Theoretical," *The Design of Computer Simulation Experiments*, T. H. Naylor (ed.), Duke University Press, Durham, N. C., 1969.

17. HILLIER, J. M., and G. S. LIEBERMAN, *Introduction to Operations Research*, Holden-Day, Inc., San Francisco, 1967.

18. KAHN, H., "Applications of Monte Carlo," *The RAND Corporation, Report RM-1237-AEC*, Santa Monica, Calif., Apr. 27, 1956.

19. KAHN, H., "Use of Different Monte Carlo Sampling Techniques," *Symposium on Monte Carlo Methods*, H. A. Meyer (ed.), John Wiley & Sons, Inc., New York, 1956.

20. KAHN, H., and A. W. MARSHALL, "Methods of Reducing Sample Size in Monte Carlo Computations," *Operations Research*, Vol. 1, No. 5, Sept.-Oct., 1953.

21. MAXWELL, W. L., "Variance Reduction Techniques," *The RAND Corporation, Report P-3139*, Santa Monica, Calif., May 6, 1965.

22. MIHRAM, G. A., "On Limiting Distributional Forms Arising in Simulation Encounters," *The Mathematics of Large-Scale Simulation*, P. Brock (ed.), *Simulation Councils Proceedings Series*, Vol. 2, No. 1, Simulation Councils, Inc., La Jolla, Calif., Jun., 1972.

23. MIZE, J. H., and J. G. COX, *Essentials of Simulation*, Prentice-Hall, Inc., Englewood Cliffs, N. J., 1968.

24. MOY, W. A., "Monte Carlo Techniques: Practical," *The Design of Computer Experiments*, T. H. Naylor (ed.), Duke University Press, Durham, N. C., 1969.

25. PAGE, E. S., "On Monte Carlo Methods in Congestion Problems: Simulation of Queueing Systems," *Operations Research*, Vol. 13, No. 2, Mar.-Apr., 1965.

26. SPALDING, J. B., "Method to Determine the Number of Random Digits to Simulate an Object System to a Desired Accuracy," *Bulletin of the Operations Research Society of America*, Vol. 19, Suppl. 2, Fall, 1971.

27. SPANIER, J., and E. M. GEBBARD, *Monte Carlo Principles and Neutron Transport Problems*, Addison-Wesley Publishers, Ltd., London, 1969.

28. STEEL, R. G., and J. H. TORRIE, *Principles and Procedures of Statistics*, McGraw-Hill Book Company, New York, 1960.

29. TOCHER, K. D., *The Art of Simulation*, D. Van Nostrand Co., Inc., Princeton, N. J., 1963.

EXERCISES

1. Read case study A-V in Appendix A.

   a) What did the authors do about: (1) starting conditions? (2) equilibrium? (3) sample size?

b) Would any of the variance reduction techniques discussed in this chapter be of potential use in this study?

2. Assume that water usage in the metropolitan area is normally distributed. Using a simulation model we have developed, we want to estimate the mean daily usage such that the error in the estimate is less than ±2000 gal, with a probability of 0.95. We estimate that the feasible range of usage covers 40,000 gal/day. What sample size (i.e., how many days) should we use in our simulation?

3. We have a simulation model of an antiaircraft, ground-based missile system. How many engagements must we run to have an 85% or better probability that the observed relative frequency of target kills will differ from the true probability, p, of a target kill by no more than 0.02?

4. Suppose that in Problem 2 we wish to estimate the variance of water usage within 1% with a probability of 0.95. What size sample must we use?

5. The following stream of numbers was produced by a computer program for the generation of Poisson distributed random numbers with lambda equal to 5.0. Calculate the lag coefficents of autocorrelation for lags up to 4 digits: 4, 6, 3, 4, 5, 4, 6, 6, 4, 2, 5, 2, 7, 4, 1, 5, 3, 6, 7, 2, 7, 5, 5, 7, 5, 3, 3, 6, 5, 11, 4, 6, 6, 6, 4.

6. In the first case study in Appendix A, the author is studying an information storage and retrieval system. He is interested in mean turn-around time (i.e., how long after an inquiry is placed is the answer forthcoming?). In an attempt to overcome autocorrelation, he lumped inquiries in groups of 25. He then used the mean of each group as a datum point. The mean turn-around time for 20 groups of 25 is shown. Calculate the estimates of the lag coefficents of autocorrelation for lags up to 3 groups. Based upon 500 observations, the mean is 205.74 and the variance is 101,921.54.

Group	Mean Time	Group	Mean Time
1	144.1	11	307.8
2	144.8	12	375.6
3	176.8	13	148.8
4	226.9	14	252.5
5	170.2	15	142.8
6	218.4	16	72.8
7	277.4	17	160.4
8	197.3	18	167.6
9	277.7	19	198.7
10	256.4	20	198.0

7. Section 5.2 discusses starting conditions and equilibrium. For the data shown in Problem 6, how much of the results would you exclude before analysis?

chapter six

VALIDATION AND ANALYSIS

Nothing will ever be attempted if all possible objections must be first overcome.

Samuel Johnson

6.1 Model validity

A model should only be created for a specific purpose, and its adequacy or validity evaluated only in terms of that purpose. Our goal is to generate a model that creates the same problems and behavior characteristics as the process or system being studied. To evaluate a model means to develop an acceptable level of confidence that inferences drawn from the performance of the model are correct and applicable to the real world system. The concept of validation should be considered one of degree and not one of an either-or notion; it is not a binary decision variable where the model is valid or invalid. It is not at all certain that it is ever theoretically possible to establish if we have an absolutely valid model; even if we could, few managers would be willing to pay the price. We must consider validity of a model on a scale from 0 to 1, where 0 represents absolutely invalid and 1 represents absolutely valid. As the degree of validity of the model increases, so too will its development cost. At the same time, its value to the decisionmaker will increase, but most likely at a decreasing rate. Anshoff and Hayes [4] have suggested that the curves would look like Fig. 6.1, in most cases. Thus, a measure of the benefit to cost ratio would presumably peak at something less than the most valid model that money can buy.

Most computer simulations of stochastic processes utilize random numbers, random variates, and replications to arrive at expected outcomes. The model often involves chains of probabi-

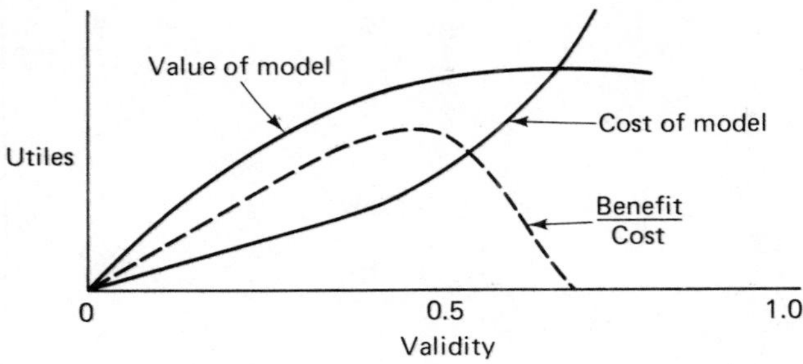

Figure 6.1 Validity.

listic events that interact in complex ways. We can demonstrate that if we take enough replications or samples, the results of the simulation *can be made as accurate as is desired.* The problem is that these *accurate* results are those which would occur if the probabilities used for each of the events were exactly the same in the real world system as they are in the model. If these probabilities are not known accurately (usually the case), there exists an inherent inaccuracy in the results, which no amount of replication or statistical analysis can remove. The following passage from Koopman and Berger [17] further demonstrates the point:

> To illustrate the issue of the piling up of errors which occurs when the mean values of the probabilities involved are not well known, I shall present an example which tremendously oversimplifies the problem: Suppose there are n independent events which occur in the operation, the ith event having a probability of success equal to p_i. In this case we get terms like $p_1 p_2 \cdots p_n$. To simplify the problem, I shall use only this one product. (Whether computers are used or not, the answer is determined by products of this form.) If we let P represent this product, then the percentage error in the answer is given by
>
> $$\frac{\delta P}{P} = \frac{\delta (p_1 p_2 \cdots p_n)}{p_1 p_2 \cdots p_n} = \delta \log(p_1 p_2 \cdots p_n)$$
>
> Thus,
>
> $$\frac{\delta P}{P} = \sum_{i=1}^{n} \frac{\delta p_i}{p_i}$$
>
> If the errors are not correlated and the percentage error in each probability is about the same, this can be approximated by
>
> $$\frac{\delta P}{P} = \sqrt{n}\left(\frac{\delta p}{p}\right)$$

If

$$\frac{\delta p}{p} = 0.3 \quad \text{and} \quad n = 10$$

it follows that

$$\frac{\delta P}{P} \cong 1$$

Thus, 30 percent error in the individual probabilities can add up to a factor of two difference in the result. Alternatively, 50 percent error in four probabilities also adds up to a factor of two difference in the result. The results of the analysis may still be useful but the analyst and the decision maker should be aware that such errors are inherent in the analysis and should not be mislead by the apparent accuracy of the results of the simulation.

Fishman and Kiviat [11] divide the process of evaluation into three categories: (1) *verification*, to insure that the model behaves as the experimenter intends; (2) *validation*, to test the agreement between the behavior of the model and that of the real system; and (3) *problem analysis*, which deals with the analysis and interpretation of the data generated by the experiments. In other words, we are concerned with the internal consistency of the model, its correspondence with the real system, and the correct interpretation of the resulting data. The second aspect immediately raises the important philosophical question, what is the value of true correspondence between the model and the real life system? How isomorphic need the model be to be considered valid? If the model consistently predicts results or reproduces problems that are borne out in practice, how important is it that the model be a true and isomorphic reflection of the real system? Can a model be grossly homomorphic and still valid? Elmaghraby [10], for example, raises the hypothetical example of an economic model for the prediction of business or economic cycles, based upon the high correlation between sunspots and these cycles. If, in fact, such a model could consistently predict better than other models (which it cannot), even though there is no reason to suspect a relationship exists between sunspots and economic cycles, should it be considered valid? After all, we can argue that it is the usefulness of the model which is important. The truth of its structure should be irrelevant.

This question is not minor or easily answered. It is a debate that has raged throughout the history of science and continues today. Belief induced from observation has been shown to be gravely lacking time and time again. Newton's theory of the world with its absolute measures of time and space appeared to satisfy what our natural senses perceived; however, Einstein's theory of general relativity and subsequent events showed us how erroneous the obvious can be. Einstein's work suggested that the best path might not be that of observation followed by induction of general laws, but rather the totally different process of postulating a theory and then discovering whether or not the facts fit it.

In general, a simulation model is a theory describing the structure and interrelationships of a system. The theory may be useful or useless, depending upon whether it is soundly conceived, inadequate, or wrong. Correctness can only be measured relative to the real system. Since all models contain both simplifications and abstractions of the real world system, no model is absolutely correct in the sense of a one-to-one correspondence between itself and real life. Consequently, there are various degrees of correctness and validity. Simulation modeling is probably not a search for absolute truth or correctness but rather a succession of theories that will progressively approach the truth. The problem of validating a simulation model is no different from the question of validating or proving any hypothesis or theory in any field of science. The basic problems are twofold: (1) What does it mean to validate a hypothesis? (2) What criteria should be used? Since scientists and philosophers have been unable to agree on answers to these questions, we would be presumptuous in assuming that we can answer them here. We can, however, look at some of the tools available and leave the problem in the hands of the model's builder and user.

Despite extensive literature dealing with validation procedures, the problem of validating simulation models remains as difficult and elusive as ever. But it is an issue we cannot avoid or push lightly aside. Unless our modeling efforts are to be pure exercises in science fiction, the need to show that a model's output does indeed bear some meaningful relationship to what can be expected as behavior from the real world remains one of the most crucial aspects of a simulation study. Furthermore, we believe that the processes by which a model is constructed, validated, and implemented are inseparable. Churchman [7] maintains that one's theories of modelbuilding and model testing are inseparable aspects of one's theory of scientific inquiry. If this is true, then it would perhaps be useful to look briefly at some different theories of scientific inquiry.

6.2 Subjective vs. objective methods

There is an apparent conflict when we are designing and validating simulation models between the need to be objective and the need to make constructive and intelligent use of our subjective beliefs.

By subjective beliefs we mean those insights, intuitions, opinions, hunches, guesses, or impressions which we have gained about how a system of interest operates. Objectivity means laying aside or ignoring our subjective beliefs or prejudgments and considering only the experimental evidence. As C. R. Blyth [5] points out, "This is just the familiar dilemma of the scientific method generally: The scientist needs to be objective, but the way he makes progress is through following up subjective insights."

The apparent conflict can be resolved readily if we recognize that the design and validation of any model involves a continuous shifting back and forth between the subjective and objective. Blyth [5] suggests that the scientific method consists in the two separate processes of construction and verification. Construction is the process of building a model to show how the system works based upon intuition, insight, observation, opinion, etc. Since the number of potential models is almost infinite, this process is largely subjective. On the other hand, verification is almost exclusively objective. Its purpose is to determine whether our model can survive objective tests against reality with no support from our prejudices.

Thus, we have the alternating steps of construction-verification-construction-verification, etc. where construction is the model-building process and verification is the process of checking this model against reality to determine their agreement.

6.3 Rationalist vs. empiricist

Throughout the history of science, rationalists and empiricists have battled over the correct method to conduct scientific inquiry. Both agree that science begins with the observation of selected parts of nature (or, for our purposes, observation of the system under study), but there the agreement ends. Should the scientist, after observation, postulate the way elements of the system interact and then discover whether the facts fit the hypothesis, or should he only include those interactions which can and have been tested empirically? Should he try to go from general to specific or from specific to general?

Rationalists are closely associated with mathematics and logic. Their effort is usually directed toward developing a mathematically expressible hypothesis devised to fit the observed facts, subsequently applying the rules of formal logic to deduce various consequences. Rationalism holds that a model is simply a system of logical deductions from a set of premises, which may or may not be open to empirical verification or appeal to objective experience. Carried to its purest form, rationalism ultimately rests upon the base of what Immanuel Kant called *synthetic a priori premises*—premises of unquestionable truth: "We do not need controlled experiments to establish their validity: they are so much the stuff of our everyday experience that they have only to be stated to be recognized as obvious." [24]

Thus, for example, Forrester's model of urban areas is based upon certain premises that he assumes are proved or validated by their mere statement [12]:

1 If conditions in the urban area are more favorable than those outside, people and industry will move in; of course, the reverse is true.

2 The changes in housing, population, and industry are the central processes involved in growth and stagnation.

3 The more spent per capita, the greater is the public service offered by the city.

4 The more houses constructed per year in the city, the greater is the city's attractiveness.

5 The larger the percentage of underemployed workers who move up to the labor class per year, the more attractive is the city to outsiders.

Forrester's urban and world simulation models are beautiful examples of models based upon the rationalist philosophy [12, 13]. *If one accepts the basic premises (which are unproved) and the logic with which they are connected, then one accepts the validity of the model.* Forrester dismisses the lack of proof of the premises and the lack of experimental measurement of the parameters with the following argument:

> Much of the behavior of systems rests on relationships and interactions that are believed, and probably correctly so, to be important but that for a long time will evade quantitative measure. Unless we take our best estimates of these relationships and include them in a system model, we are in effect saying that they make no difference and can be omitted. It is far more serious to omit a relationship that is believed to be important than to include it at a low level accuracy that fits the plausible range of uncertainty.
>
> If one believes a relationship to be important, he acts accordingly and makes the best use he can of the information available. He is willing to let his reputation rest on his keenness of perception and interpretation. [12, p. 144]

Unfortunately, any attempt to spell out explicitly all the basic assumptions and premises underlying a particular model based upon rationalism soon reveals the questionable nature of their obviousness. For example, the premise that if a city spends more tax money per capita the public service rendered is greater is not at all obvious, since it ignores the question of efficiency as well as effectiveness. This has led some authors, such as Reichenbach [23, p. 136], to go so far as to deny the very existence of a synthetic a priori premise, and others to denounce rationalist investigators as first cousins to the alchemists of the Middle Ages.

The empiricist is at the other end of the philosophical spectrum. Empiricism refuses to admit any premises or assumptions that cannot be verified independently by experiment or analysis of empirical data. Thus, Reichenbach [23, p. 256] agrues that "Verifiability is a necessary constituent of the theory of meaning. A sentence the truth of which cannot be determined from possible observations is meaningless." In commenting about rationalist simulation models of management control systems, Clay Sprowls said, "I am prepared to look at each of them as an interesting isolated case which can be described to me but from which I shall draw no conclusions" [28, p. 148]. Naylor and Finger state, "Although the construction and analysis of a simulation model, the validity of which has not been ascertained by empirical observation, may prove to be of interest for expository or pedagogical purposes (e.g., to illustrate particular simulation techniques), such a model contributes nothing to the understanding of the system being simulated" [21]. In short, in its purest form, empiricism asks that we begin with proven or verifiable facts, not assumptions.

6.4 Absolute pragmatists

We might consider the simulation model to be a black box, which transforms the exogeneous input variables into a set of output variables. Both the pure rationalist and the pure empiricist are primarily concerned with the internal structure of the black box or model. They disagree over the nature of internal relationships that are valid and admissible. A third philosophy is, for want of a better term, absolute pragmatism. The absolute pragmatist, in the pure form, says basically, "I am building the model for a specific purpose or use. If it fulfills that purpose, then it is a valid model." He does not overly concern himself with the validity of the internal structure of the black box or model, but looks only at the input-output relationships.

Schrank and Holt [25] propose "that the criterion of the *usefulness* of the model be adopted as the key to its validation, thereby shifting the emphasis from a conception of its abstract truth or falsity to the question of whether the errors in the model render it too weak to serve the intended purposes." Under such a criterion, Elmaghraby's [10] hypothetical economic model based upon sunspots should be accepted if it indeed predicts economic cycles better than any other model. Mitroff [19] has an interesting discussion of the conflicts possible between an empiricist and a pragmatist in reporting a study of modeling an engineering design process: "To me, to validate meant to substantiate in an abstract and impersonal way; for example, to select one of the test procedures listed towards the end of Naylor and Fingers article [21]. To the engineer, to validate meant to show him that the program had some direct relevance for his job, that, in short, why he should put his design 'faith' in it." He goes on to say that each had a different purpose in mind and a different goal. Mitroff's study was for the academic community to complete his Ph.D. dissertation, whereas the engineer was trying to improve his job performance. Hence, we have two distinct approaches to validation of the same model.

6.5 Utilitarian approach

Very seldom does one come across a pure rationalist, empiricist, or absolute pragmatist. Arguments and controversy are usually over matters of emphasis and degree. Most experimenters find themselves in the position of being willing to use and be concerned with all three points of view. Thus, as Naylor and Finger suggest [21], a multistage verification process usually results. This approach to validation incorporates the viewpoints of the rationalist, empiricist, and absolute pragmatist into what we might call a utilitarian approach.

The first stage is to seek face validity of the internal structure of the model based upon a priori knowledge, past research, and existing theory. In general, most complex simulation models consist in modeling a large number of simple processes. These simple processes are usually well defined and understood, but when they are combined, the large number of possible interactions makes understanding the behavior of the total system impossible. Thus, the first stage of validation entails looking at each of the simple processes modeled to ensure that the building blocks, so to speak, are the best possible. Any hypothesis that can be rejected upon the basis of a priori knowledge or past research should be so rejected until additional research or experience modifies our belief. It makes no sense to incorporate hypotheses that we have reason to believe are erroneous into our model. Some representations or hypotheses will be well known in the sense that previous validation has taken place, whereas others will have a strong theoretical base deriving from previous research. When a process is easy to observe and measure, the confidence in its representation will be high; with others, we feel less at ease and less dogmatic. We do not insist that each hypothesis be empirically testable, but we do require that each be reasonable, based upon our best knowledge of the system. Stage one is thus a modified rationalists approach. We do not insist upon Kant's synthetic a priori assumptions, but do require that the assumptions make sense.

The second stage is also concerned with the validation of the internal structure of the model, and consists in empirically testing, wherever possible, the hypothesis used. In this stage, we attempt to verify as many as possible of the assumptions that survived stage one, by subjecting them to vigorous empirical testing. This idea of testing our assumptions, parameters, and distributions appears wholly reasonable. The theory of statistics, as it pertains to estimation and hypothesis testing, provides the framework for this stage of validation based upon the empiricist's viewpoint.

The third stage attempts virgorously to verify the model's ability to predict the behavior of the real world system. Here we are faced with convincing the user that our model does what we claim it will do, i.e., that it is useful. In general, it entails comparing the input-output transformations generated by the model with those generated by the real world system. This third stage is highly critical to gaining the user's acceptance and implementation. The tools available range from highly technical mathematical techniques, such as spectral analysis as well as other goodness of fit tests, to behaviorally oriented ones, such as the "Turing" test, to the running of practical demonstrations, such as prototype and field tests. In this third stage, a model that predicts successfully is usually considered more valid than one that merely reproduces observations already known to the model builder when he developed the model.

The three stages occur in an iterative manner throughout the model development and implementation process. Even if the model or one of its hypotheses is contradicted by empirical data, it is usually not rejected unless a better model or hypothesis is available. Generally, if the model or hypothesis has any reasonable body of supporting data or theory, finding facts that do not fit it usually leads to refinement or redefinition of the hypothesis rather than its complete rejection. Therefore, there exists a continuing interplay among rationalist, empiricist, and pragmatist philosophies throughout the process.

In using simulation to study a complex system, we find several different types or classes of error, any one of which can lead us to draw erroneous conclusions:

1 errors in design;

2 errors in programming;

3 errors in the data used;

4 errors in procedure or use of the model;

5 errors in interpretation of results.

We must remember these potential errors throughout the modeling and simulation process, and do everything possible to avoid them. Since the process of creating such a model requires the integrated inputs of so many different specialists and transpires over a period of time, the potential always exists that the final result will not be exactly what was orginally desired.

In summary then, we build confidence into a model by stages and throughout the process of development by:

1 Constructing a set of hypothesis about the manner in which the subelements interact based upon all available information including: observations, previous research, relevant theory, and intuition.

2 Attempting to verify the assumptions of the model whenever possible by statistical testing.

3 Comparing the input-output transformations of the model to those of the real world system.

6.6 Tests of internal structure and hypothesis

As discussed in Chapter 2, even the most restricted portions of the real world are too complex to be studied in complete and exact detail. Consequently, we must ignore most of the actual features of a system under study and abstract from the real system certain features or aspects that together make up an idealized and simplified version of the real system. This abstract simplification, if successful, provides a useful approximation to the real world. Usually, it is convenient to break up the idealization into a number of parts for separate treatment or analysis. The possibility of doing so rests upon the existence of elements that are approximately independent of each other or mutually interact in simple ways.

Once we have selected the elements of a system, we construct a hypothesis, or trial idea concerning the nature and form of the interactions. In many cases, hypotheses about how an element functions or about the interaction between two or more elements are so simple and their consequences so obvious that we can test them directly through statistical analysis or some other form of empirical testing. For example, if we suspect that the service times for an operation are exponentially distributed or that the customer arrivals are Poisson distributed, we can either test this hypothesis if historical data exist or we can observe the system in operation.

As we have mentioned several times, validation is a continuous process that takes place throughout the modeling process. A model may suffer from several kinds of deficiencies, stated by Ackoff and Sasieni as follows [1]:

1 It may include irrelevant variables.

2 It may exclude relevant variables.

3 One or more relevant variables may be evaluated or represented inaccurately.

4 Its structure (*i.e.*, the function that relates performance to controlled and uncontrolled variables) may be in error.

The desirability of subjecting assumptions, parameter values, and probability distributions to empirical testing to avoid these deficiencies is obvious. Techniques usable for such empirical testing are based upon the statistical theory of estimation and hypothesis testing. Complete coverage of all the statistical techniques available for estimating and hypothesis testing is beyond the scope of this book; however, we discuss briefly some of the more useful and important ones, and suggest references describing them in detail.

Statistical tests used for testing the assumptions of a model and also the conclusions or inferences drawn from experiments run with the model, are in general, those which deal with:

A Hypothesis testing:

1 Tests on estimates of population parameters assuming an underlying probability distribution (parametric tests such as F, t, and Z).

2 Tests on estimates of population parameters not dependent on the assumption of an underlying population distribution (nonparametric tests, such as Mann-Whitney test of means).

3 Tests to establish the probability distribution from which a sample comes (goodness of fit tests, such as chi square or Kolmogorov-Smirnov).

4 Test on the degree of relationship among two or more variables (correlation analysis).

B Estimation:

1 Calculation of point and interval estimates of population parameters.

2 Determination of quantitative equations relating two or more variables (regression analysis).

Statistical methods used for hypothesis testing and estimating are, therefore, mainly:

1 tests of means;

2 analysis of variance and covariance;

3 goodness of fit tests;

4 regression and correlation analysis.

6.7 Tests of means

When testing means, we are generally interested in whether two sets of sample data could have come from the same underlying population. For example, we might want to test whether the average service time at work station A is the same as that at work station B, or whether customers prefer to use certain checkout stands to others at a supermarket. Two forms of hypothesis testing of means are possible. First, the means of the populations from which the two samples were drawn are equal; if either population has a larger mean, the hypothesis will be rejected. This is called a two-tailed test. Second, population A has a larger mean than population B; this hypothesis will be rejected if the mean of B is larger than that of A, but it will also be rejected if they are equal. This second form of the hypothesis calls for a one-tailed test. We must consider the hypothesis carefully to determine which test is appropriate. Choosing the appropriate test depends upon our previous knowledge and/or willingness to make assumptions about the underlying populations. If we are willing to assume normality of the underlying populations based upon the central limit theorem, previous goodness of fit tests, or our knowledge of the process, we can use one of the parametric tests. Figure 6.2 shows us that which test to use depends upon the particular situation:

1 Whether the two populations have a common variance σ^2.

2 Whether the values of the variances (or the common σ^2) are known or estimated.

3 Whether we are testing a sample against a population for which the true mean is known.

4 Whether the two samples are of the same size.

5 The size of the two samples.

We might first consider the problem of determining whether the two populations have a common variance σ^2. What we wish to

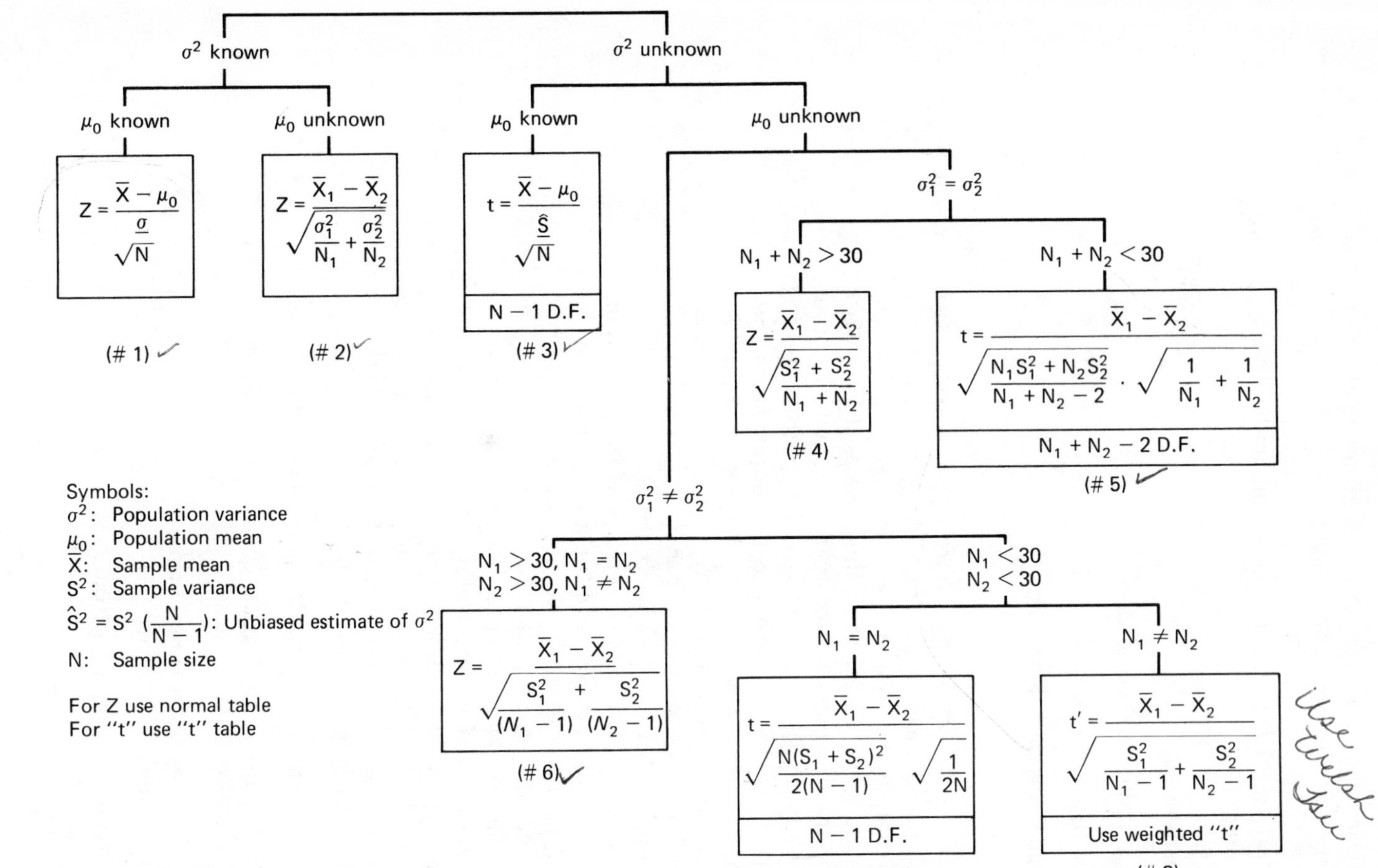

Figure 6.2 Tests of means.

test is the null hypothesis of $\sigma_i^2 = \sigma_j^2$ against the alternate hypothesis $\sigma_i^2 > \sigma_j^2$. To do so, we determine the sample variances s_1^2 and s_2^2 and calculate an F statistic by

$$F = \frac{\text{the larger } s^2}{\text{the smaller } s^2}$$

We then compare F to the tabulated values in Table C-VI, where the degrees of freedom (sample size minus one) for the larger s^2 are given across the top of the table and those for the smaller along the side. If calculated F is larger than F_α, we would reject the H_0.

Hopefully the reader has some experience or previous exposure to the testing of means, but, in any event, Fig. 6.2 is self-explanatory. Remember that all tables in this book are set up for one-tailed tests, and that the probabilities or significance levels must be doubled for two-tailed tests. Also remember that if the calculated statistic is greater than the tabulated statistic for the given level of significance, we reject the null hypothesis.

One block of Fig. 6.2 that is not self-explanatory and may be new to the reader is the use of t' or the weighted t test (block 8). To determine a significant level of t' for a given significance level, we use the Cochran and Cox [8] approximation. The value of t' such that larger calculated values are to be judged significant is calculated by

$$t' = \frac{w_1 t_1 + w_2 t_2}{w_1 + w_2}$$

where $w_1 = s_1^2/N_1$, $w_2 = s_2^2/N_2$, and t_1 and t_2 are the appropriate values of student's t for $N_1 - 1$ and $N_2 - 1$ degrees of freedom, respectively, at the chosen level of significance.

EXAMPLE 6.7.1

In a study of a chemical processing plant we have 5 weeks of past performance data for one of the subsystems. The subprogram that simulates this part of the plant is run to simulate 7 weeks' performance. The results in tons of output per week are

Subsystem: 22.0, 22.5, 22.5, 24.0, and 23.5

Model: 24.5, 19.5, 25.5, 20.0, 18.0, 21.5, and 21.5

Using an $\alpha = 0.05$, test to determine if there is a significant difference between the means of the subsystem and that of the simulator.

$$N_1 = 5 \quad \bar{X}_1 = 22.9 \quad s_1^2 = 0.68$$

$$N_2 = 7 \quad \bar{X}_2 = 21.5 \quad s_2^2 = 7.25$$

To solve, we first test $H_0: \sigma_1^2 = \sigma_2^2$,

$$F = \frac{7.25}{0.68} = 10.66$$

Looking up the critical F value in Table C-VI for $\alpha = 0.05$ with $(7-1)$ and $(5-1)$ degrees of freedom, we find $F_{\text{tab}} = 6.16$. Since $F_{\text{cal}} > F_{\text{tab}}$, reject H_0 of $\sigma_1^2 = \sigma_2^2$.

Therefore, we need to use the weighted t test from block 8, Fig. 6.2:

$$t'_{\text{cal}} = \frac{\bar{X}_1 - \bar{X}_2}{\sqrt{\dfrac{s_1^2}{N_1-1} + \dfrac{s_2^2}{N_2-1}}} = \frac{22.9 - 21.5}{\sqrt{\dfrac{0.68}{4} + \dfrac{7.25}{6}}}$$

$$= \frac{1.4}{1.17} = 1.2$$

$$w_1 = \frac{0.68}{5} = 0.14, \qquad w_2 = \frac{7.25}{7} = 1.04$$

$$t_1 = 2.78, \quad t_2 = 2.45$$

$$t_{\text{crit}} = \frac{w_1 t_1 + w_2 t_2}{w_1 + w_2} = \frac{0.14(2.78) + 1.04(2.45)}{0.14 + 1.04}$$

$$= \frac{0.39 + 2.55}{1.18} = 2.49$$

Since the calculated t' is less than the critical t', we do not reject the H_0 of $\bar{X}_1 = \bar{X}_2$. The fact that our simulator has a statistically significantly higher variance should cause us to reexamine our model, however. Even though the model produces values with the same mean, the greater variability of the model might give us trouble later, when its output becomes part of the overall model.

6.8 Paired observations

As discussed in Section 5.16, it is sometimes possible and desirable to use correlated sampling techniques, i.e., use the same sequence of random numbers for each of the alternatives to be evaluated.* Using such a sampling technique results in paired, correlated observations, which increase the ability of the experiment to detect either small differences or the same difference with a smaller sample size. The use of paired or correlated observations allows us to eliminate a source of extraneous variance—that existing from pair-to-pair. We do so by calculating the variance of

*Rereading Section 5.16 would be helpful at this point.

the difference (D) rather than the variance among the individuals within each sample. The null hypothesis tested is that the mean of the population of differences is zero. The number of degrees of freedom of the test statistic is based upon one less than the number of pairs. The test statistic used is t, except we now calculate it as

$$t = \frac{\bar{X}_1 - \bar{X}_2}{\sqrt{\dfrac{\sum_j D_j^2 - (\sum_j D_j)^2/N}{N(N-1)}}}$$

where D_j is the difference $(X_{1j} - X_{2j})$ between members of the jth pair and N is the number of sample differences or pairs.

EXAMPLE 6.8.1

A simulation model has been run using correlated sampling (same random number sequences) for 10 time periods under two different policies. The results are shown in Table 6.1. Test the hypothesis of no difference between the population means against the alternative that policy A results in higher values. Use $\alpha = 0.05$.

Table 6.1 *Simulation results after correlated sampling.*

Time Period	Policy A	Policy B	Difference (D)	D^2
1	4.0	4.4	-0.4	0.16
2	5.2	3.7	1.5	2.25
3	5.7	4.7	1.0	1.00
4	4.2	2.8	1.4	1.96
5	4.8	4.2	0.6	0.36
6	3.9	4.3	-0.4	0.16
7	4.1	3.5	0.6	0.36
8	3.0	3.7	-0.7	0.49
9	4.6	3.1	1.5	2.25
10	6.8	1.9	4.9	24.01
Sum	46.3	36.3	10.0	33.00
$\bar{X}$	4.63	3.63		

Using the analysis for paired comparisons, we now calculate

$$t_{cal} = \frac{\bar{X}_1 - \bar{X}_2}{\sqrt{\dfrac{\Sigma D^2 - (\Sigma D)^2/N}{N(N-1)}}}$$

$$= \frac{4.63 - 3.63}{\sqrt{\dfrac{33 - (10)^2/10}{10(10-1)}}} \qquad = \frac{1}{\sqrt{\dfrac{23}{90}}}$$

$$= \frac{1}{\sqrt{0.25}} = 2$$

$$t_{tab} = 1.83$$

$$\therefore\ t_{cal} > t_{tab}; \quad \text{reject } H_0 \text{ of no difference}$$

6.9 Mann-Whitney test

If the assumptions of normality are suspect, we can use the nonparametric Mann-Whitney test [26, 34], one of the most powerful of the nonparametric tests. Mood [20] showed that its relative power efficiency is close to 95% compared with the t test, even if the populations were, in fact, normally distributed. Although we cannot give a complete presentation of the Mann-Whitney test, a limited discussion is justified because of its importance. This test can be used to test whether two independent groups or samples have been drawn from the same population. It can be employed in place of the parametric t test when the analyst is unwilling to make the assumptions necessary for using the t test.

Given two samples of size M and N, where $M \leqslant N$, we rank all the observations (from both groups) from 1 to P where $P = M + N$. Assign the rank of 1 to the observation with the lowest algebraic value and the rest in ascending order. Assign tied observations the average of the tied ranks. Now, if R is the sum of the ranks assigned to the observations in the sample of size M, we can calculate the Mann-Whitney U statistic by

$$U = MN + \frac{M(M + 1)}{2} - R$$

How we determine or test the significance of U depends upon the size of N. There are three possibilities:

1 If N is 8 or less, we must also calculate $U' = MN - U$ and determine which is smaller, U or U'. We use the smaller of these two values to look up the exact probability associated with a value as small as the calculated U or U' in special tables, which can be found in references [26, 34].

2 If N is between 8 and 21, we again calculate both U and U', but now we can look up in special tables critical values for U or U' (whichever is smaller) for the α level of interest, and if $U_{cal} > U_{tab}$ reject the H_0 of no difference. Appropriate tables are found in the references [26, 34].

3 If N is larger than 20, we may determine the probability associated with a value as extreme as the observed value of U by calculating a Z value for our data and looking it up in our normal table (Table C-IV). The Z value in this case is calculated as

Section 6.9
Mann-Whitney test

$$Z = \frac{U - \dfrac{MN}{2}}{\sqrt{\dfrac{(MN)(M + N + 1)}{12}}}$$

EXAMPLE 6.9.1

Assume that we have historical data from 18 weeks of operation of a real world production system and data from 22 weeks of simulated operation of the system from our simulation model. The data are given in Table 6.2. Test to see if there is a significant difference at $\alpha = 0.01$, using a Mann-Whitney U test. In this case

$$M = 18$$

$$N = 22$$

$$R = 251.5$$

$$\begin{aligned} U &= MN + \frac{M(M+1)}{2} - R \\ &= 18(22) + \frac{18(18+1)}{2} - 251.5 \\ &= 315.5 \end{aligned}$$

$$\begin{aligned} Z &= \frac{U - \dfrac{MN}{2}}{\sqrt{\dfrac{(MN)(M+N+1)}{12}}} \\ &= \frac{315.5 - \dfrac{18(22)}{2}}{\sqrt{\dfrac{[18(22)]\,[18 + 22 + 1]}{12}}} \\ &= 3.19 \end{aligned}$$

Reference to Table C-IV shows that $Z = 2.33$ for a one-tailed test. Since this is smaller than our calculated F, our decision is to reject the H_0 of no significant difference.

The Mann-Whitney test assumes that the data are from a continuous distribution. With very precise measurement of a continuous variable, the probability of a tie is zero. However, with

Table 6.2 *Weekly production figures for real and simulated systems.*

	Real World System		Simulation Model	
Week	*Production*	*Rank*	*Production*	*Rank*
1	80	19	90	29.5
2	70	1	91	31
3	79	17.5	95	35.5
4	74	5	90	29.5
5	85	24	93	33
6	89	28	83	22
7	76	9	97	38
8	82	21	72	3
9	76	9	95	35.5
10	77	13.5	84	23
11	76	9	76	9
12	71	2	77	13.5
13	73	4	79	17.5
14	94	34	92	32
15	75	6	96	37
16	77	13.5	87	26
17	78	16	98	39
18	81	20	86	25
19			99	40
20			76	9
21			88	27
22			77	13.5
Totals	1413	251.5	1921	568.5
$\bar{X}$	78.5		87.3	

the type of measurement we typically use, ties may often occur. When ties occur between data points, we give each of them the average of the ranks they would have had if no ties had occurred. If ties occur between two or more data points in the same group, the value of U is not affected; however, if the ties involve data from both groups, U is affected, although usually negligibly. We can correct for the effect of ties as follows:

$$Z = \frac{U - \frac{MN}{2}}{\sqrt{\frac{MN}{P(P-1)}\left(\frac{P^3 - P}{12} - \Sigma T\right)}}$$

where

$$T = \frac{t^3 - t}{12}$$

t = number of data points tied for a given rank

ΣT = summing the T's over all groups of tied observations

$P = M + N$

We can demonstrate this correction for our example by noting that we have the following tied groups:

5 scores of 76

4 scores of 77

2 scores of 79

2 scores of 90

2 scores of 95

To find $\Sigma\ T$, we sum the values of $(t^3 - t)/12$ for each of the tied groups:

$$
\begin{aligned}
\Sigma\ T &= \frac{5^3 - 5}{12} + \frac{4^3 - 4}{12} + \frac{2^3 - 2}{12} + \frac{2^2 - 2}{12} + \frac{2^3 - 2}{12} \\
&= 10 + 5 + 0.5 + 0.5 + 0.5 \\
&= 16.5
\end{aligned}
$$

As the reader can easily see, this correction would be almost undetectable in the result of our example. When the correction is employed, it tends to increase the calculated Z value slightly. Thus, if we do not correct for ties, our test is conservative. Siegel [26] recommends that we correct for ties only if the proportion of ties is very large, some of the t's are large, or if the Z obtained without the correction is very close to the critical value.

6.10 Comparison of input-output transformations

A computer simulation model is in the simplest sense an input-output transformation device. One of the most obvious approaches to helping validate a model of an existing system is to compare the outputs of the real world system and the model, using (if possible) identical inputs. By employing an appropriate two-sample test, we could then determine statistically if the two samples are actually from different populations or "practically" from the same population.

If the measure of effectiveness yields only one response characteristic, then the solution is concerned with the univariate two-sample problem. As we have seen, numerous such tests may be considered for the univariate case. We can test means by the rather robust t test or the nonparametric Mann-Whitney test. We can test variances using the F test or we can test for goodness of fit using the chi-square, Kolmogorov-Smirnov, or the Cramer-Von Mises tests (see Sections 2.12 and 2.13).

Since most simulation models generate a time series as an output, and since this time series is often autocorrelated, spectral analysis has been suggested to test the conformance of the two time streams (one from the model and one from the real system). The procedures to test the equivalence of two spectra have been discussed by Jenkins [16], Fishman and Kiviat [11], and others. Spectral analysis for comparison of the time series has both advantages and disadvantages. Fishman and Kiviat [11] give the advantages as follows:

1 The problems of autocorrelation are automatically taken into account.

2 Spectral analysis considers the average level of activity, deviation from this level, and how long these deviations last once they occur.

3 Spectral analysis provides a means of objectively comparing two time series and constructing confidence bands.

The disadvantages of spectral analysis, as given by Van Horn [32], are as follows:

1 If the spectra are equivalent, the analyst increases his confidence in the model, but if they do not pass the test for equivalence, the interpretation is less clear.

2 Comparison of spectra requires a large number of observations.

3 Spectral analysis assumes that the time series are covariance stationary, whereas many systems we study either are nonstationary or never reach steady state.

Although spectral analysis is potentially very useful, its shortcomings are presently such that it has only limited use with simulation models. For additional discussion of some of the limitations, see Watts [35], as well as Howey and Kelejian [14].

Ideally, we would like to have a test that would work with small samples, evaluate multivariate responses (see Section 6.11), and handle nonstationary or non-steady-state situations. If we want to use the term "test" in a broad sense, there is just one: the so-called Turing test. Turing [31] was actually trying to provide a way for comparing human and machine intelligence when he suggested the procedure. The test hinges on the notion of an imitation game. To run it, we find a person or persons who are intimately familiar with the operations of the system being studied. We then present them with one or more sets of input-output data from the real world system and one or more data sets from the model and ask them to differentiate between them. If they succeed, we ask them how they were able to do it. This then gives us a clue as to what might be wrong with our model. Not much can be said about the power of such a test, but it is an intuitively appealing approach and much better than simply relying upon the modeler's intuition. Mitroff [19] presents a very interesting discussion of the practical and philosophical problems associated with the use of Turing-type tests.

Other tests have also been proposed, such as Theil's inequality coefficient [29] and the use of regression analysis [9]. Aigner [2] has raised some serious questions as to the validity of the assumptions underlying the use of regression analysis for input-output comparisons as proposed by Cohen and Cyert. There is also some question as to the utility of Theil's inequality coefficient for other than certain economic models. Hunt [15] demonstrates its use for trying to decide which of several models is the better predicter of system behavior when humans are an important part of the system.

6.11 The multiple response problem

As we have already seen, the usual approach to the validation of a model of an existing system is basically simple. Two independent random samples are obtained—one from the model and the other from the real world operational system. By employing an appropriate two-sample test, we can determine statistically if the two samples are from similar or different populations. If we are interested in only one response characteristic, we can use numerous statistical tests. Unfortunately, a more serious problem exists with the validation of multivariate response models (i.e., models yielding more than one response characteristic of interest). Often, we can bypass the multiple response problem by treating an experiment with many responses as many experiments each with a single response (i.e., validate the model separately for each of the response variables of interest). However, this approach may leave us with a serious decision problem if the model validates satisfactorily for some response variables but not for others.

Another common approach used whenever multivariate responses are encountered is to reduce the problem to the univariate case by transforming or combining (e.g., by addition) the responses into a composite measure of "effectiveness." This composite measure can then be used to validate the model. Such a composite measure is often difficult to obtain and/or interpret in a meaningful way, since it represents an abstraction. Also, unfortunately, the validation of the model on this composite measure does *not* insure its validity on the multivariate responses. The only time one can have confidence in the results when using a composite measure technique is when the validity of the model is rejected. A rejection of the model on the basis of the composite measure should certainly be heeded by the analyst, but a failure to reject may merely mean that the true characteristics and behavior of the system and/or model are masked by the composite measure (e.g., a very high response could offset a very low response).

Let $O_n = [X_1, X_2, \ldots, X_p]$ and $O_m = [Y_1, Y_2, \ldots, Y_p]$ be two sets of p component random vectors defined on a sample space where the p components represent different characteristic responses. Each of the two sets could be the result of a particular run of the simulation model (under different policies or configurations), or O_n could be the output of the simulation model and O_m the output of the real world system under identical inputs. The decision we are asked to make is, "Are the sets O_n and O_m drawn from the same population?" The potential difficulties faced by the analyst in a multivariate response problem such as this are readily evident if we consider the simplistic two-dimensional results shown in Fig. 6.3.

In Fig. 6.3(a), if we ran a univariate, two-sample test (such as for means), we would probably find that we would reject the null hypothesis (H_0) of no difference when we tested for response characteristic R_1 but would fail to reject the H_0 for R_2. Considering Fig. 6.3(b), we would probably not be able to reject the H_0 for either characteristic. In both cases, we have deliberately made it clear that a difference exists between O_n and O_m. Imagine now the problems when the dimension p goes from 2 to 4 or 6, and/or the area or volume of responses are not mutually exclusive.

What we need is an efficient and widely applicable method for multivariate two-sample significance tests, which allows some or all of the response characteristics to be noncontinuous and is preferably nonparametric in nature. Even parametric multivariate two-sample tests are of very limited availability. If all the components in the response are continuous, and if the usual parametric assumptions of normality, equivalent scatter matrices,

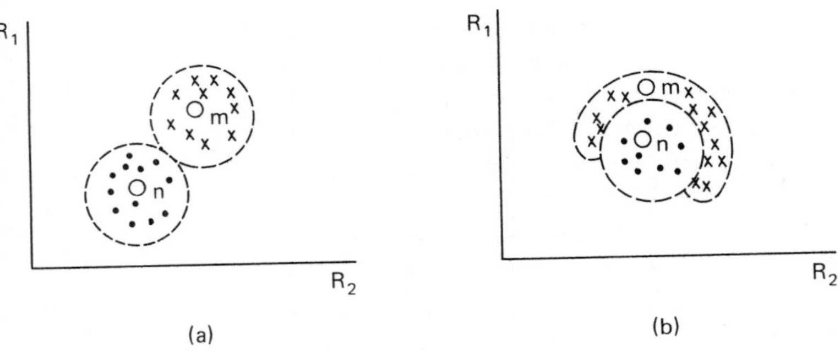

Figure 6.3

etc., are conscientiously justifiable, then perhaps Hotelling's T^2 test would be appropriate, or, if we are dealing with only two response variables, we might run correlation tests. Beyond that possibility there is an almost complete void. Fortunately, several multivariate, nonparametric, two-sample tests have been proposed.

Presently, the most promising class of multivariate two-sample nonparametric tests utilize the concept of tolerance regions. Under this concept, a set of mutually exclusive and exhaustive tolerance regions are defined on one of the two samples of observations. These tolerance regions form a partition on the multivariate sample space. The number of observations from each of the two samples lying within each region are then counted. These frequency counts then form the basis for determining the outcome of the comparison test.

The most promising methods for forming the multivariate tolerance regions would appear to be some variation on the methods propsed by Wilks [36, 37, 38], Wald [33], Tukey [30], and Boase [6]. These methods have the following advantages: (1) the two sets of observed data vectors need not be random and/or independent (a more general constraint of joint null distribution symmetry is imposed); (2) the data need only be partially continuous (at least one component must be continuous); (3) once the partitions are derived and the frequency counts completed, a number of well developed nonparametric tests can be utilized; and (4) the partition construction process may terminate at any stage whenever sufficient knowledge of the test statistic value is obtained. Other procedures, such as those presented by Anderson [3], or Puri and Sen [22], also show some promise.

6.12 Field tests

The strongest verification of any simulation model occurs when we demonstrate that the model can successfully predict events which have not yet transpired. Even though we have succeeded in developing a model that appears to be reasonable and adequately reproduces the past performance of the system, the question still remains, "Will it faithfully predict the future?" It is fair to say that in most studies it is the future which concerns us. We need answers to questions such as, "What happens if I change this?" If the study is sufficiently significant, we can sometimes afford to conduct field verification tests to explore the model's ability to predict. An example of this type of approach is the Small Independent Action Force (SIAF) simulation study conducted for the U.S. Army's Advanced Research Projects Agency in the early 1970's [18].

As a direct result of experiences with guerrilla warfare in Vietnam, the U.S. Army became intensely interested in the use of

Small Independent Action Forces (SIAF), which are defined as small military patrols, ranging generally from 5 to 10 members, that operate independently of their parent units, but that may continue to receive firepower, logistical support, and communications from them. The SIAF's missions are those of reconnaissance and combat patrol in insurgency environments. In 1968, the U.S. Army sought to improve the effectiveness of these small infantry elements. They were particularly concerned with two aspects: (1) the development and testing of new equipment, including sensors, patrol location devices, weapons, and communication devices; and (2) the study of critical patrol capabilities, including communications, movement, insertion and extraction, and target detection. It was determined that a simulation model could be an invaluable aid in studying potential changes in either equipment or operational policies, and development of such a model began. The model developed is extremely interesting from an analyst's viewpoint; the manner in which terrain and movement were represented is particularly innovative. However, our concern is not with the model but rather its verification.

The model was designed in modular fashion (see Section 7.9), in that submodels were developed to represent the various processes and logical decisions required. To provide confidence in the various elements of the model, its developers initiated a separate experimental task in 1970 to run simultaneously with the model development. These tests were conducted on a military reservation in California, where the terrain and conditions were similar to those in Vietnam. The model predicted outcomes based upon various scenarios set in the California locale. Mock patrols were conducted under the leadership of experienced patrol leaders just back from Vietnam, and results were compared. These field tests allowed the analysts to verify some of the assumptions made during the model development and to develop a great deal of confidence in the ability of the model to predict future outcomes.

The use of designed field tests wherever possible helps both analyst and client-user to develop confidence in the model. Obviously, the demonstration of the ability of a model to predict some future event is a much stronger test than merely duplicating past events upon which the design of the model was predicated. Unfortunately, it is usually difficult, expensive, and sometimes impossible to devise meaningful field tests for complex systems. We should not, however, overlook the possibility of perhaps devising field tests to check out at least parts of the system or some of the submodels and assumptions. For example, if we have developed a model of police patrolling in a city to explore various manning and deployment policies, why can't we take a small part of the city and use it for field tests? This has long been a practice of the military with their maneuvers and war games. In my opinion, it could and should be more widely practiced in the industrial and social sectors.

6.13 Confidence intervals

We often use results of simulation models to obtain estimates of population parameters of the response variable. Because they are estimates, we must be in a position to say something about their accuracy, which we usually do by placing confidence bands or intervals about our estimates. For example, if we have an estimate ($\bar{X}$) of the true mean (μ), we would like to establish an upper limit (UL) and lower limit (LL), such that the probability of the true mean being between the upper and lower limits is equal to a specified exact probability. That is,

$$P_{(LL \leqslant \mu \leqslant UL)} = 1 - \alpha$$

In this case, it is relatively easy to establish LL and UL if we can assume a normally distributed distribution. The sample statistic t is defined as

$$t = \frac{\bar{X} - \mu}{s_{\bar{x}}}$$

with μ the only unknown, and where $s_{\bar{x}} = s/\sqrt{N-1}$, s = sample standard deviation, and N = sample size. Therefore, in general we can represent confidence limits for a population mean estimated by a small sample ($N < 30$) as

why? How about $\sigma_{\bar{X}} = \frac{\sigma}{\sqrt{n}}$

$$\bar{X} \pm t_{\text{crit}} \frac{s}{\sqrt{N-1}}$$

where the values $\pm\ t_{\text{crit}}$ are the critical values of t as given in Table C-V for $N - 1$ degrees of freedom and the chosen α level. If by chance we know the true standard deviation of the sampled population, or if the sample size is large ($N \geqslant 30$), then the confidence intervals would be set as

$$\bar{X} \pm Z_c \frac{\sigma}{\sqrt{N}}$$

In this case, Z_c is the normal statistic from Table C-IV, and we would use the unbiased estimator $s^2\ [N/(N-1)]$ for σ^2.

EXAMPLE 6.13.1

A sample of 10 measurements of the turn-around time in a model of a computer center gave a mean of $\bar{X} = 43.8$ min. and a standard deviation of 6 min. Find (a) 95% and (b) 99% confidence intervals for the actual turn-around time.

a) The 95% confidence interval is given by

$$\bar{X} \pm t_{0.995}\left(\frac{s}{\sqrt{N-1}}\right)$$

since degrees of freedom are $10 - 1 = 9$, then $t_{0.975} = 2.26$, and

$$43.8 \pm 2.26\left(\frac{6}{\sqrt{9}}\right) = 43.8 \pm 4.52$$

We can be 95% confident that the true mean lies between $(43.8 - 4.52) = 39.28$ min and $(43.8 + 4.52) = 48.32$ min.

b) For the 99% confidence interval, $t_c = t_{0.995} = 3.25$. Then the 99% confidence interval $= 43.8 \pm 3.25(6/\sqrt{9}) = 43.8 \pm 6.50$, and the interval is 37.3 to 50.3.

EXAMPLE 6.13.2

A sample of 100 measurements of the processing time of a simulated machine showed a mean of 82.4 sec and a standard deviation of 4 sec. Find (a) 95% and (b) 99% confidence intervals for the mean processing time.

a) For our problem $\bar{X} = 82.4$ sec. and the unbiased estimator of $\sigma^2 = s^2\ (N/N - 1) = 16(100/99) = 16.16$ and $\sigma = 4.02$. Since $Z_{\alpha/2} = 1.96$, the 95% confidence interval is given by $X \pm 1.96\ \sigma/\sqrt{N} = 82.4 \pm 1.96\ (4.02/\sqrt{100}) = 82.4 \pm 0.79$. Thus, we are 95% confident that the true mean processing time lies between $(82.4 - 0.79) = 81.61$ and $(82.4 + 0.79) = 83.19$.

b) For the 99% confidence interval, the only change is the critical Z value, which is now 2.58. Thus, $\bar{X} \pm 2.58\ \sigma/\sqrt{N} = 82.4 \pm 2.58(4.02/\sqrt{100}) = 82.4 \pm 1.04$, and the 99% interval is 81.36 to 83.44.

The assumption of normality is not really as restrictive as it might at first appear owing to the central limit theorem (see Section 5.3). As we have pointed out, $\bar{X}$ is usually normally distributed for large samples, and, in addition, the characteristics of most complex simulation models cause the central limit theorem to hold for their outputs. We should mention the interpretation of confidence intervals. It is incorrect to state that the true mean μ lies between the limits with a 0.95 probability. We should say, the procedure which leads us to state that μ lies between the two numbers will result on the average in our making 95 correct statements out of 100. We make this distinction because the interval and not the population parameter is the random variable.

If the statistic of interest is the proportion of successes in a sample of size N drawn from a binomial distribution (see Section 5.6), the procedure is similar. If we let P equal the proportion of successes in the sample of size N, and p the true probability of success in a single trial, then $\mu = p$ and $\sigma^2 = pq = p(1 - p)$. For large samples ($N \geqslant 30$), a good approximation of the confidence interval can be calculated as

$$P \pm Z_c \sqrt{\frac{p(1 - p)}{N}}$$

To calculate these confidence intervals, we can use the sample proportion P as an estimate of p. This will generally prove satisfactory, so long as $N \geqslant 30$. Another criterion sometimes used to decide whether this approximation is valid is to require $Np > 5$ and $Nq > 5$. A more exact method for obtaining these confidence limits can be found in most texts on probability and statistics [38].

EXAMPLE 6.13.3

Suppose that we have 100 simulation data points, with the result that 60 are classified as successes and 40 as failures. Find the 95% confidence interval.

$$p = P = \frac{60}{100} = 0.6 \quad \text{and} \quad Z_c = 1.96$$

$$P \pm Z_c \sqrt{\frac{P(1 - P)}{N}} =$$

$$0.6 \pm 1.96 \sqrt{\frac{0.6(1 - 0.6)}{100}} = 0.6 \pm 1.96 \sqrt{0.0024} =$$

$$0.6 \pm 0.096$$

Therefore, the 95% confidence interval is approximately from 0.5 to 0.7.

6.14 Sensitivity analysis

As we mentioned in Section 1.16, one of the real advantages of simulation is the ability to perform sensitivity analysis readily. Sensitivity analysis usually consists of systematically varying the values of the parameters and/or the input variables over some range of interest and observing the effect upon the model's response. Such experimentation can help tremendously in building our confidence in the results of the study.

First, it can show us how sensitive the results or outputs of the model may be to the parameter values used. Since much (if not most) of the data upon which we may have based the design of the model is often questionable, we need to know how far off the data could be without changing our basic conclusions. If the model's response is relatively insensitive to fairly wide-range changes in a certain parameter, then we need not concern ourselves too much about its accuracy. On the other hand, if the model proves to be highly sensitive to a certain parameter, we might want to consider seriously spending the time and resources necessary to get a more realistic and valid measurement.

Second, sensitivity analysis in the form of varying the input values over a wide range can give us some indication of the impact of our decisions if the external environment changes. Again, if the results of the model prove to be insensitive to broad changes in the input variables, we are more comfortable in our decision to proceed to the implementation stage. On the other hand, an indication that slight changes in the input would result in our making different decisions might cause us to proceed with some caution in implementation. For example, we might want to proceed by stages, trying always to leave ourselves with a viable alternative, or we might want to build-in some controls, which would quickly warn us of changing conditions.

Third, sensitivity analysis can provide us with valuable clues for possible modifications to be made to the model later. In practice, if a model really proves useful in decisionmaking, it will usually be updated and modified continuously. Unless we are very careful, the model will continue to grow and become more and more complex with each revision. Sensitivity analysis can tell us what parts of the model we might simplify by changing from a Monte-Carlo routine to the use of a mean value, or by dropping a whole subsystem from the model. Sensitivity analysis can likewise show us where it might be fruitful to model the system in more detail.

Presenting management and/or the client-user with the results of our sensitivity analysis will usually have a very positive psychological effect insofar as obtaining acceptance of our results is concerned. The mere fact that we have explored the sensitivity of the model results to changes, errors, etc., will help reassure the decisionmaker or ultimate user of the thoroughness of our study and the validity of our results.

6.15 Summary

The final and by far the most important test for the validity of our model and the results obtained should be the answer to the question, "Does it make sense?" I firmly believe that the professional judgment of the people most intimately familiar with the design and operation of a system is more valuable and valid than any statistical test yet devised. Unquestionably, we hope our model will give us insights into the operation and performance of the system that we would not have obtained otherwise, but if those insights fly in the face of everything we know and have experienced with the system, they are suspect and should be examined carefully before we accept them.

This "caveat" is probably unnecessary, since we have yet to hear a manager say, "Well, your results don't make sense to me, but if that is what your model says, that is what we'll do." In fact, quite the opposite is true. If the results are reasonable, if they appear to fit our previous experiences (even if we didn't understand them), then we tend to minimize concerns about the details of the model or the way in which the study was conducted. On the other hand, if the results or recommendations do not make common sense, all the statistical tests and analyses ever devised will not convince the decisionmaker to accept them.

We might summarize this chapter by saying that the validation and analysis of a simulation study constitute a continuous process, beginning with the start of the study. Confidence is built into the model as we go along; it is not something we develop solely at the end. The greatest possible validity is achieved by:

1 using common sense and logic;

2 taking maximum advantage of the knowledge and insight of those most familiar with the system under study;

3 empirically testing, by the use of appropriate statistical techniques, all the assumptions, hypotheses, etc., possible;

4 paying close attention to details and checking and rechecking each step of the model building process;

5 assuring ourselves that the model performs the way we intend it to by using test data, etc., during the debugging phase;

6 comparing the input-output transformation of the model and the real world system (wherever possible), using statistical and Turing-type tests;

7 running field tests or peripheral research where feasible;

8 performing sensitivity analysis on input variables, parameters, etc.;

9 checking carefully the predictions of the model and actual results achieved with the real world system.

REFERENCES

1. ACKOFF, R. L., and M. W. SASIENI, *Fundamentals of Operations Research*, John Wiley & Sons, Inc., New York, 1968.

2. AIGNER, D. J., "A Note on Verification of Computer Simulation Models," Management Science, Vol. 18, No. 11, Jul., 1972.

3. ANDERSON, T. W., *An Introduction to Multivariate Statistical Analysis*, John Wiley & Sons, Inc., New York, 1958.

4. ANSHOFF, H. I., and R. L. HAYES, "Role of Models in Corporate Decision Making," *Proceedings of IFORS Sixth International Conference*, Dublin, Ireland, Aug., 1972.

5. BLYTH, C. R., "Subjective vs. Objective Methods in Statistics," *The American Statistician*, Vol. 26, No. 3, Jun., 1973.

6. BOASE, R. L., and J. E. WALSH, "Sequential Multivariate Quality Control Tests Using Tolerance Regions," *Southern Methodist University THEMIS Report* No. 110, Dallas, Tex., 1971.

7. CHURCHMAN, C. W., *Challenge to Reason*, McGraw-Hill Book Company, New York, 1968.

8. COCHRAN, W. G., and G. M. COX, *Experimental Designs*, 2nd Ed., John Wiley & Sons, Inc., New York, 1957.

9. COHEN, K., and R. CYERT, "Computer Models in Dynamic Economics," *Quarterly Journal of Economics*, Vol. 75, No. 1, Feb., 1961.

10. ELMAGHRABY, S. E., "The Role of Modeling in I. E. Design," *The Journal of Industrial Engineering*, Vol. XIX, No. 6, Jun., 1968.

11. FISHMAN, G. S., and P. J. KIVIAT, "The Analysis of Simulation-Generated Time Series," *Management Science*, Vol. 13, No. 7, Mar., 1967.

12. FORRESTER, J. W., *Urban Dynamics*, The Massachusetts Institute of Technology Press, Cambridge, Mass., 1969.

13. FORRESTER, J. W., *World Dynamics*, Wright-Allen Press, Cambridge, Mass., 1971.

14. HOWREY, P., and H. H. KELEJIAN, "Simulation versus Analytical Solutions," *The Design of Computer Simulation Experiments*, T. H. Naylor (ed.), Duke University Press, Durham, N. C., 1969.

15. HUNT, E. B., "The Evaluation of Somewhat Parallel Models," *Mathematical Explorations in Behavioral Science*, F. Massauh and P. Ratoosh (eds.), Richard D. Irwin, Inc., and The Dorsey Press, Homewood, Ill., 1965.

16. JENKINS, G. M., "General Considerations in the Analysis of Sprectra," *Technometrics*, Vol. 3, No. 2, May, 1961.

17. KOOPMAN, B. O., and H. N. BERGER, "Use and Misuse of Simulation," presented at NATO Conference on The Role of Digital Simulation in O. R., Hamburg, Germany, Sept., 1965.

18. MILES, R. H., "Historical Background of the SIAF Model," *Transactions of the Seventeenth Conference of Army Mathematicians*, ARO-D Report 72-1, Army Research Office, Durham, N. C., Mar., 1972.

19. MITROFF, I. I., "Fundamental Issues in the Simulation of Human

Behavior: A Case in the Strategy of Behavioral Science," *Management Science*, Vol. 15, No. 12, Aug., 1969.

20. MOOD, A. M., "On the Asymptotic Efficiency of Certain Non-parametric Two-Sample Tests," *Annals of Mathematical Statistics*, Vol. 25, 1954.

21. NAYLOR, T. H., and J. M. FINGER, "Verification of Computer Simulation Models," *Management Science*, Vo. 14, No. 2, Oct., 1967.

22. PURI, M. L., and P. K. SEN, *Nonparametric Methods in Multivariate Analysis*, John Wiley & Sons, Inc., New York, 1971.

23. REICHENBACH, H., *The Rise of Scientific Philosophy*, University of California Press, Berkeley, Calif., 1951.

24. ROBBINS, L., *An Essay on the Nature and Significance of Economic Science*, Macmillan & Co., Ltd., London, 1935.

25. SCHRANK, W. E., and C. C. HOLT, "Critique of: Verification of Computer Simulation Models," *Management Science*, Vol. 14, No. 2, Oct., 1967.

26. SIEGEL, S., *Nonparametric Statistics for the Behavioral Sciences*, McGraw-Hill Book Company, New York, 1956.

27. SPIEGEL, M. R., *Theory and Problems of Statistics*, Schaum Publishing Company, New York, 1961.

28. SPROWLS, C., "Simulation and Management Control," *Management Controls: New Directions in Basic Research*, C. P. Bonini *et al.* (eds.), McGraw-Hill Book Company New York, 1964.

29. THEIL, H., *Economic Forecasts and Policy*, North-Holland Publishing Co., Amsterdam, 1961.

30. TUKEY, J. W., "Nonparametric Estimation II. Statistically Equivalent Blocks and Tolerance Regions—The Continuous Case," *Annals of Mathematical Statistics*, Vol. 18, No. 4, 1947, pp. 529-539.

31. TURING, A. M., "Computing Machinery and Intelligence," *Mind*, Vol. LIX, Oct., 1950, pp. 433-60.

32. VAN HORN, R. L., "Validation of Simulation Results," *Management Science*, Vol. 17, No. 5, Jan., 1971.

33. WALD, A., "An Extension of Wilk's Method for Setting Tolerance Limits," *Annals of Mathematical Statistics*, Vol. 14, No. 1, 1943, pp. 45-55.

34. WALSH, J. E., *Handbook of Non-Parametric Statistics I and II*, D. Van Nostrand Co., Inc., Princeton, N. J., 1962, 1965.

35. WATTS, D., "Time Series Analysis," *The Design of Computer Simulation Experiments*, T. H. Naylor, (ed.), Duke University Press, Durham, N. C., 1969.

36. WILKS, S. S., "On the Determination of Sample Sizes for Setting Tolerance Limits," *Annals of Mathematical Statistics*, Vol. 12, 1941.

37. WILKS, S. S., "Statistical Prediction with Special Reference to the Problem of Tolerance Limits," *Annals of Mathematical Statistics*, Vol. 13, No. 3, 1942, pp. 400-409.

38. WILKS, S. S., *Mathematical Statistics*, John Wiley & Sons, Inc., New York, 1962.

EXERCISES

1. Read case study A-VI in Appendix A.
   a) What other actions could the author have pursued to validate his model?
   b) What philosophic viewpoint did the author pursue in his validation process?

2. A researcher makes two runs of his model using different values for one of his multipliers in a rate equation. Using $\alpha = 0.01$, determine if there is a significant difference in the resulting means.

   Run 1: 3, 4, 8, 7, 4, 3, 2, 0, 5, 5, 5, 7, 7, 1, 4, 4

   Run 2: 1, 4, 5, 4, 5, 2, 1, 5, 8, 4, 9, 8, 8, 0, 3, 3

3. The government specifications for a shoulder-launched rocket called for a range of 1000 yd with a variance of 100 yd. A sample of 50 launchings from a simulation model yields a mean range of 950. Using $\alpha = 0.05$, test the hypothesis that the mean of the model output is equal to 1000 yd.

4. Two runs of a simulation model under different decision policies yield the following results. Test to see if there is significant difference between the means at $\alpha = 0.05$.

$$N_1 = 31 \qquad \bar{X}_2 = 4$$
$$N_2 = 35 \qquad s_1^2 = 0.05$$
$$\bar{X}_1 = 3 \qquad s_2^2 = 0.07$$

5. Two police patrolling policies are to be compared under identical circumstances using a simulation model. The measure of effectiveness is the number of calls answered within 5 min per shift. The same sequence of demands was generated under both policies for 12 shifts. Using $\alpha = 0.05$, test the hypothesis of no difference.

Policy												
A	31	34	29	26	32	35	38	34	30	29	32	31
B	26	24	28	29	30	29	32	26	31	29	32	28

6. Two preliminary runs have been made of a simulation model for the operations in a large computer center under two scheduling policies, A and B. The output of the model is the number of jobs not completed per 8-hr shift. It was hoped that policy A would result in fewer uncompleted jobs. Given the following data, run a Mann-Whitney U test to determine at $\alpha = 0.05$ whether policy A does appear better.

Policy	
A	7, 29, 15, 5, 31, 35, 24, 19, 31, 26, 39, 12, 41, 3, 1, 4, 8, 44, 14, 36, 37, 25
B	36, 40, 8, 13, 37, 37, 34, 16, 18, 30, 17, 38, 31, 27, 9, 42

7. For the data shown in Problem 6, assume that the analyst feels justified in believing that the number of late jobs is normally distributed. Under this assumption, retest the data, with $\alpha = 0.01$.

8. The output of an airport simulation model is used to compare the average delay time that a customer will experience after landing, in unloading and obtaining his luggage. Output data for two airlines are given. Test to determine if there is a significant difference in means, using $\alpha = 0.05$.

Golden Carpet:	No. flights = 20 Average delay = 5.5 min Variance in delay = 6.2 min
Regional Airlines:	No. flights = 25 Average delay = 10.9 min Variance in delay = 40.1 min

9. Set 95% and 99% confidence intervals for the two means of the data in Problem 4.

10. Set 95% confidence intervals on the means calculated for both policies of the data in Problem 6.

11. Suppose that in 1000 engagements of a ground-to-air missile defense system there are 485 kills. Find the 99% confidence interval for the calculated kill probability.

12. Set 95% confidence intervals on the mean delay times calculated for two airlines in Problem 8.

13.* Reconsider Problem 3, Chapter 1. Assume that Mr. Chance now plays until one of three things happens: (1) he runs out of money; (2) after each game, he checks the length of the line waiting for the bus and if it is greater than 10 he quits to be assured of a seat; (3) it is time for the bus. The time between arrivals for people wishing to board the bus is exponentially distributed with mean equal to 0.5 people per minute. The amount of time he has available after finishing breakfast is

*These two problems will require a computer model to be programmed. Appendix B gives random variate generators.

uniformly distributed from 5 to 15 min. It takes him ½ min per toss of the coin. He is considering two strategies in terms of how much money to start out with each day: $10 and $30. Develop, program, and run a computer simulation of this situation to determine his expected daily gain or loss under the two strategies. Use a sample size of 100 days for each strategy. Set 95% confidence intervals on the mean for each strategy. At the $\alpha = 0.05$ level, is there a significant difference in his expectations?

14.* We have a small clinic with two doctors and two nurses. Patients arrive randomly at an average rate of five every half-hour. Patients are served in the order of their arrival, but cannot specify which doctor or nurse they wish to see. There are three types of patients, each requiring different services and service times. Pertinent information for each patient type is as follows:

Type of Patient	Services Required	Probability	Service Time
A	Doctor alone	0.45	$\mu = 10$ min, $\sigma = 2$
B	Nurse alone	0.20	$\mu = 5$ min, $\sigma = 1$
C	Both	0.35	$\mu = 12$ min, $\sigma = 3$

Service times for all three types of patient are normally distributed with the means and standard deviations as shown. If a patient requires the services of both a doctor and a nurse but no nurse is presently available, one of the nurses who is waiting on a patient alone is called to assist the doctor while her patient waits. The service time for a patient is the same with or without interruptions. There are four examination rooms. The clinic is open from 8 A.M. to 5 P.M. No additional patients are admitted after closing, but all patients waiting at closing are seen. Develop, program, and run a simulation of this clinic to:

a) Estimate the mean time a patient will spend in the clinic for each type of patient. Set 95% confidence intervals on your estimate.

b) Estimate the percent utilization of both doctors and nurses.

chapter seven

MANAGEMENT ASPECTS

The big problem with management science models is that managers practically never use them.

John D. C. Little

7.1 Previous experiences

To get some idea of the process we are trying to manage, we should examine the past experience of others. I am unaware of any studies directed specifically and exclusively at simulation studies; however, a number of studies have been aimed at operations research activities [7, 15, 20]. Since they have shown that anywhere from 25 to 50% of the activities of operations research (O.R.) departments are devoted to simulation studies, the statistics reported in these studies should be relevant to our interests.

Turban [20] surveyed the "top 500" largest corporations in the United States, as listed by *Fortune Magazine*; 107 companies replied, with useful results. Of those replying, only statistical analysis ranked above simulation as the most widely used tool, whereas linear programming ranked third. These three techniques constituted 73% of the efforts of these O.R. teams. The average duration of the projects reported was 10.1 months. The duration of individual projects varied from one (1) month to eighty (80) months (see Table 7.1).

As we would expect, these operations research projects generally involve teamwork rather than individual effort. Contrary to the recommendation of many O.R. textbooks, however, the size of the teams working on individual projects is relatively small. On the average, 2.5 researchers were used per project. The development of a computer simulation of a complex system requires the services of a variety of skilled personnel, such as operations research analysts, statisticians, computer programmers,

Table 7.1 *Duration of O.R. projects [20].*

Length of Project (months)	Percent
Less than 3	9
4–6	25
7–12	30
13–18	24
19 or more	12
Total	100

systems analysts, and subject-matter specialists. However, they are not all needed at the same time, and each specialist may join the team as needed and move on to other jobs when his input is completed. In the early stages, only operations analysts and subject-matter specialists are needed. Expert statistical advice will be needed at various stages of the development, whereas programmers and systems analysts will not be required until the first rough models are being converted into computer programs. Table 7.2 shows the average number of researchers per project as reported by Turban. Notice that with an average duration of 10.1 months, utilizing an average of 2.5 professionals, we can expect the average simulation project to be a 25.25 man-month effort not counting the support personnel.

Table 7.2 *Number of researchers per project [20].*

No. Researchers/Project	% of Projects
1	7
2–5	60
6–10	20
11 or more	13
Total	100

Table 7.3 *Degree of implementation [20].*

Degree of Implementation	Answer to General Question (%)	Compiled Answer—Individual Projects (%)
Completely implemented	48.1	35.4
Mostly implemented	18.0	21.5
Some parts implemented	13.7	20.3
Very few parts implemented	6.6	4.4
Not implemented	13.6	11.4
Not clear yet		7.0
Total	100.0	100.0

Last but not least is the question of how many projects are actually implemented. To study the extent of successful implementation, Turban built a cross-check into his questionnaire. First he asked a general question about the percent of projects that were implemented, and then he asked about the fate of each individual project reported. Table 7.3 summarizes the results. They agree rather closely with other studies [13, 17], which show that anywhere from 35 to 45% of all studies are something less than total successes—not a very good track record. However, from these and other studies, we can hopefully learn how to improve this rate.

7.2 Decisionmaking process

As we have stated several times throughout this book, the primary aim of a simulation study is to help improve the quality of managerial decisions. If we are going to provide management with the right data, in the right form, and at the right time, to assist in making the right decisions, we must understand the decision-making process and decisionmakers clearly. As soon as we begin to read about decision theory and decisionmaking, we find that there are two distinct types of literature in this field. William T. Morris [11] has categorized these as:

1 normative decision theory, which describes how managers *ought* to decide in an optimum fashion;

2 descriptive decision theory, which attempts to describe how managers *actually* decide.

In this chapter, we try to deal with some of the management aspects of planning, carrying out, and implementing the results of simulation studies. In doing so, we deal with both normative and descriptive decision theories and their possible effects upon our success or failure. We assume that the types of decisions with which we are concerned derive from and are an integral part of an organizational context or setting. Thus, we are concerned with organizational decisionmaking and not necessarily with how each individual within the group reaches his own decision.

Reduced to its essentials, the decisionmaking process appears to be simple and straightforward. According to Churchman, Ackoff, and Arnoff [5], the minimal necessary and sufficient conditions for a decisionmaking situation are the following:

1 an individual who perceives that a problem exists and a decision is required;

2 an outcome or objective that the decisionmaker desires;

3 at least two alternative courses of action that appear to have some chance of achieving the desired objective;

4 the decisionmaker's state of doubt as to which alternative is best;

5 an environment or context within which the problem exists.

Given these minimal, necessary, and sufficient conditions for a problem to exist, the decisionmaker

1 perceives that a decision is required;

2 searches for and evaluates alternative courses of action;

3 selects one or more of the alternatives that he believes will satisfy the needs.

Although this sounds very simple and straightforward, our understanding of the dynamics of this process, particularly in an organizational setting, is presently very poor, which may be one reason why so many operations research, management science, and management information system projects are unsuccessful. We cannot successfully design models or systems to aid a process we do not completely understand. I propose a tentative model of this process that appears to fit observations of the real world. Figure 7.1 depicts this proposed decisionmaking process in organizational contexts, and it is used in the following discussion.

A decision can involve either taking action or accepting an idea. It is easy to assume that the executive knows he has a problem, knows what the problem is, and has set out to solve it. Usually, however, the perception of a need or problem is a gradual event, which slowly dawns on the executive after he observes the system for which he is responsible and discusses it over a period of time with his subordinates and colleagues. As his recognition of the problem thereby becomes clearer, someone eventually assumes, or is assigned, the responsibility of looking into the matter. The responsible person can be either a staff person, a subordinate, or one of the executives. This constitutes the first decision point.

The preliminary investigator begins tentatively to clarify and define the problem. Originally, he identifies symptoms, such as schedule delays, frequent shortages, insufficient profit, numerous customer complaints, etc. As the investigation proceeds, the nature and severity of the problem becomes clearer. Again, informal discussions are held and a joint decision is made as to whether a more formal investigation is justified. This constitutes the second decision point in the process. If the decision (usually arrived at by concensus of a group of managers and the preliminary investigator) is to commit resources to a more formal study, the next step is how to proceed. This may entail deciding whether to assign the project to the operations research-management science group, to bring in consultants, or to assign it to one

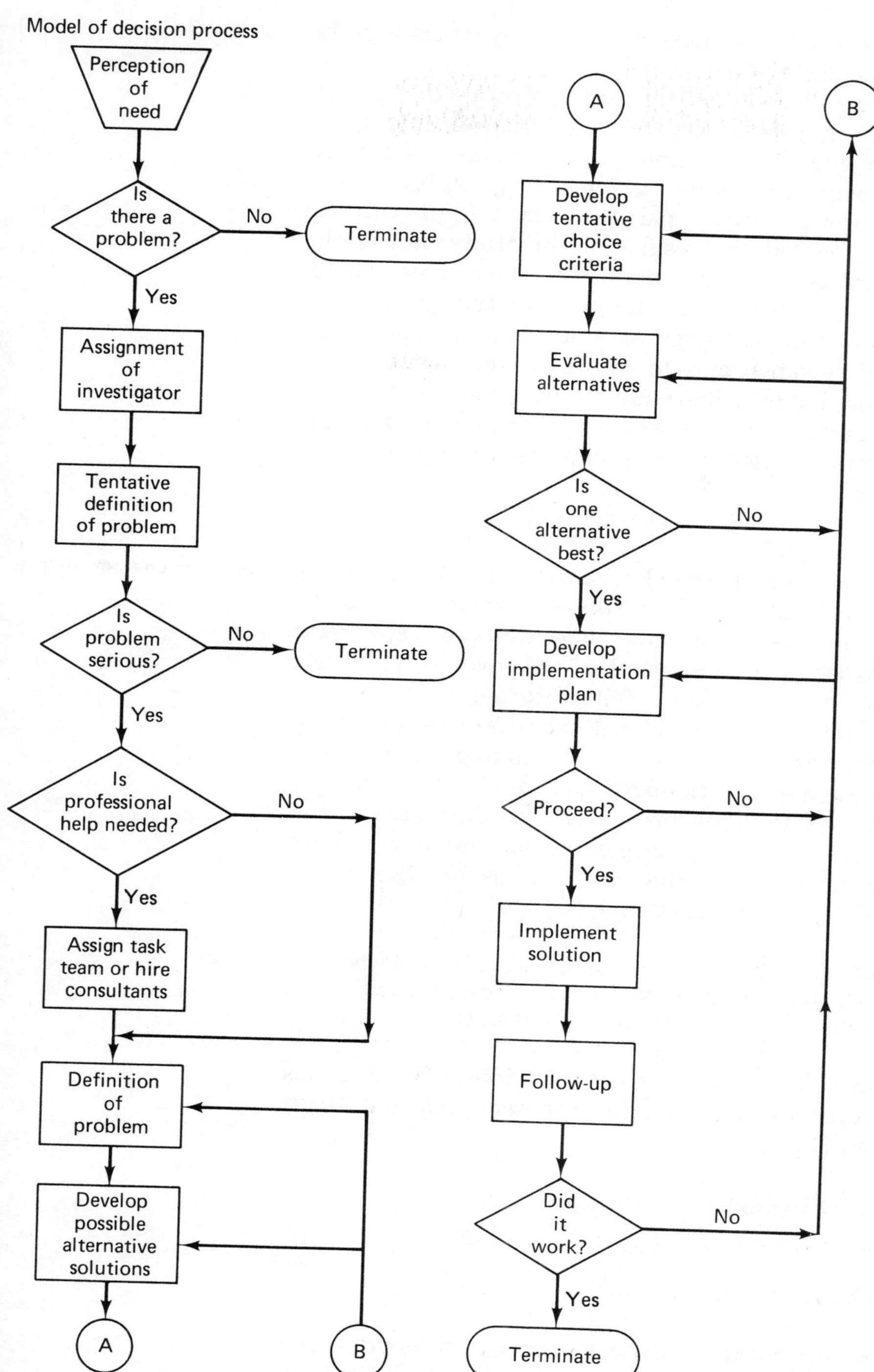

Figure 7.1 Decision process.

of the line organizations. Decisions at this early stage can be critical to the future success of the project.

Once the task team is assigned, the process proceeds through steps 7 to 12. If the team decides to use a simulation model, the process depicted in Fig. 1.3, Chapter 1, is utilized. We must recognize that the decision required at step 12 is generally arrived at by a consensus of all affected parties. Very seldom will the responsible executive accept the results of the study and implement the findings unless there is general consensus. This usually requires several iterations through the loops. Note that each of the affected managers will seek to perpetuate and increase their own prestige, power, and rewards, and to minimize threats to themselves. It is rarely possible to achieve unanimous agreement on the part of all affected parties, but the wise executive recognizes that unless there is general consensus, the implementation will not succeed.

7.3 Decision characteristics

The executive learns to be skeptical regarding the necessity for imposed decisions and is, therefore, biased toward failing to perceive problems if they are not presented in clear and unmistakable terms. Beyond the element of risk involved are other factors that tend to bias him toward failure to perceive the need for a decision. One of these is training. Executives are taught, or very soon learn the hard way, to avoid making unnecessary decisions. Since decisions (particularly at high levels) tend to disrupt operations, create additional work, and interrupt the current flow of events, the manager avoids invoking these consequences over trivia. The following advice by Barnard is typical of that usually given the executive [3]:

> The fine art of executive decision consists in not deciding questions that are not now pertinent, in not deciding prematurely, in not making decisions that others should make.

We all tend to avoid self-imposed decisions. Barnard has categorized the source of requirements for decisions into three categories, as follows [3]:

1 those imposed by higher authorities;

2 those required by subordinates;

3 those coming from self-initiative.

The first type of requirement virtually always results in a decision by the manager; the second type also always results in a decision, even though it may be to make no decision, i.e., continue the status quo; however, the third category lacks a forcing mechanism and hence can be ignored for long periods of time with no immediate penalty. If an immediate decision will or may involve adverse tradeoffs, the manager may well elect to postpone investigating the problem. He is gambling that the situation will correct itself in time to avoid future "management pain." Albers [1] puts it this way: "The fear of wrong decision is frequently more compelling than the possible adverse consequences of no decision."

There is also a very strong tendency to persist with the present course of action. Programs and organizations tend to have a built-in inertia. In the case of a large system development program, this inertia can become quite large. Once plans are made, budgets established, hardware allocated, contracts let, etc., it is difficult to change. The larger the program, the more difficult become changes of direction. This fact reinforces the need for good planning and analysis in the earliest phases of program development. In view of the difficulties inherent in reorienting such a program, it is not surprising that managers have a built-in bias not to "rock the boat."

In virtually all cases, even if they perceive the need for a decision, managers are concerned with finding satisfactory, not optimum, courses of action. March and Simon [9] state that, "Most human decision making, whether individual or organizational, is concerned with the discovery and selection of satisfactory alternatives." Textbooks speak very glibly of "optimum solutions," but even the most comprehensive models portray only a small portion of reality. It is also a rather obvious, although often forgotten, fact that the alternative selected can only be "optimum" in comparison with alternatives considered. There may very well be (and probably is) an alternative that was not considered which is superior.

Finally, we must consider another aspect of decisionmaking discussed by Herbert A. Simon [19], who points out that every decision involves elements of two kinds—*factual* and *value* elements. Factual elements are statements about the observable world and the way in which it operates. They should be capable of being tested to determine whether they are true or false. Value elements, on the other hand, are ethical or moral statements, whose validity cannot be tested. The same dichotomy is true of objectives. Objectives such as weight, speed, range, etc., are factual elements. Objectives such as deterrent effect, national prestige, customer goodwill, morale, etc., are value elements. Most decision problems contain both objective and subjective elements.

Unfortunately, we have not made as much progress in being able to handle the subjective aspects of a problem as we have the objective. The mathematical tools for handling the factual or objective parts are highly developed and sophisticated, but only during the last few years has there been any attention given to the subjective or value aspects. Unquestionably, value or subjective aspects of a problem must somehow be handled in the decision-making process, and only the affected decisionmakers can make these judgments. In some cases, this can be accomplished through close cooperation between the decisionmakers and the modeler, thus integrating these value judgments into the model. In other cases, we must be content to make our model as realistic as possible so far as the factual or objective aspects are concerned and let the decisionmaker then integrate the output of the model with his value judgments to arrive at a final decision. In either case, there must be a close working relationship between the decisionmaker and the analyst. The analyst is an extension of the decisionmaker, not his substitute. Therefore, it is extremely important to recognize the fact that good system simulation studies cannot exist in a vacuum. Too many of the value judgments that make the difference between success and failure must come from the decisionmaker himself. The only way that systems simulation can help the decisionmaker is to provide appropriate division of labor.

Our purpose for this discussion of the decisionmaking process and the characteristics of the decisionmaker are twofold: (1) to support our contention that meaningful simulation studies can only be conducted directly for the pertinent decisionmaker(s); (2) to argue that the final result and rationale behind the recommendations must be capable of being fully understood and appreciated by the decisionmakers. If the decisionmaker refuses to acknowledge that he has a problem, he is not likely to implement the results of any study. Therefore, if they wish to survive and prosper, modeling groups must work on those problems which the decisionmaker recognizes. This point takes on added significance when we note that of the 217 projects reported by 107 different industrial corporations in one study, 51% of the projects were initiated by the operations research department itself [20].

This natural reluctance on the part of managers leads to the second important consideration to remember throughout the study. If the consequences of change are potentially significant, we cannot expect the decisionmaker, who must take the ultimate responsibility for his decisions, to base his future upon blind faith in the project team. He must fully understand every aspect of what was done, including the assumptions upon which the results are based and the degree of validity of the data inputs. Fortunately, the more carefully the problem is defined and the whole project planned and carried out, the easier it will be to explain and sell the end result.

7.4 Factors of success

Since about 1960, a group at Northwestern University under the leadership of Albert H. Rubenstein has been conducting a series of in-depth studies on the life histories of industrial management science groups. Based upon the data from these investigations, they identified 10 factors that appeared to determine the success or failure of O.R./M.S. groups [13]:

1. level of managerial support including the extent of managerial understanding and acceptance of the need for such activities;
2. organizational location of the activity or where it reported in the organization;
3. adequacy of the resources allocated to the activity;
4. client receptivity, i.e., freedom of the O.R./M.S. group to select projects, gather data, and implement results;
5. level of opposition to the O.R./M.S. activity within the organization;
6. reputation of the activity within the organization;
7. general perception of the level of success of the O.R./M.S. activity within the organization;
8. organizational and technical capability of the O.R./M.S. person or group;
9. relevance of the projects undertaken to the needs of the organization;
10. influence which the group and its leadership could bring to bear in the organization.

Relationships among the project team, management, and the client or clients are therefore most critical to the success or failure of any O.R./M.S. activity. Rubenstein and his co-workers distinguished between management and clients. In their discussion, management appears to refer to top management or that part of the hierarchy above the level to which the group reports; clients are those members of the organization for whom projects and services are proposed or performed by the O.R./M.S. activity. Although Rubenstein and others stress the importance of executive exposure and personal relationships, they also point out that, ultimately, success or failure is determined by whether favorable results are forthcoming. Since large simulation projects are costly, management will naturally hesitate to proceed with a new project if it has experienced previous failure or only mediocre success. This is one reason that new O.R./M.S. groups are always wise if at the outset they select a few, manageable projects with a high probability of being successfully completed in a short time period. A contribution in the early stages that is significant in the opinion of both management and clients will serve to improve the reputation and acceptance of the group and lead to improved support.

7.5 Conflicts of model criteria

Models in general and simulation models in particular are used for two different purposes. Researchers who apply the scientific method use modeling as one of the crucial steps in the sequence of hypothesis formulation, model building, prediction, and validation. Anshoff and Hayes point out that researchers tend to judge their models by their quality [2]. They define a high-quality model as one that is *nontrivial*, *powerful*, and *elegant*. A model is considered nontrivial if it leads to insights into the system not readily perceivable by direct observation; powerful if it provides a large number of nontrivial insights; and elegant if the structure can be kept simple and if it runs efficiently on the computer. The scientific reputations, personal self-fulfillment, and, to some extent, chances for promotion of the model builders are generally tied closely to the quality of their models.

Managers, on the other hand, use models to gain insight into the value of possible decision alternatives. They are concerned with pragmatic action decisions and tend to judge a model by its applicability. From their viewpoint, an applicable model is *relevant*, *valid*, *usable*, and *cost-effective*. A model is *relevant* if it deals with problems of importance to the manager; *valid* if a high degree of confidence can be placed in any inferences drawn from it; *usable* if it provides acceptable solutions that can be implemented; and *cost-effective* if the improvements it makes possible exceed the expense of developing and applying the model [2].

Anshoff and Hayes [2] contend that much of the conflict between model builders and managers derives from the fact that they use different criteria for judging the modeling effort. The manager's objective is to solve organizational problems, whereas the model builder typically seeks model quality. The model builder is not held accountable if a particularly pressing problem does not receive the proper attention. On the other hand, the manager is responsible and his career may, in fact, be jeopardized by his failure to solve a critical problem. In addition, he receives very little reward for selecting an elegant solution to a problem of secondary importance. This leads to two potential areas of conflict: (1) problem choice; and (2) the 20% syndrome.

The model builder wishes to work on problems that will give him the greatest professional recognition and allow him the widest latitude for applying his skills. However, such problems are not necessarily the most urgent and consequential from the manager's viewpoint. Thus, the modeler's tendency is to select projects in which he feels he can foresee the development of a high quality model to display his technical skills. Professional challenge rather than importance becomes the choice criterion, and the researcher may, in fact, be more interested in methodological and technological advancement of the state of the art than in working on management's problems.

A second difficulty derives from the so-called 20% syndrome. Anyone who has worked in any aspect of applied operations research or management science knows that the greatest payoff comes during the first 80% of the study. The return on invested resources may be 10:1 or higher during this portion. On the other hand, the payoff between 80 and 100% completion may be as low or lower than 1:1. Most managers are concerned with making money or solving problems rather than coming up with an elegant and complete package; therefore, they often believe that the resources needed to go from 80 to 100% completion are wasted and could better be used in starting other projects. The researcher may insist on working on this last 20% because it injures his professional and analytical pride to do otherwise.

Many managers state that this is the most exasperating and infuriating problem they have in trying to work with operations research-management science personnel. Many organizations have attempted to alleviate some of these problems by establishing a board of advisors, who decide what projects are to be worked on, when to stop projects, and when to move on to something else. This board is usually made up of top executives who represent various parts of the organization and who can adopt an overview of what is most important to the organization as a whole.

7.6 Communications

One of the most difficult problems in trying to gain acceptance of the modeling effort is the modeler's inability (or lack of interest) to communicate effectively with anyone other than his technical colleagues. People trained in the sciences, mathematics, and statistics are often inarticulate when it comes to selling or explaining their ideas, their work, or themselves to people outside their field.

One of the most confusing aspects of operations research-management science to the neophyte and the manager is the practitioners' vocabulary. When the analyst speaks of algorithms, stochastic systems, and heuristic approaches, instead of procedures, random systems, and intuitive approaches, he confuses rather than enlightens. This insistance upon using unfamiliar terminology is unfortunate although understandable. This vocabulary, which appears to be some foreign tongue to the uninitiated, has come about from two sources. The first results from the interdisciplinary nature of the endeavor; the tools and concepts originate from many different academic fields—mathematics, economics, engineering, physics, etc. Therefore, the modeler's vocabulary contains terminology from many fields.

The second source of the vocabulary is less justifiable today; it derives from the ego needs of the practicioners. The mathematician, economist, etc., who becomes interested in management applications soon finds that he may no longer be professionally acceptable to colleagues in his field. Since he is interested in practical solutions to practical problems, he finds himself a victim of academic and professional snobbery. One of the unwritten rules of some segments of the scientific community appears to be that if what you are doing is practical, lucrative and easily understood, it must not be very profound or scientific. All too often, for example, practical, easily understood papers are not acceptable for publication. Since this sort of activity must remain practical and solution-oriented in order to survive, there is one other alternative open to achieve acceptability in the scientific community: make it uncomprehensible to the uninitiated. Hence, one never uses intuition but rather a heuristic; one never follows a procedure but rather an algorithm; etc. Although this process has contributed to scientific acceptability, it has unfortunate side effects. The primary difficulty is in communicating with managers and clients; the second is that it is an open invitation to charlatans and frauds.

We have stated several times that a simulation study is a failure if its results are not used, despite how well it is done. It is a well-known psychological fact that no manager will accept a recommendation, the implementation of which may have significant impact, on blind faith. He must understand the derivation of results, the underlying assumptions, etc. This problem necessitates clear, concise, easily understandable communications between the modeling group and the clients. A study that is not understood by the decisionmakers will not be implemented. Therefore, the good analyst must operate with and develop two vocabularies: one for communicating with his colleagues and a second for dealing with the layman.

This necessity to communicate with users and clients presents also another problem in simulation studies. Most managers (even those with technical backgrounds) have a low level of mathematical sophistication. Communications problems therefore arise when the analyst tries to explain a complex mathematical, statistical, or computer programming technique to the decisionmaker. The manager who does possess a technical background presents an especially touchy situation; he has had training in mathematics, but has usually forgotten it owing to disuse and will be reluctant to admit it, even to himself. A very common error among modeling groups is to assume the decisionmaker possesses a higher degree of mathematical sophistication than he does.

It is equally true that the user or client organization also has its own specialized vocabulary, which is often just as incomprehensible to the management scientists. Thus, two groups, each speaking a different language, must communicate some rather complex concepts to each other if the project is to be successful.

Rubenstein *et al.* [13] found a very close correlation between successful groups and the presence of a dynamic, articulate group leader. In many cases, when this leader left the group rapidly declined in influence and often disappeared altogether. One can readily hypothesize that the key factor to the leader's success was his ability to bridge the communication gap. Bean's very interesting study [4] shows a positive influence on success by using intermediaries to bridge the communication gap between clients and modeling group. This study indicates that reliance upon direct communications between the management science group and the client-users resulted in dissatisfaction, strained relationships, and a poor record of successful implementations of results. The study further shows that the most successful intermediary was one trained in the functional area being studied. Such studies serve to reinforce strongly the notion that the forms of communication between the modeling group and the user are critical to success.

7.7 Skills mix

There is considerable discussion in the literature concerning the desirability of the interdisciplinary team approach to O.R. studies. In actual practice, however, the typical operations research department is very small with an average staff of five professionals and one or two clerks and secretaries [20]. Therefore, it does not make much sense to talk about the desirability of interdisciplinary teams and then list 10 or 12 different desirable skills. Based upon existing practice, of the five professionals in a typical department, two are trained in mathematics, statistics, or some physical science, two are engineers (mainly industrial, mechanical, or electrical), and the fifth is trained in either business administration or economics. Usually at least one has his doctoral degree, and two others their master's degrees. Several, and perhaps all, the members are competent with the computer and are expert programmers.

When we speak of the desirability of a mixed team approach,

we are not necessarily speaking of interdisciplinary in the traditional sense. Rather, we mean a variety of backgrounds consisting in a mixture of analysts, operating personnel, and others who are intimately concerned with the problem area. Harvey N. Shycon [18] has described the ideal team as consisting of:

1 *One or more management scientists.* These will be persons with skills and experience in applying management science techniques to real management decision problems. The orginal discipline background of the individuals may vary, possibly, the physical sciences, or more recent graduates would have degrees directly in operations research. In either case, they should be basically quantitative analysts with the capability to abstract the problem, model, consider uncertainties where appropriate, and execute the kind of analysis required to suit the situation.

   Qualities here should be practical applications orientation, with the ability to orient the techniques to fit the problem requirements (not the other way around), and a willingness to analyze the problem in depth in order to determine what the *real* problem is.

2 *One or more computer scientists.* A computer scientist, in my definition, is not simply a computer programmer. He is himself an analyst, a model builder, but whose primary orientation is toward the use of computing power to manipulate data and develop information. He will implement the mathematical techniques toward a meaningful result.

   The computer scientist may in some cases be the same individual as or replace the management science specialist mentioned above. They may be very much the same animal, though one having great facility with the computer, the other having great facility in defining the problem and building models.

3 *One or more well-rounded, nontechnical individuals* who know the company and the organization, and particularly the problem area. These individuals can perhaps best understand the problem in the operating sense, can help interpret and frame it for the management scientists, and can assist in formulating the logic of the method of solution. Further, they can interpret back to operating personnel the language and logic of the management scientist, can perhaps open doors which would not otherwise be open to the project team, and may even know where to find certain information concerning the problem area.

   These individuals can be among the most valuable on the project team. They are in a position to know the organization better than the analysts themselves, to know where to obtain information, to assist in getting the cooperation of operating people, and finally, are perhaps better equipped to help the project team "sell" the project solutions than anyone else.

4 *One or more representatives from the problem area itself,* those concerned with the operating functions being studied and who perhaps know the problems involved better than anyone else. These individuals, working under the careful guidance of the management scientists, can be among the greatest contributors to the project and its solution simply because of their local knowledge of the problem area.

When we examine Fig. 1.3, Chapter 1, and Fig. 7.1 in this chapter, we see clearly that the development of a simulation model requires diverse expertise. No single person has all the expertise necessary to develop and implement such models. Although the management or computer scientist can develop models that are internally consistent, they usually lack the information and knowledge of goals and value judgments required to ensure that the model reflects the real world. Not only are the aid and support of others required in a simulation task, but also the efforts of all the experts must be integrated into a single effort, which all will support and understand. This suggests that we must devise a management scheme, which will do the following:

1 integrate the activities of people with diverse backgrounds;

2 translate the specialized knowledge of various organizational segments;

3 control resources.

7.8 Feasible organization schemes

In deciding what is necessary in the way of organizational setups, we should consider the following characteristics of all simulation projects, suggested by Olsen and Sasser [12]:

1 A specified task is to be done.

2 The task has a specified or desired date of completion.

3 There is a high degree of uncertainty about the direction and results of the study.

4 Stringent resources constraints exist.

5 A number of diverse activities are involved.

6 The nature of the task requires the efforts of diverse individuals from various organizational elements.

7 A high degree of coordination of activities is required.

8 A high degree or level of commitment is required.

9 At some point in the project, the task results and/or model are turned over to a user group, and the development organization disbands or goes on to another task.

These characteristics are common to most any research and development project, despite its goal, which suggests that many of the concepts and techniques developed over the years for research and development project management are equally applicable to the management of a simulation development.

We can use at least three methods of team organization and team interrelationships in developing and implementing a simulation study [16]:

1 The autonomous self-contained management science project team. This is a team consisting entirely of management scientists, related computer science people, systems analysts, and statisticians.

2 The semiautonomous self-contained project team with communication to management. This team consists largely of the people in method 1, but with a frequent reporting relationship to middle and upper management from user-client organizations and with ready access to their advice and counsel.

3 The integrated management science-management project team. This team consists in a true working partnership among members of the management science group, members from the user-client organization(s), and user-client management.

Each of these forms of team organization has been employed successfully in the past and will be in the future. Each has distinct advantages and disadvantages that must be evaluated for each particular study project. All three schemes employ the team concept, but vary as to the degree of participation of personnel and management from the user-client organizations. Harvey Shycon [16] has described the advantages and disadvantages of each as follows:

> In the first, the wholly management science team, the advantages are having a close working knowledge of the techniques and the thinking of the group among the team members, requiring the least amount of formal communications and explanation, and running the least risk of misinterpretation by one member of the team to another. A team of this composition can generally define its problem as it sees it with dispatch and achieve a result efficiently. Disadvantages of this arrangement are that while the

team has excellent communication within itself, it may tend to overlook aspects of the problem which are important to management though not obvious to those outside the management group.

The second method of constituting the team, the management science team with communication to middle management, has the advantage of greater communication with the "customers" of the services rendered. Customers are those who have the problem being addressed and those, therefore, who are most familiar with it. This method, to a large degree, avoids the problem of extensive internal team communications. Disadvantages of this method are those generated by the need to keep a greater number of persons informed on study progress, methods, indications, and results. Management people with whom communication is maintained are generally not familiar with management science, and it may require, therefore, lengthy explanations of the analysis being performed and the reasons for performing the study in this manner. As a consequence, the cost of the project may be somewhat greater than method #1.

The third method of team organization is that which involves a working partnership of members of the management science organization and members of the management group. It involves frequent meeting at both middle management and upper management levels. We call this the "inter-participative" team. Its advantages are clearly those of communication at all levels of the organization and direct contact with the problem area and those who know the problem best. Further, knowledge of the analysis and its results are disseminated widely in the organization, and a true management involvement in the progress of the study is achieved. Acceptance of ultimate study results is frequently more easily obtained. Disadvantages of this method of organization are the extensive communications necessary and the sometimes cumbersome nature of team deliberations. The team so constituted is generally larger and more time is spent in active communication up and down the line.

As we mentioned earlier, each of these schemes has been used successfully and each has its time and place. Which is most effective in a given case depends upon the given circumstances. In some cases there will be no choice. For example, if the development of the model is contracted out to another organization that is geographically apart from it, one has no choice other than to use the self-contained approach. On the other hand, if the decision process is indeed as we have portrayed it in Fig. 7.1, one would suspect that the third method would be more successful—i.e., have a better track record of successful implementations. Based upon personal observations and what little formal analysis has been done, such indeed appears to be true. One of the papers by Shycon [17] supports the contention that the percentage of successes is directly proportional to the degree of participation by client personnel and management. Careful examination of the reports of Rubenstein's group also provides further evidence [13]. Unfortunately, Shycon's data also show that the cost of projects run under the type 2 scheme ran about 100% higher than those under type 1, and that the costs of projects under the type 3 scheme were 70% higher than those of type 2. Thus, we pay a rather steep price to increase our chances of successful implementation; however, we must remember that success breeds success, and that the long-term survival of the O.R./M.S. group is greatly enhanced if we can use the interparticipative team approach.

7.9 Modular design and documentation

A very critical factor in the success of designing, programming, testing, and using a simulation model is the method used to write the computer program—not the choice of language but rather the "building" method used to produce the program. Any researcher who has been involved with computers and computer programmers for any length of time can tell you all kinds of horror stories related to programming problems; most are related to the basic way in which program was designed, coded, and tested.

In the early days of computer programming, a single programmer would sit down with the specification or logic flow diagram and proceed to write the code instruction-by-instruction until the complete program was coded. Only then would testing and validation begin. The resulting model or program is referred to as a *monolithic* or *integrated* model. Although such an approach was perhaps acceptable when computer programs were small and relatively uncomplicated, it is very undesirable today, particularly for simulation models. The strongly recommended alternative is modular programming, defined by Maynard [10] as, "a system of developing programs as a set of interrelated individual units (called modules) which can later be linked together to form a complete program."

Before we talk about the advantages of modular programming or design, we should perhaps first discuss the kinds of problems inherent in monolithic or integrated designs. First, the complete testing of monolithic programs is virtually impossible, owing to the large number of total logic paths, complex interrelationships, and the inability to isolate key portions of the program for testing purposes. Second, the logic of the program cannot usually be changed or modified readily without a major rewrite of the program. Third, the organization and research team become very dependent upon the continued availability and cooperation of the original programmer.

Computer programmers are very individualistic people, who like to demonstrate this attribute in their programs; this is particularly true when programs are being written in machine or assembly language where more individuality is possible. But the individualistic tendency is also evident in a simple experiment, such as asking 10 programmers to write a simple FORTRAN program to generate and sum the numbers 1 to 25. The result will be 10 distinct, different programs. Programmers (like everyone else) tend to get bored, and they will devise a great many methods of coding simple routines to keep their minds active during long coding jobs. For example, it is not unusual for a programmer to use his own, or a girl friend's, or his childrens' names for variable names. If the programming job takes some time, this individuality often results in the program author having forgotton how he intended a particular section of coding to work when he gets ready to begin testing it. It also means that probably no one except the original author will be able to modify the program.

A far more serious problem arises if the original author quits, is injured, or becomes ill before the program is completed and tested. Before a replacement can complete the job, he must find out how the program functions, what testing has been completed, and what unusual or nonstandard coding techniques and variable names the author used when he originally wrote the program. In many if not most cases, the replacement will find it easier and faster to start over, rewriting the entire program. Anyone who has ever tried to figure out someone else's integrated program will recognize this profound truth, and so will any manager who has ever asked a programmer to modify someone else's program.

On the other hand, modular programming reduces the overall job into small manageable units, such that each unit or module can be developed as an independent unit. We cannot provide here a complete discussion of the process of performing modular programming; it is found, however, in the book by Maynard [10], devoted entirely to this subject. Suffice it to say that each module will perform a single logical function or a number of small related logical functions. An analysis of the model is performed to determine what subsystems and logical functions are required. Once these logical functions have been identified and isolated, each module can then be coded and tested on its own.

The critical stage of modularization is the development of an initial module linkage chart showing the basic split into modules. When linked together, the individual modules of a program communicate by passing data from one to another. Thus, after all the modules have been coded and tested, and are known to work correctly, the development of a working model requires only that the modules be linked correctly and called in the correct sequence.

One of the great advantages of modular programming is that it produces small sections of easily reusable and/or replaceable coding. A single module can very simply be "lifted" from an existing program and used elsewhere, or replaced with another module. For example, if one of the purposes of our model is to examine different policies, then that module representing the logic function of the policy can be replaced readily with different modules representing different policies. In an integrated program, the policies usually end up so imbedded in the program that a major rewrite is required for each policy change.

Closely related to this advantage is our ability to simplify, elaborate, or modify our model at will. We can pull out a module that generates stochastic variables via Monte Carlo and substitute a constant value such as the mean (or vice versa). We can pull out a normal variate generating module and plug in a Poisson variate generating module. We can pull out a module that chooses customers on a first-come-first-served basis and plug in one that chooses customers on some priority basis. No integrated or monolithic design can approach the modular design for rapid change and flexibility.

The next factor we must discuss is documentation. Most programmers detest spending the time necessary to document each module and the overall model design properly. Complete documentation requires the following:

1 flow diagrams of each module and the overall model;

2 description of inputs necessary for executing the program, including: input card number, symbol, definition, whether integer or real, and the field;

3 definition of program variables not used as input to the program;

4 verbal descriptions of all modules as to purpose and function;

5 input deck setup to run on the computer of interest;

6 listing of the program.

I have found that a good rule of thumb is to prohibit any module from containing more than fifty (50) programming steps and to insist upon the liberal use in the program of comment cards; which are statements of definition or explanation inserted into the program code.

Therefore, management can do a great deal to aid the successful design, programming, testing, verification, and implementation of computer simulation models, if it will insist upon:

1 modular design;

2 no main program or subroutine to exceed approximately 50 lines;

3 each main program and subroutine to begin with a statement of what it does and a dictionary of the variable names used in the routine (done by including comment cards in the program);

4 comment cards included before each major calculation, thereby explaining the program;

5 complete documentation, including how to input data.

Although these requirements may sound simple and obviously sensible, it has been my experience that most managers and researchers learn about them only after several costly mistakes. What it boils down to is that common sense does not appear to be common. Ask any experienced simulation modeler for his favorite horror stories and I believe you will invariably find that one or more of the common sense suggestions given here were ignored. I will also guarantee that the first time you see six months' or more effort go down the drain because of inadequate documentation or because the model could not be modified easily, you will become very hard nosed about insisting upon "common sense" procedures from programmers.

7.10 Implementation

The best simulation model in the world is valueless until it is sold to or accepted by the people who must implement the results. Unless implementation is achieved and favorable results are obtained, all our efforts have been in vain. The fact that we have built an elegant computer simulation model and utilized good statistical and mathematical analysis in the generation and interpretation of the results is of little or no consequence to most operating managers, who are concerned with the problem and how it will be solved. Managers could not care less about the sample sizes, levels of confidence used, and all sorts of details that concern the researcher when he prepares his paper for presentation at a technical meeting or publication in a journal.

The information generated by the simulation model must be acceptable to the client-user before it will be used. The criteria for acceptability cover several aspects of credibility and utility. Obviously, the output information must appear reasonable; that is, the model should not be capable of providing absurd answers, even if absurd or extreme values are used for the input variables. If the outputs from the model are not believable under certain sets of circumstances, they immediately become suspect even when the results do appear reasonable.

A second aspect of acceptability and implementation is that the client-user must be able to see how he can act upon or use the information generated. If he cannot use the information directly to either help himself make decisions or assist someone else in making them, then he will ignore the information and the modeling effort will be wasted. The success of the effort and the utility of the results are thus obviously closely tied to the goal or objective set for the study—the reason we have stressed and stressed again the importance of realizing always the goal for which we are building the model. By this point the reader is sick and tired of hearing, over and over again, "Which way you go or what you do depends upon your goal or purpose." But the fact remains that if the model is to have any utility and if the results are to be used, the goal or purpose must be realized.

Very closely related to goal achievement is the requirement that the data needed to utilize or exercise the model must be realistic. To be realistic, the data required must be (1) available, (2) of acceptable quality, and (3) obtainable at reasonable cost. As obvious as that sounds, it is amazing how many times great effort is expended to develop a model for which input data are unavailable. For example, I am familiar with a very interesting simulation model of a hospital that took almost a year to develop. Unfortunately, it will probably never be used, for the simple reason that a qualified physician would have to examine each patient's records to provide a proper classification for him, and the hospital would have to revise its whole bookkeeping procedure.

Finally, any simulation model used must permit a manager to evaluate both the decisions that satisfy his own concepts of rationality and the likely results of decision strategies that he has formulated. The key point here is personal involvement. Ultimately, the manager, not the modeler, must take responsibility for the results of any decisions model. The manager assumes (probably correctly) that he is the person most likely to be able to devise satisfactory solutions. He looks to the analyst for help, not replacement. The reader should read the article "What is a Decision," by S. Eilon [6], a paper discussing various aspects of rationality in decisionmaking and the concept of personalistic vs. nonpersonalistic choice.

Thus, to maximize the chances for implementation of the results of a simulation project, we should have ended up with a model that is,

1 understandable to the client-user;

2 capable of giving reasonable answers;

3 capable of giving implementable answers;
4 realistic in data requirements;
5 capable of giving answers to "what if"-type questions;
6 easily modified;
7 cost-effective in use.

7.11 Presentation of results

Hopefully, the modeling team will have the opportunity to keep management or the client-user abreast of development all during the study. There is no doubt that the more familiar they are with the project and its details, the more likely they are to accept and implement the results. Such an approach also gives them the opportunity to make inputs all through the process and thus feel that they have participated in the end result.

It is extremely important that all presentations are made in a reasonable and plausible form at a level that is readily comprehensible. Obviously, a presentation that accomplishes this requirement must be planned carefully and prepared imaginatively. This whole idea of selling the results is as important as the study itself, and it should receive the same creative thought and attention.

In preparing to present results to management or to client-user, the analyst should remember and/or emphasize several points:

1 Always try to make verbal presentations using well-thought-out visual aids rather than written reports. The evidence is overwhelming that written reports are seldom read. In addition, the feedback of questions and answers is immeasurably helpful in determining whether you are getting your points across. Have handouts available for the participants to refer to and take with them.

2 Try to make a series of informal presentations instead of one big formal one. What you are trying to get across may be simple and readily understandable to you since you have lived with the project for some time and participated in or made all the decisions, but it is probably all new to the listener. New, complex ideas take a certain time to absorb and are learned better in small doses.

3 Don't assume that the listener knows all about computers, simulation, statistics, and mathematics. He probably does not. Concepts such as "standard deviation" or "confidence level" are basic to the analyst but usually completely foreign to the manager.

4 Stress the logic of the approach you have used to the problem and its solution. A long explanation of the merits and techniques of simulation, the problems of analysis encountered, the heroic methods of solution used, etc., will generally fall on deaf ears. What the listener wants to know is, do you understand the problem? Was your analysis and approach logical? Are the solutions you are proposing reasonable?

5 In the initial presentations, clearly and carefully spell out the assumptions and constraints of the study. This allows the listener to raise any objections early, and they can be taken into account during later analysis and/or presentations.

6 Spell out carefully the objective function or response variable used in the study. It is important that the listener understand what the analyst was attempting to optimize on the basis of comparing the different alternatives studied.

7 You must explain carefully and in the most nontechnical terms possible the conceptual approach used, the principal relationships and variables considered, and the reasons for interpreting the results in the manner that you have.

8 The burden of proof is upon us to interpret our work to the manager or client-user. We should therefore present our results in a format that meets their expectations and is understandable to them.

9 Summarize clearly the alternatives being considered and the predicted results in terms of both pros and cons. Also indicate your confidence or lack thereof in each prediction.

10 Finally, don't deliver a testimonial to the techniques used or the professional abilities of the modeling team. Remember that the home buyer is not interested in the brand name of the hammer the carpenter used or the number of nails driven without bending. In general, management assumes the best or they wouldn't have brought you on board in the first place.

7.12 Follow-up

It goes without saying that the modeling team should help in the implementation phase to the highest degree possible and continue to follow up for as long as the model is used. In most organizations, it is not sufficient to develop the model, test it, run the study, and present the results in a report, only to sit and expect acceptance and implementation of the results. The sad truth is that the implementation phase can also make or break the project and thus should be planned and carried out as carefully as possible.

The nature of the implementation problem depends upon whether the result of the project is a one-time decision or a tool for aiding in repetitive decisionmaking. In either case, the team needs not only to sell the model and its results but also to remain active during the implementation and follow-up. This is necessary because:

1 No matter how careful the researchers have been in designing, validating and running the model, shortcomings may not become apparent until the results are implemented.

2 The means of implementing the solution or the continued running of the model may not be as obvious as the research team thinks.

3 If the client-user tries to modify the model himself without thoroughly understanding it, the results may be disastrous.

4 The real world system may change with time, so that the model will no longer be a valid representation.

Researchers naturally tend to lose interest in a project as soon as they have developed, checked out, and run the model. They are anxious to get on to the next challenging problem and to leave the follow-up and implementation to someone else. This is one reason why so many projects that look good on paper are never implemented, or are implemented with very disappointing results. The Army has a saying that they tack up all over their training facilities: "If the student hasn't learned, the teacher hasn't taught." We might change that to read, "If the user hasn't used the result, the researcher hasn't finished his task."

7.13 Check list for managers and analysts

In summarizing some of the ideas and activities that are important to the success of a simulation project, we might consider a couple of check lists to jog our memories. First, the manager responsible for the modeling activity should:

1 Be sure the problem is one of more than casual interest to the client-user.

2 Define the problem carefully in conjunction with the client-user, including the proposed scope of the study (i.e., what is and is not proposed).

3 Establish an implementation plan, including what personnel will be used, how the client-user will be involved in the design and carrying out of the project, what resources will be required, and a time schedule for implementation, including clear-cut milestones.

4 Establish the organization for implementing the study, including the reporting relationships to management and the client-user.

5 Obtain the concurrence of management and the client-user of the plans, schedules, budget, and the total scope of the project.

6 Establish and maintain close relations and liaison with management, client-user, and the project team.

7 Regularly compare progress and resource expenditures against plans and budgets.

8 Insure that consideration is given to each and every step of the check list for modelers.

9 Insist upon a modular design as well as adequate and complete documentation as the model is programmed and checked out.

10 As the project nears the end, give serious thought and careful planning to the methods and approach to be used to sell and implement the results.

11 Play the devil's advocate by anticipating all the weaknesses and shortcomings of the model, results, and plans for implementation that may be raised so that possible objections can be answered and/or overcome.

12 Follow up during the implementation phase and keep in close contact with the client-user.

13 Objectively evaluate the entire project in terms of its successes, failures, errors, etc., so as to learn for the next project.

14 Write up a project summary and critique and insure that all aspects of the model and the project are well documented.

We can likewise summarize some of the other things we have

discussed in this book by establishing a check list for simulation modelers, which must include the following:

Section 7.13 Check list for managers and analysts

1 Be sure you understand the problem and the system to be studied. Check and recheck your understanding with those most familiar with the system.

2 Define carefully the objective and scope of the study. What are the specific goals? If you can't write them in just a few sentences, you haven't defined them.

3 Spell out the boundaries of the system to be studied. What is and is not to be considered a part of the system for this study?

4 Establish a specific plan of attack. Is simulation the correct approach to the problem? If so, what specific manpower and other resources are required? Set up a schedule, budget, etc., for the study.

5 Obtain concurrences of management and client-users of the goals and objectives, approach to be taken, plans, and budget.

6 Document the subsystems, interrelationships, etc., of the system. Find out what usable data are available.

7 Construct a logic diagram of the model as you conceive it.

8 Run empirical tests (goodness of fit, etc.) where possible to determine the parameters and characteristics of the interrelationships, inputs, etc.

9 Decide upon the programming language and timekeeping mechanism to be used.

10 Construct and program the model using a modular design, liberal use of comment cards, careful documentation and checkout of each stage.

11 Test to ensure that all subprograms and the overall model run as designed and intended.

12 Check all random variate generator subprograms to ensure that they are indeed random and give the statistical distribution desired.

13 Carefully plan how the model is to be used. This may include preliminary runs and includes consideration of:

(a) choice and definition of experimental variables and the levels to be used for input variables;

(b) presence or absence of autocorrelation;

(c) design of the experiment;

(d) starting conditions and when to start taking measurements;

(e) possibility of using variance reduction techniques;

(f) sample sizes to be run;

(g) how the model will be validated.

14 Run and validate the model. Continuously check the results with the client-user for reasonableness.

15 Analyze the results, including the performance of sensitivity tests and the establishment of confidence intervals, etc.

16 Carefully plan how the results and recommendations will be presented to management and the client-users. Be sure to include a specific plan for implementing the results. Make liberal use of good visual aids in all presentations.

17 Sell, sell, sell the results and recommendations!

18 Follow up continuously during the implementation phase and participate, if at all possible.

19 Be sure every phase of the model and project are documented.

20 Learn from the errors and mistakes made!

7.14 What of the future?

Even a casual examination of the applications of simulation cannot help but impress us with the pervasive and important impact it has had and will continue to have upon our lives. One is hard pressed to find any area of human concern in which at least tentative steps have not been taken to utilize simulation. It might therefore be interesting to examine briefly where we have been, where we are, and where we are likely to go.

In a keynote address at the 1972 Summer Computer Simulation Conference held in San Diego, California, Karplus [8] discussed the "Four Ages" of technology and simulation. His presentation was based upon the premise that all technological advances go through a similar, irreversible evolution consisting of four basic phases characterized by the types of individuals who play leading roles at that time. The four basic phases of development can be characterized as follows:

1 the inventor;

2 the engineer;

3 the theorist;

4 the ultimate user.

Throughout the following discussion we are talking about the "center of gravity" of the activity, even though progress and research continues in all the preceding phases. Also, that the history of simulation and that of computers are intimately interconnected.

During the early years of any field, the inventor plays a paramount role. For computer simulation, the phase of the inventor can be considered to have covered the period from 1938, when the first suggestion for operational amplifiers was made (leading to analog simulators), until approximately 1953, when the first commercially available electronic analog computers were introduced. During this time also, the work of von Neumann and Ulan appeared, suggesting Monte-Carlo sampling, the design of digital computers was making great strides, and the first UNIVACs appeared. The introduction of the IBM 650 in 1953 launched a period of intense competition in computer design.

In the engineer phase, the emphasis shifts to reduction to practice. Useful inventions continue to be made, but there is a concerted effort to make inventions more usable. For the field of simulation, this period lasted from about 1953 to perhaps 1962. During this period, the development of analog, digital, and hybrid computers far outdistanced users' abilities to employ their capabilities effectively. Starting in the late fifties and continuing through the early sixties, we saw the rapid development of simulation lanaguages designed to help the modeler translate his ideas to the computer in a rapid fashion. The programmer and numerical analyst came into great demand. Computer software firms and service companies began to appear, and relatively simple systems simulations appeared in the literature.

As more time elapses, the emphasis shifts toward the theorist. This phase for simulation lasted from about 1962 to 1970, during which time the scientists and mathematicians began to examine the concept of simulation and to put the whole field on a sounder, more formal, scientific basis. Articles began to appear dealing with the finer points of simulation, such as sample sizes, starting conditions, statistical analysis of output data, etc. This emphasis on theory led to simulation becoming a respectable academic pursuit, with the emergence of formal course work and the formation of technical societies (e.g., Simulation Councils, Inc.) devoted to the field.

We can summarize the status of simulation at the end of the theorist phase as follows. By 1970, the literature contained many excellent articles dealing with simulations of queueing situations, materials handling systems, transportation systems, aerospace hardware, logistic and inventory systems, etc. We could do a very good job of simulating mechancial systems that do not require man as an integral part of the system. We were also successful at solving management problems through simulation, if they had the following characteristics:

1 concerned with operating problems of the firm;

2 occurred at middle management levels;

3 had quantifiable variables;

4 dealt primarily with economic variables;

5 had scalar payoff functions.

It is interesting to compare the proceedings of the Summer and Winter Computer Simulation Conferences held in 1970 vs. those held in 1973 as to the types of papers presented. By 1973, sessions are devoted to urban problems, national economic policy, world and global simulation, political science, public policy, the environment, and many other areas of public concern. It would be accurate to say that the majority of papers presented in 1973 dealt with social and political science-oriented areas—strong evidence that the ultimate user phase is already well under way.

The ultimate user phase is characterized by a greater concern for the ultimate impact of the study than for its technical details. The ultimate user may be an economist, political scientist, lawyer, doctor, or humanist. There tends to be much less sophistication both mathematically and in the details of computation as the user relies upon the power of the computer and simulation languages to handle these concerns for him. On the other hand, there is a greater concern with the ultimate utilization of the simulation and the importance of social, economic, and political considerations. One result is less concern for "optimizing a design" and more concern for exploring the probable outcomes or pursuing various alternatives. In fact, the results of many of these simulation models will not actually be adopted in practice but will be used as ammunition in political battles fought over vital societal issues.

What of the future? It is safe to say that we shall continue to see advances from the inventors and engineers in terms of hardware, software, and more powerful languages. For example, advances in computer graphics and on-line computing techniques will play an ever-increasing role in simulation.

Likewise, we shall see a strengthening of the theoretical and scientific base of simulation as the scientists and mathematicians attack the remaining problems of multiple response designs, validation, autocorrelation, etc. But the real action and activity will be concerned with those problems which:

1 occur at the top levels of management;

2 are strategic and integrative in nature;

3 have multidimensional payoff functions;

4 contain nonquantifiable variables;

5 require the inclusion of human values and judgments.

Simulation is such a powerful and flexible technique for attacking complex systems problems that there is no doubt of its continuing importance. I suspect that we could easily list 300 or more applications today. As the activity picks up in social, political, and economic areas, the list will grow into the thousands. I once heard a speaker say that there was no problem he couldn't solve using simulation. Although that is certainly an exaggeration of the power of good simulations, there is no doubt that simulation modeling has already had a profound impact upon each of our lives and will continue to do so for a long time to come.

REFERENCES

1. ALBERS, H. H., *Organized Executive Action*, John Wiley & Sons, Inc., New York, 1961, p. 229.

2. ANSHOFF, H. I., and R. L. HAYES, "Role of Models in Corporate Decision Making," *Proceedings of IFORS Sixth International Conference*, Dublin, Ireland, 1972.

3. BARNARD, C. I., *The Functions of the Executive*, Harvard University Press, Cambridge, Mass., 1938, p. 194.

4. BEAN, A. S., "Client Receptivity and the Structure of the MS-Client Interface," Presented at Joint National Meeting of TIMS, ORSA, and AIIE, Atlantic City, N. J., Nov., 1972.

5. CHURCHMAN, C. W., R. L. ACKOFF, and E. L. ARNOFF, *Introduction to Operations Research*, John Wiley & Sons, Inc., New York, 1957, p. 132.

6. EILON, S., "What is a Decision?" *Management Science*, Vol. 16, No. 4, Dec., 1969.

7. HOVEY, R. W., and H. M. WAGNER, "A Sample Survey of Industrial

Operations—Research Activities," *Operations Research*, Vol. 6, 1958, pp. 878-881.

8. KARPLUS, W. J., "The Four Ages of Technology—and Simulation," *Simulation*, Vol. 19, No. 3, Sept., 1972.

9. MARCH, J. S., and H. SIMON, *Organizations*, John Wiley & Sons, Inc., New York, 1958.

10. MAYNARD, J., *Modular Programming*, Auerbach Publishers, Princeton, N.J., 1972.

11. MORRIS, W. T., *The Analysis of Management Decisions*, Richard D. Irwin, Inc., Homewood, Ill., 1964, p. 502.

12. OLSEN, R. P., and W. E. SASSER, "Managing the Development of a Large-Scale Simulation," *Harvard Business School Working Paper 71-7*, Presented at 39th National ORSA Meeting, *Operations Research*, Vol. 19, Suppl. 1, Spring, 1971.

13. RUBENSTEIN, A. H., *et al.*, "Some Organization Factors Related to the Effectiveness of Management Science Groups," *Management Science*, Vol. 13, No. 8, Apr., 1967.

14. SCHUMACHER, C. E., and B. E. SMITH, "A Sample Survey of Industrial Operations—Research Activities II," *Operations Research*, Vol. 13, No. 6, Nov.-Dec., 1965.

15. SHANNON, R. E., and W. E. BILES, "The Utility of Certain Curriculum Topics to Operations Research Practitioners," *Operations Research*, Vol. 18, No. 4, Jul.-Aug., 1970.

16. SHYCON, H. N., "All Around the Model," *Interfaces*, Vol. 1, No. 4, Jun. 1971.

17. SHYCON, H. N., "All Around the Model," *Interfaces*, Vol. 1, No. 6, Aug., 1971.

18. SHYCON, H. N., "All Around the Model," *Interfaces*, Vol. 2, No. 3, May, 1972.

19. SIMON, H. A., *Administrative Behavior*, The MacMillian Company, New York, 1961, p. 203.

20. TURBAN, E., "A Sample Survey of Operations-Research Activities at the Corporate Level," *Operations Research*, Vol. 20, No. 3, May-Jun., 1972.

appendix A

CASE STUDIES

Case Study A-1

SIMULATION OF AN INFORMATION CENTER'S INQUIRY ANSWERING SERVICE*

Bernard J. Schroer†

This case study was selected to provide the reader with an overview of the types of issues the modeler must face and solve. Particularly note the discussion of data collection and tactical planning.

Introduction

The Apollo Parts Information Center [4, 5, 6] is an information center for collecting, storing, and disseminating parts and materials information for the Apollo Program. Specifically, the center is concerned with parts and materials information such as item identification, requirements information, qualification data, test data, failure data, environment and environmental extreme data, usage data, and inspection data.

Figure A.1.1 outlines the general procedures for entering information in the center. The information required for retrieval is abstracted from the various input documents and stored on magnetic tape, which comprise the bibliographic files. After the abstracted information has been coded, the documents are recorded onto video magnetic tape, which comprise the document files. The original documents are then stored in the manual file, which consists of standard filing cabinets.

The center offers two major services to its users; an index and

* Condensed from, "A Simulation Model for Evaluating the Projected Expansion of the APOLLO Parts Information Center's Inquiry Answering Service," unpublished Master's Thesis, University of Alabama, Huntsville, Ala., 1967.

† Research Associate, Center for Environmental Studies, University of Alabama in Huntsville.

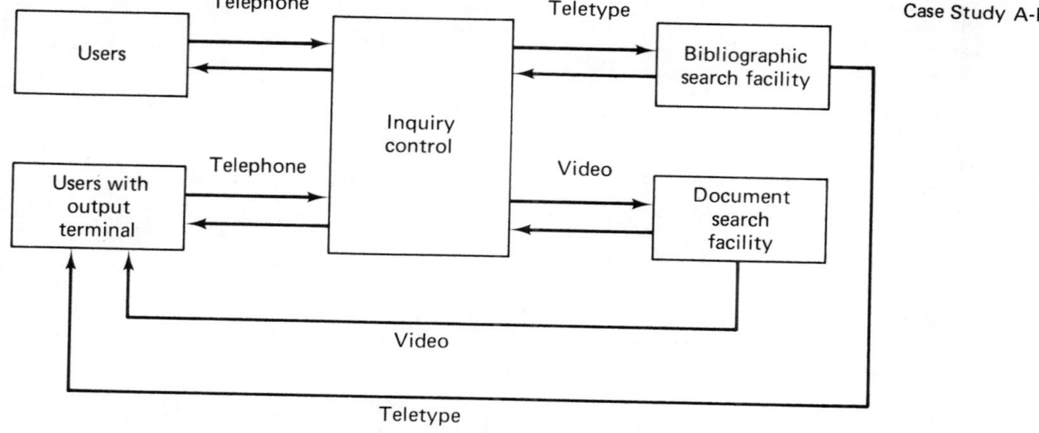

Figure A.1.1 Generalized inquiry response flow diagram.

an inquiry capability. The index is a periodic publication that presents in abstract form information for a specific part or material. The primary function of the index is to provide the user with a summary of the information available from the center. The inquiry capability is the major service offered by the center, allowing the user to submit telephone inquiries to the center and receive a response by either telephone or mail.

Inquiry processing flow

All inquiries are received at inquiry control via telephone, where there are two rotary lines with two telephones per line. Because of the complexity of the types of inquiries, no capabilities exist for the user to query the center directly. Instead, the center employs technical personnel for receiving, standardizing, and coding inquiries. There are three processing routes or classes of inquiries: special manual search, bibliographic search, and document search.

A *manual search* is conducted for inquiries that cannot be answered by searching the bibliographic files. Figure A.1.2 presents the general procedures for answering this class of inquiries. After receiving an inquiry, one of inquiry control's experienced personnel is assigned to locate the answer. The answer is documented and the user is notified by telephone to determine if a copy is desired of any of the available documents. If the user does not require copies, the response is prepared for mailing; if the user desires a copy, the documents are reproduced on a copier and prepared for mailing.

A *bibliographic search* is an automated search of the bibliographic files to determine, in summary form, the information that is available in the center. Figure A.1.3 presents the general procedures for answering this class of inquiry.

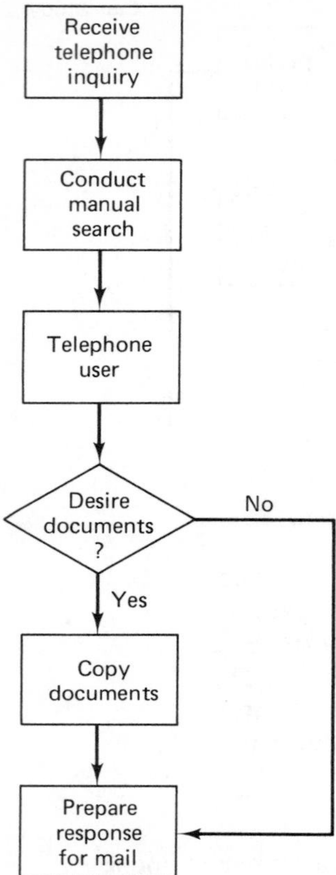

Figure A.1.2 Processing the manual search class of inquiries.

After an inquiry is received, it is manually entered onto a teletype. The center has one teletype both for transmitting and for receiving. If the teletype is in use, the inquiry is punched on cards and the cards entered in the teletype card reader.

The inquiry is relayed to a communications processor, which is very similar to a telephone exchange for the switching of telephone calls. At the switching processor, the inquiries are stored and relayed to the central processor on a first-arrive first-service basis. Since the central processor is shared by many organizations, inquiries may queue until the processor is available.

The inquiry is answered by the central processor searching the bibliographic files. The located information is then returned to the switching processor, where it enters a queue until the receiving facility is available. If the inquiry was received from a user who has access to an output terminal and who desires his response via teletype, the answer in the switching processor is sent directly to the terminal. If the user desires a copy of any of the documents contained in the teletype response, he must submit another inquiry.

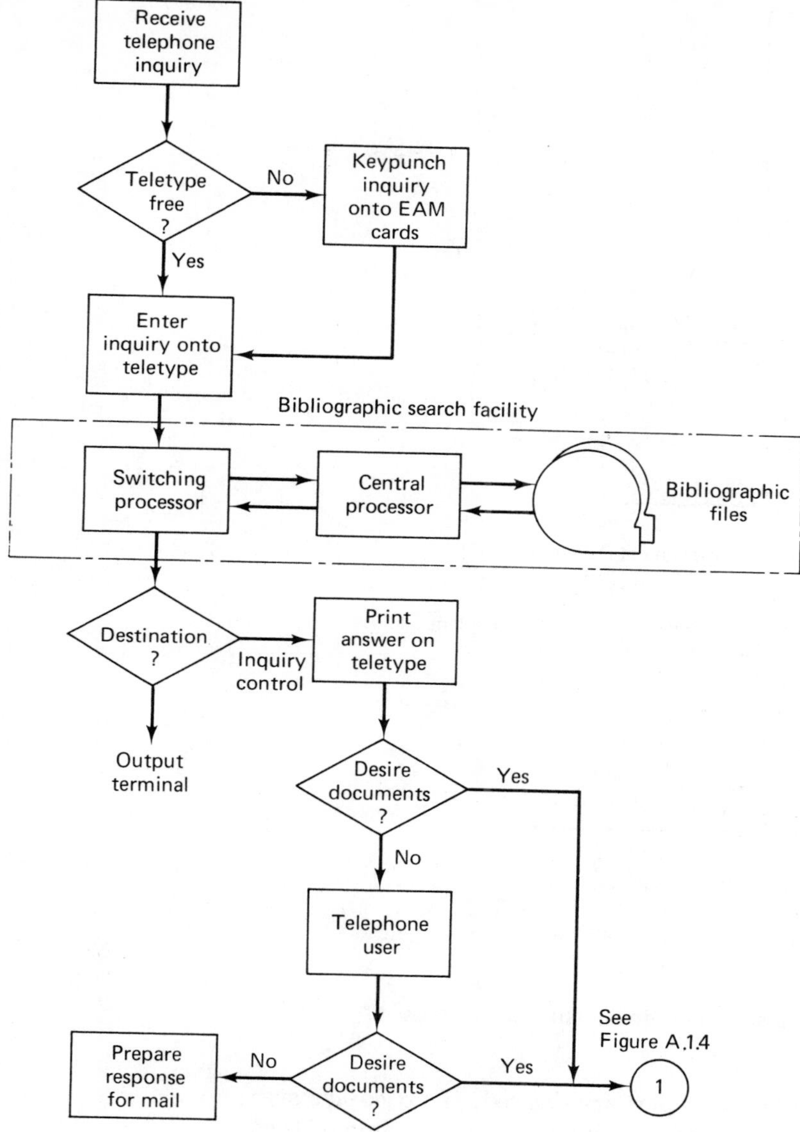

Figure A.1.3 Processing the bibliographic search class of inquiries.

For inquiries from users not having access to an output terminal, the answer in the switching processor is printed on the inquiry control teletype. If at the time of the telephone call the user stated that copies of all documents located are desired, the inquiry is then routed to the document search facility. However, if the user did not state that such copies were wanted, he is notified

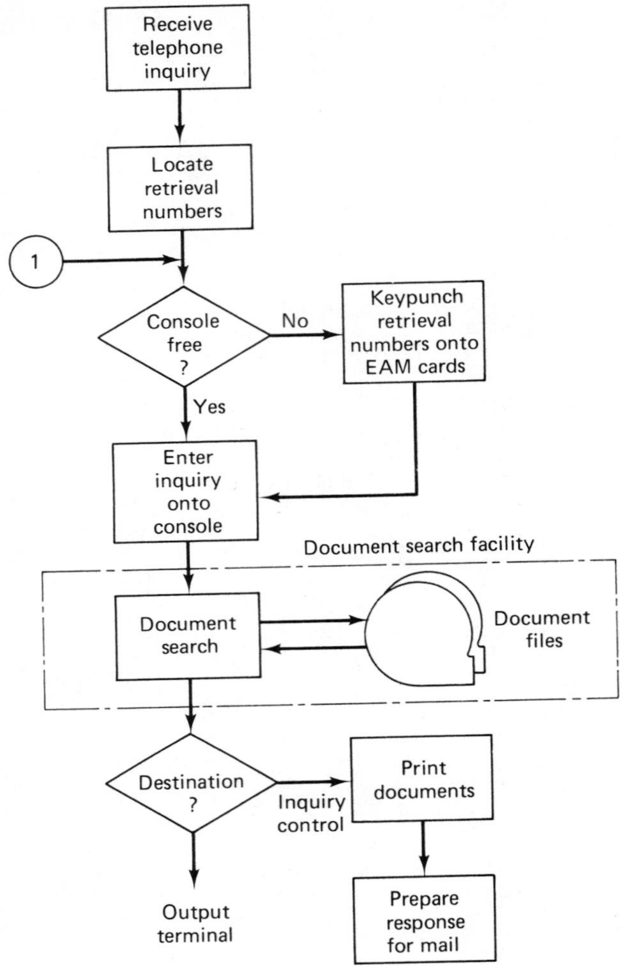

Figure A.1.4 Processing the document search class of inquiries.

by telephone to determine if copies are desired. If no copies are required, the response is prepared for mailing. If at the time of the telephone call the user states that he wants a copy of available documents, the inquiry is routed to the document search facility.

The *automated document search* is a proposed addition to the center. This class of inquiry is a request for specific copies of the documents stored in the document files. Figure A.1.4 presents the procedures for answering this class of inquiry. The graphic index that contains the document numbers is manually searched to locate the retrieval numbers, which are then entered onto the document console keyboard. If the console is in use, each retrieval number is keypunched onto a card and entered in the console card reader. If the user has access to an output terminal and desires the documents sent to his terminal, the document images are relayed directly to the output terminal. At the terminal the requestor has the alternative of viewing the retrieved images on his monitor and/or receiving a copy from the terminal printer. If the user does not have access to an output terminal, the document images are sent to the high-speed printer at inquiry control.

Development of the model

The information center is currently operational; however, an evaluation was essential to assure that the center could respond to the estimated increase in the number of inquiries. In addition, numerous equipment studies had resulted in recommendations concerning the types of hardware for expanding the center. However, a void existed in integrating these recommendations into the present center, and then evaluating the operation of the center.

The objective of this study was to evaluate the present center, including the proposed additions, to determine the capabilities of the center's facilities for receiving and answering inquiries. The selected approach in evaluating the center was to determine the turn around time of the proposed inquiry answering service. Turn-around time is defined as the total time that elapses once a requestor has made connection with inquiry control, through the processing of the inquiry, to the time that the response has been forwarded to the requestor.

A series of simple numerical statements was used to describe the answering of an inquiry. Categorizing these statements as operating components and delay components and summing these components resulted in expressions of the turn-around times. Once the model had been constructed based on the expressions for the turn-around times, the center was simulated under a variety of conditions. Since these conditions were controllable, the center could be tested by varying one parameter or a combination of parameters.

The model was based on the following set of definitions:

1 Processing an inquiry requires the utilization of a defined set of facilities ($F1, F2, \ldots, FM$).

2 Between the facilities there may exist a queue ($Q1, Q2, \ldots, QN$).

3 A component ($C1, C2, \ldots, CK$) is defined as an expression of time. The time to process an inquiry at a facility is defined as an operating component ($t_1(F1)$,

$t_2(F2), \ldots, t_i(FM)$). The time that may elapse in a queue is defined as a delay component ($t_1(Q1), t_2(Q2), \ldots, t_j(QN)$).

4 The turn-around time for each type of inquiry is a function of a set of components ($C1, C2, \ldots, CK$) and can be expressed as

$$T_q = f(C1, C2, \ldots, CK) \tag{1}$$

The set of operating and delay components associated with each T_q can be expressed as a function of

$$T_q = f(t_1(F1), t_2(F2), \ldots, t_i(FM) \text{ and } t_1(Q1), t_2(Q2), \ldots, t_j(QN)) \tag{2}$$

Assuming that the operating and delay components are independent of each other, we can express T_q as

$$T_q = \sum_{i=1}^{M} t_i(FM) + \sum_{j=1}^{N} t_j(QN) \tag{3}$$

where T_q is the turn-around time for each type of inquiry ($q = 1, 2, \ldots, 10$), $t_i(FM)$ is the time required at facility M to process an inquiry, and $t_j(QN)$ is the time that may elapse at queue N before the following facility is available to accept the inquiry.

The Univac General Purpose Systems Simulator II (GPSS) [3] was selected as the simulation language. The block structure of GPSS, along with its use of transactions, made it an ideal language. The model consisted of 250 blocks, 38 functions, 12 facilities, 13 queues, and 24 output tables.

Data collection

The data collection effort was oriented toward the collection of the data represented by the operating components, $t_i(FM)$, and by the additional inputs required for the simulation program. The data collection was divided into two phases: the collection of available empirical data, and the generation of data for which no empirical data were available and which were associated with the proposed additions to the center.

The collection of empirical data was made from January, 1966, through September, 1966. Primary data sources were inquiry log records, telephone tape recordings, and internal weekly and monthly status reports. The arrival distribution of incoming inquiries was determined through statistical testing to be Poisson.

Since the arrivals were Poisson, the time between inquiries followed the negative exponential distribution. Therefore, from projected inquiry growth rates, the appropriate exponential distributions were generated. A number of the service time distributions were determined to be Erlang, such as the time on the telephone and the time to conduct a bibliographic search. Several of the distributions could not be approximated closely by any of the standard theoretical distributions. Some of these were the number of questions per a class 1 inquiry, the number of documents per a response, and the number of pages per document. In these instances, the actual sampled data were used as input.

The method of forwarding responses for class 2 and class 3 inquiries was a proposed modification to the center; therefore, no empirical data were available for the frequency of occurrence of the types of inquiries within each of these classes. The assumption was made that the remote inquiry terminals would be initially installed within a 30-mile radius of the center. By using the data collected from inquiry log records, an estimate was made of the inquiries.

No empirical data were available concerning the delay that may occur because the requestor's telephone is in use at the time the call is placed for verbally relaying the response. Therefore, it was assumed that an arbitrary percentage of responses would be delayed because the requestor's telephone was in use. It was also assumed that a constant amount of time would elapse before an attempt would be made to replace the call. The time to place a telephone call is dependent on the type of call—such as a person-to-person, station-to-station, or direct dial call—and on the distance of the call. A personal conversation with a Southern Bell Telephone representative resulted in a distribution of the mean times to place different types of calls.

During the data collection, it was noticed that the printing time of the response to a question on the teletype was changing because of the computer programming modifications concerning the output formats and the quantity of output being printed. Therefore, accurate empirical data were not available. Other operating components for which data were not readily obtainable were the time per question to conduct a manual search, the time to prepare a response for mailing, and the time to search the index to locate a retrieval number. Through discussion with inquiry control personnel, an estimate of the mean service times was obtained for each of the components.

Tactical planning

In testing the inquiry answering capabilities of the center, we can consider a combination of parameters as input variables into the simulation, such as projected inquiry arrival rates. Therefore, the facilities and queues of the center are evaluated as a function of the various arrival rates. Before executing the simulation model, we had to determine the starting conditions, definition of equilibrium, stopping conditions, and the sample size.

Since the simulation model requires an amount of time to elapse for the system to approach a state of *equilibrium*, or stabilization, an initial portion of the run must be omitted from the output results. This stabilization period is a function of the number of inquiries in the center and the number of inquiries that have been processed by the center. According to R. L. Conway [1], there are no objective criteria for determining when measurements should be collected; however, he does recommend truncating a series of measurements until a measurement is neither a maximum nor a minimum of the previous measurements.

The time required for the system to approach equilibrium depends upon the initial starting condition. Since the model begins with all its facilities and queues idle and empty, any starting condition other than idle and empty would decrease the time required for the center to reach equilibrium. Ideally, the starting condition should be selected to correspond with the condition when the center has reached equilibrium. However, since such knowledge of the center is not known, the selection of the starting condition must be subjective. The selected starting condition for the center was initially to generate 10 arrivals at time zero.

After these arrivals are generated, the selected input arrival rates are used. The approach taken by Conway is used in determining the stabilization point of the center. The inquiries are grouped into sets of 25 in the order of their termination from the center. The mean turn-around time for each set is computed and presented in Fig. A.1.5. The results in the figure also include the previously stated starting conditions. The first mean, neither a maximum nor a minimum of the previous series, is at the fifth set. Therefore, data for the first 125 inquiries were not included in the output results. This stabilization point was constant for all runs.

The *stopping condition* of the model is to stop the run after a given number of inquiries have been processed by the center. No maximum is assigned to the number of inquiries inputed into the model; therefore, after the stopping condition is obtained, there may still be inquiries that are being serviced in facilities and delayed at queues.

Many methods are used for determining *sample sizes.* One technique frequently used is to determine the sample size needed to achieve a given confidence that a facility or queue is in a given state or condition. Since the center has many facilities and queues, and since some of these are infrequently used, such an approach would require a very large sample size. Therefore, the inquiry

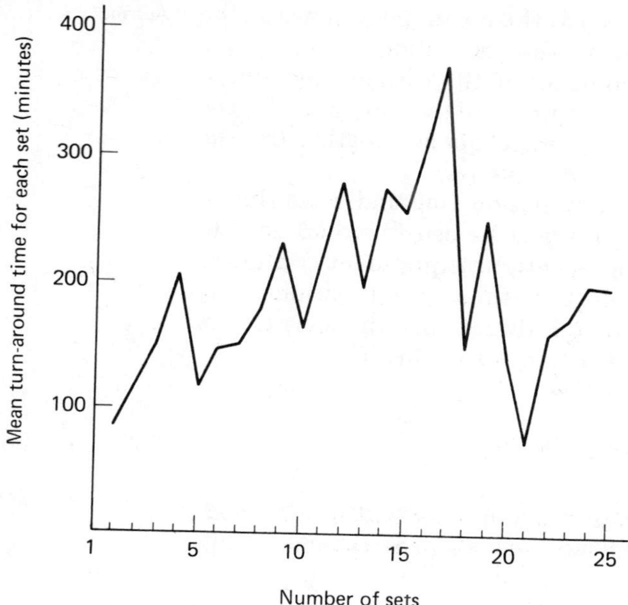

Figure A.1.5 Mean inquiry turn around time for each set of 25 inquiries.

turn-around time was used to determine the sample size. A relative comparison of alternatives was then used between the results of the facilities and queues.

To determine sample size, the inquiries were grouped into sets of 25 in the order of their termination from the center. By computing the mean turn-around time for each set, a group of n measurements, $x_1, x_2, x_3, \ldots, x_n$, was obtained. By the central limit theorem, we can assume that the $\bar{x}$ of each group is normally distributed. By Tchebycheff's inequality, n_x can be chosen such that for a 95% level of confidence, where $\bar{x}$ is the estimate of the true mean,

$$(1.96)\,\sigma_{\bar{x}} \leqslant \bar{x} \tag{4}$$

Since the n measurements are related to time, we cannot assume that the consecutive measurements are independent. Therefore, the variance of the mean of an autocorrelated series was used [2]. A minimum sample size of 500 resulted from the preceding calculations.

Model outputs

Outputs of the model provide information on the following aspects of the center: (1) the mean time to answer the various types of inquiries (i.e., the turn-around time); (2) the average utilization of the center's facilities; (3) the mean delay times at the center's facilities. To complement the preceding outputs, the model also outputs (1) the distribution of the turn-around times; (2) the distribution of the delay times in the queues; (3) the maximum queue lengths; (4) the average queue lengths; (5) the total entries that spent zero time in the queues.

Analysis of the results of the simulation runs indicates that if the projected arrival rates were achieved, the center would need to make some minor modifications concerning equipment configuration and personnel allocation. Most of these modifications only require an additional piece of hardware or an increase or reallocation of personnel assigned to various facilities.

REFERENCES

1. CONWAY, R. W., *Some Tactical Problems in Simulation Methods*, Memorandum RM-3244-PR, The Rand Corporation, Santa Monica, Calif., 1962.
2. GEISLER, M. A., *The Sizes of Simulation Samples Required to Compute Certain Inventory Characteristics with Stated Precision and Confidence*, RM-3242-PR, The Rand Corporation, Santa Monica, Calif., 1962.
3. *General Purpose Systems Simulator II on the Univac 1107 Users Manual*, Univac Division of Sperry Rand Corporation, New York, 1965.
4. *Operation Program Plan*, Astrionics Laboratory, George C. Marshall Space Flight Center, Huntsville, Ala., 1965.
5. *Specifications for Supplementary Hardware to a Mass Storage Device System*, Astrionics Laboratory, George C. Marshall Space Flight Center, Huntsville, Ala., 1965.
6. The "Videofile System," *Datamation*, Jan., 1965.

Case Study A-II

AIR DEFENSE MODEL*

Robert E. Shannon

This case study was selected to show how logic and knowledge of simple physical and time relationships can be used to develop the key structure of situations for modeling purposes. Many real world systems and situations are so simple and transparent that those familiar with the system can understand them by inspection and discussion.

Introduction

The purpose of this paper is to describe a simple computer simulation model useful in defensive weapon system parameter sensitivity analysis. The utilization of war-gaming simulation models to aid in evaluating proposed systems is an approach well accepted by systems analysts. Detailed simulation models have been developed and are constantly being improved to study the interaction of related components of complex defensive systems. These models have continued to increase in size and complexity owing to the desire to approach as near as practically possible an exact model of the real life situation. Unfortunately, as these models increase in size and complexity, it becomes more and more difficult to separate out the influence of individual system parameters upon the overall system's effectiveness. Therefore, it is becoming increasingly apparent that to evaluate a system's effectiveness adequately and to choose between alternatives, we must utilize a whole family of simulation models of various degrees of complexity, rather than only one big, complex one.

The development of any model has two primary goals or purposes. First, the model must be simple enough so that some kind of meaningful answers can be obtained fairly rapidly. Secondly, the simplifying assumptions necessary in creating the

*Condensed from "Air Defense Engagement Duel-Model I," *UARI Report No. 53*, University of Alabama Research Institute, Huntsville, Ala., Sept., 1968. This research was supported by the U. S. Army Missile Command under Contract DA-AH01-67-01630.

model should not obscure the critical factors involved in the operation of the system. Thus, the developer of a model is immediately faced with the necessity of certain tradeoffs. The model must approximate the critical aspects of the situation in which the defensive system will be used, but at the same time it must be simple enough to provide rapid answers.

The model described in this paper is intended to be but one in a family of models for evaluating air defense missile systems. The primary purpose is to aid in the evaluation of certain parameters of such defensive systems against low-altitude attack by enemy aircraft.

The purpose of low-altitude air defense systems is to protect civilian populations, supply depots, vehicle concentrations, artillery positions, troop concentrations, etc. The purpose of any offensive weapon system is to destroy a given set of targets (value units) with the least possible cost. It is generally accepted that a defensive system cannot prevent a determined and powerful offense from destroying a given number of these targets if the offense is willing to pay the price. The defense objective, then, is to try to extract a high cost from the offense. In gaming theory terms, the goal of the defense is to maximize the offensive cost while minimizing the defensive cost. The defensive units themselves are high-value target units, whose destruction is sought by the offense. In fact, the defense system itself will generally be the first or highest-priority target at the beginning of any enemy attack.

In general, the offense will try to destroy the defensive units early in an engagement by attacking them with missiles fired from low-flying aircraft. Attack by missiles as opposed to that by bombs allows the enemy aircraft to strike from greater range, thereby minimizing the chances of its own destruction. Also, by attacking from low altitude, a missile attack minimizes the chances of detection prior to being within range to fire offensive weapons, thus minimizing its own exposure time.

Model assumptions and inputs

The basic purpose of this particular model is to provide a means of studying the effect of certain critical parameters in a one-on-one duel between a defensive missile ground battery and an attacking enemy aircraft armed with air-launched missiles. Each is trying to destroy the other. Only four outcomes to the engagement duel are possible:

1 The aircraft is destroyed but the ground battery survives.

2 The aircraft survives but destroys the ground battery.

3 Both are destroyed.

4 Both survive.

In addition to including the critical parameters that characterize both the offense and the defense, the model must include certain "rules" of the game or encounter, which for this model are as follows:

1 Neither the aircraft nor the ground battery can be wounded. Each is either completely destroyed or completely functional at all times.

2 The ground defense battery fires only at the aircraft not at the airborne missiles.

3 Both the aircraft and the ground battery will fire their first missiles when the opponent is at a distance equal to the respective threshold range.

4 Each will continue firing until either one of the first three permissible outcomes specified in the preceding paragraph occurs, or it has expended its available missiles.

5 The aircraft flies at constant velocity directly at the target until either one of the first three permissible outcomes occurs or it has expended all its missiles; it then reverses its direction and exits at the same velocity.

The basic parameters for the attacking aircraft and its missiles are as follows:

V = velocity of aircraft

D_1 = delay times between firing of air missiles

M = number of missiles to be fired by aircraft

R_1 = threshold or maximum range of the air missiles

$U_1(R)$ = average velocity of an air missile as a function of firing range

$P_1(R)$ = probability of a single-shot kill by an air-fired missile as a function of firing range

The basic parameters for the ground defense battery and its missiles are as follows:

D_2 = delay time between firing of ground missiles

N = number of ground missiles available for firing by the battery

R_2 = threshold or maximum range of the ground missiles

$U_2(R)$ = average velocity of a ground-fired missile as a function of firing range

$P_2(R)$ = probability of a single-shot kill by a ground-fired missile as a function of firing range

Model outputs

As stated earlier, only four outcomes to each engagement duel are possible. The usefulness of the Monte-Carlo outcome of a single or few engagements is meaningless; however, since the model operates very rapidly, one can easily and quickly run sample sizes of 1000 engagements (80 sec on the Univac 1107) or more for any one set of inputs. This allows the analyst to determine the probability of each outcome with a high degree of precision and confidence.

In addition to the probability of occurrence of each of the four possible outcomes, we are also interested in the expected cost (number of missiles expended) to obtain a favorable outcome. Thus, by having generated a history of how many missiles were fired prior to obtaining a kill, the program calculates the expected number of missiles necessary to obtain a favorable outcome.

The output of the computer model for a single run can be summarized as follows:

1 a critical event history in time sequence;

2 the history event of each engagement;

3 the probability of occurrence of each possible outcome over the long run;

4 the number of times x number of missiles were required to obtain a kill for both the ground battery and aircraft;

5 the expected number of missiles required to obtain a kill for both the ground battery and aircraft.

Model operation

Once the input parameters have been selected for both the offense and defense, the computer program proceeds as follows:

1 computes the number of missiles fired by the ground battery;

2 determines when (time) each of the air missiles and ground missiles is fired;

3 calculates the appropriate range for each firing;

4 computes when (time) each of the air missiles and ground missiles reaches its respective target;

5 time sequences the critical events in the engagement;

6 computes the Monte-Carlo history samples based upon the sequence of events in step 5;

7 calculates the outcome probabilities and other output.

Derivation of equations

This model is critical event sequenced, and all the equations are predicated on relative time. If the time $t = 0$ is defined as the time the engagement begins (i.e., when the first missile is fired, despite whether it is defensive or offensive), all other events in the encounter can be referenced to that time. One of the rules or assumptions of this model, as previously given, is that both the aircraft and the ground battery will fire their first missiles when the opponent is at a distance equal to the respective threshold range as specified by the inputs. Therefore, all the equations will depend upon whether $R_1 > R_2$ or $R_2 > R_1$, since this relationship will determine who fires first and hence sets $t = 0$.

If we define t_c as being the relative time at which the aircraft fires its last missile, then

$$t_c = (M - 1)D_1 \quad \text{if} \quad R_1 > R_2 \tag{1}$$

$$t_c = \frac{R_2 - R_1}{V} + (M - 1)D_1 \quad \text{if} \quad R_2 > R_1 \tag{2}$$

According to the rules of the game as previously given, as soon as the aircraft has fired its last missile, it instantaneously reverses its direction and exits at the same speed. Therefore, we can calculate the range $R(t)$ of the aircraft from the ground battery at any given relative time t as

$$R(t) = R_1 - Vt \quad \text{when} \quad t < t_c \quad \text{if} \quad R_1 > R_2 \tag{3a}$$

or

$$R(t) = R_1 - 2Vt_c + Vt \quad \text{when} \quad t \geqslant t_c \tag{3b}$$

$$R(t) = R_2 - Vt \quad \text{when} \quad t < t_c \quad \text{if} \quad R_2 > R_1 \tag{4a}$$

or

$$R(t) = R_2 - 2Vt_c + Vt \quad \text{when} \quad t \geqslant t_c \tag{4b}$$

The maximum number of missiles that the aircraft will fire is equal to M, the number available. However, the maximum number of missiles the ground battery will fire is a function of the length of time the aircraft is within threshold range of the ground battery, up to a maximum of N, the number available. We can calculate the length of time $T(R)$ during which the aircraft is within threshold range of the ground battery as

$$T(R) = 2\left[\frac{R_2 - R_1}{V} + (M - 1)D_1\right] \tag{5}$$

We can now calculate L, the maximum number of missiles fired by the ground battery, as

$$A = \frac{T(R)}{D_2} \tag{6}$$

$L = A + \Delta$ where Δ is of a size as to make L the next highest integer value if A contains a fractional remainder (7)

Obviously, if $L > N$, the ground battery will have fired all its missiles and L is set equal to N.

Once t_c (critical time) and L (the maximum number of ground missiles to be fired) have been determined, we can calculate the relative times of the critical events in which we are interested. These are,

E_i = time at which ith air missile is fired by A/C (aircraft)

F_i = time at which ith air missile reaches target

where

$$i = 1, 2, \ldots, M$$

and

G_j = time at which jth ground missile is fired by G/B (ground battery)

H_j = time at which jth ground missile reaches A/C

where

$$j = 1, 2, \ldots, L$$

The calculation of each of the E_i, F_i, and G_j values is straightforward and presents no difficulties. If $R_1 > R_2$,

$$E_i = (i - 1)D_1, \quad i = 1, 2, \ldots, M \tag{8}$$

$$F_i = E_i + \frac{R(E_i)}{U_1[R(E_i)] + V}, \quad i = 1, 2, \ldots, M \tag{9}$$

$$G_j = \frac{R_1 - R_2}{V} + (j - 1)D_2, \quad j = 1, 2, \ldots, L \tag{10}$$

Note that $R(E_i)$ is calculated for each time E_i, using Equation 3. The average velocity of an air missile as a function of firing range is $U_1[R(E_i)]$. The function $U_1(R)$ is one of the required inputs to the model.

Returning to the calculation of the critical event times, we see that if $R_2 > R_1$,

$$E_i = \frac{R_2 - R_1}{V} + (i - 1)D_1 \tag{11}$$

$$F_i = E_i + \frac{R(E_i)}{U_1[R(E_i)] + V} \tag{12}$$

$$G_j = (j - 1)D_2 \tag{13}$$

In this case, the range $R(E_i)$ is calculated by Equation 4.

The calculation of H_j (the time when the jth ground missile reaches the aircraft) is not nearly so easy. We are faced with three possible situations:

I The aircraft is approaching the ground battery throughout the flight of the jth missile—i.e., $G_j < t_c$ and $H_j < t_c$.

II The aircraft is departing or going away from the ground battery throughout the flight of the jth missile—i.e., $G_j > t_c$ and $H_j > t_c$.

III The aircraft is approaching the ground battery when the jth missile is fired, but has turned around and is departing by the time the missile reaches it—i.e., $G_j < t_c$ but $H_j > t_c$.

We solve each of the three possible cases by finding the appropriate value of H_j subject to the constraint that $H_j > G_j$, which satisfies the appropriate equation:

Case I: $$H_j - G_j - \frac{R(H_j)}{U_2[R(H_j)] + V} = 0 \tag{14}$$

$$\textit{Case II:} \quad H_j - G_j - \frac{R(H_j)}{U_2[R(H_j)] - V} = 0 \tag{15}$$

$$\textit{Case III:} \quad H_j - t_c - \frac{R(H_j) - [R_1 - (N-1)VD_1]}{U_2[R(H_j)] - V} = 0 \tag{16}$$

We can readily identify case II by testing whether $G_j > t_c$. However, if $G_j < t_c$, the correct solution is the smallest value of H_j that satisfies either Equation 14 or 16.

To demonstrate how such a model might be used for sensitivity studies, we show an example problem. Assume first that we are interested in (1) the effect of varying the delay time between firing of the ground missiles, and (2) the effect of varying the maximum or threshold range of the ground missiles. We define five different simulation runs as follows:

Standard Case:

$N = 6$

$D_2 = 10$ sec

$R_2 = 18{,}000$ ft

$U_2(R) = 2200 + 0.2R - 0.00000833R^2$

$P_2(R) = 1.1 - 0.00005R$, except $P_2(R) = 0.7$ if $R \leqslant 8000$

$V = 600$ ft/sec

$M = 10$

$D_1 = 3$ sec

$R_1 = 20{,}000$ ft

$U_1(R) = 2000 + 0.2R - 0.00000833R^2$

$P_1(R) = 0.764 - 0.0000033R$, except $P_1(R) = 0.6$ if $R \leqslant 5000$

Case I: Same as standard, except change D_2 from 10 to 12 sec

Case II: Same as standard, except change D_2 from 10 to 8 sec

Case III: Same as standard, except change R_2 from 18,000 to 15,000 ft

Case IV: Same as standard, except change R_2 from 18,000 to 21,000 ft

The coefficients for the equations of the average velocities as a function of range [$U_1(R)$ and $U_2(R)$] would be achieved by curvilinear regression techniques of curve fitting. The single-shot kill probabilities as a function of range [$P_1(R)$ and $P_2(R)$] are based on linear regression lines up to a certain minimum range and then a constant up close, but they could be represented by any appropriately shaped curve.

A simulated sample of 1000 engagements was run of each situation with the following results:

	Standard (%)	Case I	Case II	Case III	Case IV
Ground killed—AC survived	51.5	49.5	42.3	62.8	46.2
AC killed—ground survived	26.5	27.1	29.0	20.8	35.2
Both killed	22.0	23.4	28.7	16.4	18.5
Both survived	0.0	0.0	0.0	0.0	0.1
Expected no. AC missiles/kill	3.47	3.50	3.45	3.50	3.39
Expected no. GB missiles/kill	1.37	1.38	1.51	1.17	1.73

The only surprising result is case I, where the increase in the delay time is shown as favoring the ground battery. This results from random variation and a statistical test of significance shows that the indicated difference is well within the range of statistical variability. The total computer run time for the five situations was 166 sec, including compilation time on the Univac 1108. Sample sizes of 10,000 engagements would raise this to about 5 min or less.

Summary

We readily discern that factorial experiments (those in which all levels of a given factor are combined with all levels of every other factor in the experiment) can be run rapidly and that we can learn a great deal about the probable influence of the system parameters. Improved guidance would reflect in the single-shot kill probability curves, whereas quicker acceleration would result in a steeper ascent of the average velocity curves.

The model presented can readily be adapted to more complex situations. These could include scenarios with multiple attackers and defenders; terrain effects obscuring visibility; etc. Such extensions, however, would obviously increase the time required to exercise the model. It is believed that, even in this simple form, the model can be highly useful for conducting defensive weapon system parameter sensitivity analysis quickly and cheaply.

Case Study A-III

DIGITAL SIMULATION OF AN EXISTING WATER RESOURCES SYSTEM*

Augustine J. Fredrich†

This case study was selected to demonstrate a system that can only be represented as a continuous flow model. Such models confront the modeler with very real problems of data presentation and analysis. This case study also presents a situation that would lend itself to analog, as well as continuous digital, simulation.

Introduction

Concern for environmental and sociological effects of water resources projects has led to the need for techniques to analyze and evaluate the performance of existing systems. The increased awareness of the populace and policymakers is resulting in a greatly increased need for studies to: review and update operational plans for existing systems; establish base conditions for comprehensive land and water resource planning in basins where there are existing developments; and define the economic, social, environmental, legal, and functional effects of changes in criteria or priority of service among water uses.

The rapidity with which changes occur in modern society and the diversity of interests among the various segments of society have stimulated the operating entities to expand the scope of operation objectives and consider a wider range point-of-view in making operation decisions. Consequently, the studies upon which the operation decisions are based must be more comprehensive in order to identify the degree of interaction among water uses,

*Condensed from a paper presented at the 1971 IEEE Joint National Conference on Major Systems, by permission of the author and the U.S. Army Corps of Engineers, Hydrologic Engineering Center, Davis, Calif.

†Chief, Research Branch, The Hydrologic Engineering Center, U.S. Army Corps of Engineers, Davis, Calif.

provide information on competitive and complementary aspects of the interactions, and direct attention toward problems that have not been fully considered in past studies.

During the past few years, many techniques for developing optimal operating plans have been described in the technical literature, but these analytical techniques are not completely satisfactory for use in evaluating system operation plans for a large, complex, existing system owing to the necessity for simplification, linearizations, and generalizations to make models of the existing system mathematically tractable. To avoid the problems inherent in attempting to quantify and handle explicitly some of the constraints, we can apparently use digital simulation to analyze and evaluate the operation of the system, with the idea that a satisfactory operation plan might be developed through successive incremental improvements in operation policy.

By postulating an operation plan, operating the simulation model to determine the results of the proposed plan, evaluating the results in terms of the desired operation objectives, making modifications to the proposed operation plan to rectify any errors or inconsistencies in the policy as indicated by the results of the simulation study, and repeating the process until the desired objectives are realized, we can develop an operation plan to satisfy feasible operation objectives. This paper describes some of the efforts expended thus far with respect to perception and formulation of operation objectives and evaluation of study results for one particular system.

The Arkansas-White-Red Rivers system

The existing reservoir system in the Arkansas River, White River, and Red River basins illustrates very well many of the complexities we encounter in studying the operation of an existing system. As shown in Fig. A.3.1, the system is composed of 23 reservoir projects located in three river basins. Although the river basins are hydraulically independent, they are electrically interconnected so that system power demands could theoretically be met by any one of the 19 power projects in the three basins. In addition to the generation of hydroelectric power, the projects provide flood control, water supply, and navigation, and have operation requirements to modify and enhance fish and wildlife environment, water quality, and water-based recreation. Not all these purposes are served by each project, but almost all the projects are operated to provide direct service to at least two water uses.

There are no physical facilities for diverting water between basins, so all demands for water must be met by projects within the basin where the demand occurs. Legal and institutional constraints limit the services which can be provided from some of

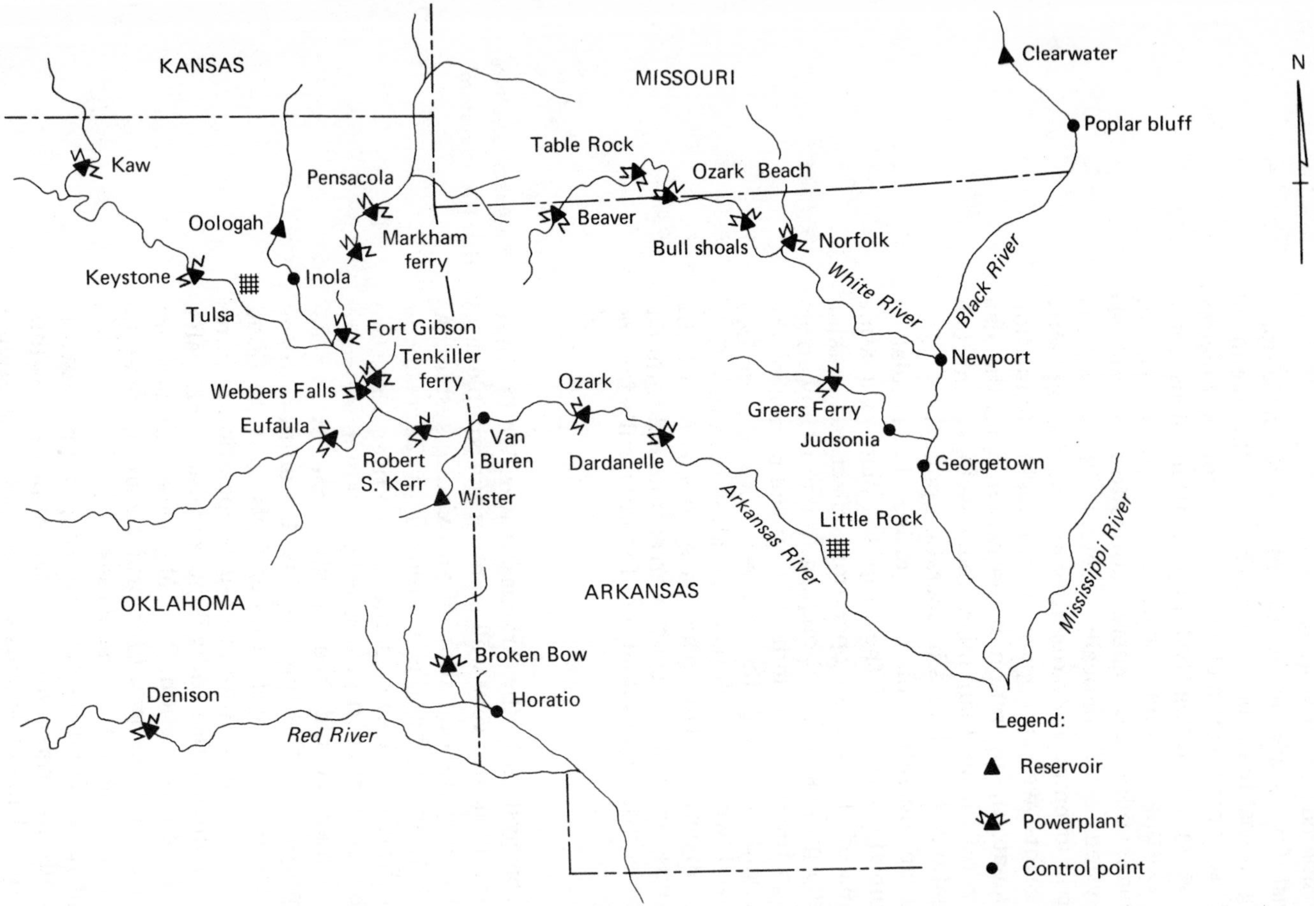

Figure A.3.1 Reservoirs and powerplants, Arkansas-White-Red River Basins system.

the reservoirs, so that it is not always possible for all projects having the physical capability to meet a given demand to do so.

The reservoir projects range in size from 4,350,000 acre-feet of usable multiple-purpose storage to 19,000 acre-feet of power pondage. The total installed capacity of the 19 hydroelectric projects in the system is almost 2 million kilowatts. Water supply, fish and wildlife, water quality, and water-based recreation are at present less important than flood control, navigation, and power production. However, the pressures for additional attention to recreation and water quality are increasing, and many operation decisions are already based, either explicitly or implicitly, on requirements for these two purposes.

Three independent entities own and operate the projects, and a fourth agency is directly involved in operation decisions because of power marketing considerations. The Federal Government (Corps of Engineers) owns and operates 20 of the 23 projects, and a separate Federal agency arranges for marketing of the power output of the 16 Federal power projects. A state agency owns and operates two of the 23 projects, both of which have power installations, and the remaining project, which also has a power installation, is owned by a private utility.

The multiple ownerships create problems in analyzing the operation for any purpose, but the problems involved in the analysis of hydropower operation are vividly illustrative of the complexity resulting from the addition of institutional and legal constraints to the physical and hydrologic constraints that exist naturally in the system. The power projects are interconnected and their outputs marketed in a way that, for purposes of an operation study, create three power subsystems, which must be analyzed separately but simultaneously. As shown in Fig. A.3.2, the Bull Shoals and Table Rock projects are electrically interconnected. The output from these two power projects is marketed by the Federal marketing agency to an area having a seasonally varying demand with a substantial peak demand during the winter. The non-Federal projects (Ozark Beach, Markham Ferry, and Pensacola) are operated by their owners, and their output is not marketed by the Federal marketing agency. Consequently, they form a system with water outputs that contribute to the Federal projects, but with power outputs that do not contribute to the Federal power supply. The remaining 14 power projects are electrically interconnected, and they comprise a third system. The output from these projects is marketed in an area with a seasonally varying demand with a substantial peak demand in the summer. Furthermore, a portion of the output of the Denison project is marketed to utilities in Texas that are not connected to the other utilities in the major marketing area. Therefore, this output must

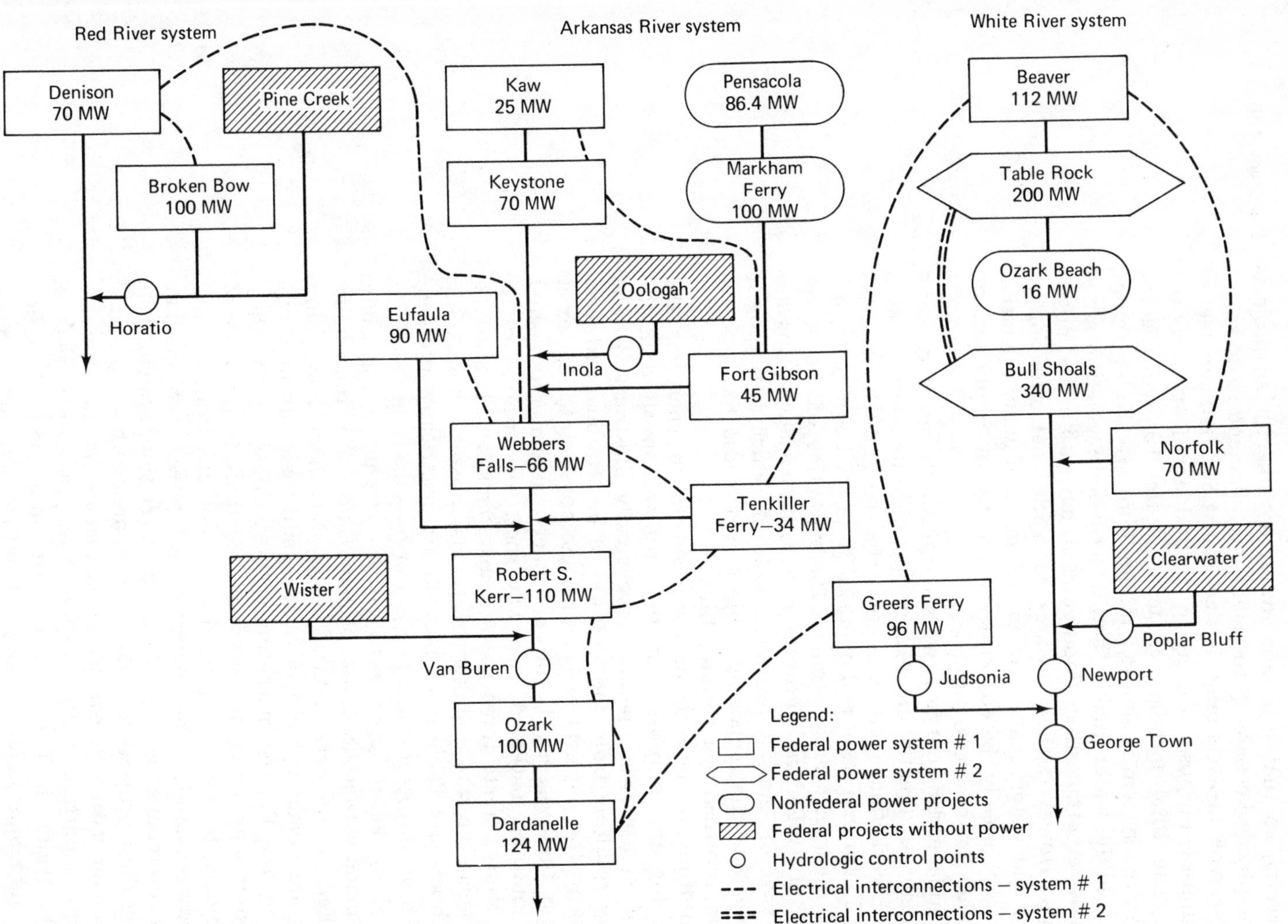

Figure A.3.2 System schematic Arkansas-White-Red River Basins system.

be deducted from the total power output of Denison before we calculate Denison's contribution to the main system.

The bulk of the power demand in the market area is met by thermal generation that is not Federally owned or operated, and the hydroelectric generation is used primarily to meet peaking demands. More than 400,000 kilowatts of the 1,112,000 kilowatts of installed capacity in the 14-plant Federal system is located at navigation lock and dam projects on the Arkansas River. The storage at these projects is only adequate to sustain peaking generation for daily, or, at most, weekly cycles. Since the storage volume upstream of these essentially "run-of-river" projects is not large with respect to the water required to provide energy to support this amount of installed capacity, and since there are no physical facilities for diverting water from the large storage projects in the White and Red River basins, the power generation allocations among the basins must be carefully planned to utilize the available streamflow fully and meet the system power demands. The development of operation criteria to accomplish this allocation effectively is a major part of the problem of operating the hydroelectric system.

In arranging to market the hydroelectric power, it is necessary to provide for the capability to purchase thermal energy to support the hydroelectric capacity during periods of deficient streamflow. Since the thermal purchases represent a cost that must be deducted from the revenues obtained from the sale of hydroelectric energy, it is not sufficient simply to maximize the hydroelectric energy production. Instead, the hydroelectric generation must be integrated with the thermal purchase in a way that minimizes the thermal purchases without endangering the capability of the hydroelectric plants.

The simulation model

The use of simulation as a tool in studying the operation of reservoir projects in the Arkansas, White, and Red River basins is not new. For at least 20 years, various simulation studies using handcrafted simulation models (i.e., manual routing studies) have been conducted by the Little Rock and Tulsa District of the Corps of Engineers to evaluate the operation of individual projects and systems of projects. In the past, these studies have been limited in both scope and objective. However, during the past few years, events have occurred that have increased the need for and feasibility of comprehensive studies of water resources systems.

The operation of the system is studied by simulating its performance through 45 years of historical hydrologic data. The range of hydrologic events during this period is believed to be such that it includes representative critical conditions for evaluating alternative operating plans and gives a reasonable approximation of the long-term average output of the system.

Basically, the computer program used in the A-W-R system study does essentially the same type of computation that has been done with handcrafted simulation models in the past. Only the degree of refinement, speed of computation, and degree of complexity have been increased. In the computer simulation model (program), it is possible to consider many more factors than could be considered in traditional routing studies, to consider each factor in much more detail than previously, and to study a much larger system than previously.

Generally, the computer program requires that the location of each component in the system (i.e., reservoir, power plant, downstream control point for flood control, etc.) be specified with respect to the other components in the system, and that operation requirements for all pertinent purposes be specified at each reservior and control point where the purpose is significant. In general, these requirements must be specified in terms of a flow rate, such as release in cubic feet per second or total river flow in cubic feet per second. However, it is also possible for requirements to be specified in terms of storage volume remaining in a reservoir and kilowatt-hours of electrical energy production. Since these three parameters do not always lend themselves to direct relationships with some operation objectives, it is sometimes necessary to develop relationships externally. For example, navigation operation may be dependent upon river stage at a certain location. Although the program will not accept river stage as an input parameter, the stage-discharge relationships at the location can be developed externally, and the operation requirement in terms of discharge can be calculated and provided as input to the program.

In addition to specifying the operation requirements for each purpose at each component of the system, it is necessary to describe the physical relationships and constraints that control the operations of the component (e.g., outlet capacity of powerplants, spillway elevation-discharge relationships, conduit invert elevations). Finally, if the system operation is to be dependent upon the relative state of components (e.g., the generation of power at a powerplant to contribute to meeting a system power requirement being dependent upon the reservoir storage at the powerplant as compared with other powerplants in the system), operation rules must be specified. All the operation rules and operation requirements, and some of the constraints, can be varied from period to period, if desired (e.g., power requirement of 2000 megawatt-hours in January, 2700 megawatt-hours in February, 2500 megawatt-hours in March, etc.). For the A-W-R study, it was also necessary for the program to simulate in a crude fashion the operation of a thermal generation system, because the relationship between the hydroelectric and thermal generations had to be considered.

The computations performed by the program are based on the principle of continuity, as expressed by the equation

$$S_i = S_{i-1} + I_i - Q_i - E_i$$

where

S_i = reservoir storage at the end of the current period, i

S_{i-1} = reservoir storage at the end of the previous period, $i-1$

I_i = inflow volume during period i

Q_i = release volume during period i

and

E_i = evaporation volume during period i

When I, Q, and E are properly defined, this basic equation is appropriate for storage accounting, where the length of the period i is long enough that the travel time through the reservoir is insignificant. Proper definition of I implies that all diversions into the reservoir and releases from upstream reservoirs must be added to the natural inflow to obtain the inflow volume; proper definition of Q implies that all diversions out of the reservoir, leakage from the reservoir, and releases for different purposes are added together to obtain the total release volume; and E must reflect the gain or loss in reservoir storage volume that would occur as a result of net evaporation (evaporation minus precipitation) over the impoundment area during the period.

Power calculations are based on the equation

$$GE_i = 0.08464\, Q_i h_i e_i$$

where

GE_i = energy in kilowatt-hours generated during period i

Q_i = flow in cubic feet per second through the generating units during period i

h_i = effective head on the turbine during period i

e_i = efficiency of the generating units during period i

This equation is appropriate for use when Q_i has been defined as only that part of the release volume which passes through the generating units; when h_i is defined as the head existing during the period i (calculated by subtracting tailwater elevation and head loss from the reservoir surface elevation); and when e_i reflects the average overall station efficiency during period i. The calculation of head is based on the elevation corresponding to mean reservoir storage for the current period (average of the beginning and ending reservoir storage), and the tailwater elevation is specified as either a constant value or a function of the mean release rate for the period.

The specific components of the entire system that contribute to meeting a system requirement (such as system power requirement, a streamflow requirement at a point downstream of several reservoirs, or a flood control limitation at a downstream point) are specified in a manner that insures only those projects which should (with respect to legal or institutional ability as opposed to physical capability) contribute to meeting the system requirements are permitted to do so. The system requirements can either override or supplement the individual project requirements, according to the study requirements. Operation rules are specified to implement the desired interaction among projects in meeting the system demand.

The simulation model operates by considering the water and power requirements at each pertinent point in the system in a sequential fashion, beginning at an upstream point and moving in a downstream direction through each river basin. The release required to meet the at-site requirements for all pertinent purposes is determined by evaluating each operation requirement and all physical and operation constraints at each site. Also, an index of the relative state of each reservoir (usually a function of reservoir storage) is determined according to the specified operation guides. After all at-site requirements have been met at all points in the system (or shortages declared if water is not available), the various system requirements are examined to determine whether additional water releases or power generations will be needed to satisfy the system demands. If they are, the additional needs are proportioned among projects that have been specified to be available for meeting that system requirement in accordance with the relative state of the projects, as evidenced by the indices previously computed. The additional releases are added to the previously computed releases for meeting at-site requirements, and the system and at-site requirements are thus met (or system and at-site shortages are declared if water is not available). This process is repeated for each period of study, with the ending state of the projects in the system for the current period being the beginning state for the next period.

Results from the successive applications of these calculations on a period-after-period basis are recorded for all points in the system (including nonreservoirs) by an accounting procedure that simply accounts for the movement of the water through the system by using the specified relative location of the reservoirs and downstream control points. By adding releases to natural streamflow to obtain total streamflow, and by adding inflows to storage volume and subtracting releases from storage volumes, we can calculate the state of any component and the flow at any point in the system. As these results are calculated, they are stored and finally printed out, on a project-by-project basis, to produce a continuous record of inflow, storage, outflow, power generation, and other pertinent data. These results may be rearranged in many ways to serve various needs in analysis or evaluation of the system operation.

Through careful examination and evaluation of the results, the response of the system to the specified operation requirements and the specified operation rules can be determined. If the response of the system is satisfactory and if there are no deficiences or inconsistencies in the operation plan, the results can be used as a basis for implementing the operation plan. If, on the other hand, the response of the system is unsatisfactory, we must examine the results further to determine the nature of changes in the operation plan to produce a satisfactory response. When the contemplated changes have been identified, they must be transformed into specific operation rules, and the entire simulation study must be repeated to determine the response of the system to the modified rules. The entire process must be repeated until a satisfactory system operation plan emerges.

The model also can relate hydrologic or physical parameters to economic returns through the use of benefit functions. Benefit functions can be specified separately for each reservoir and downstream control point and for each different purpose, if necessary. Furthermore, the benefit functions can vary seasonally if this is necessary to reflect the seasonal variations in value of water or storage for some purposes. At the end of each simulation run, we apply the parameters, such as regulated streamflow, power production, and reservoir storage, to the appropriate benefit functions to obtain an indication of the relative economic returns that we could expect from the operation plan used in the study.

Study results

The study of the operation of the A-W-R system is a continuing one, which requires that objectives and criteria be updated periodically to reflect the changes in priority of water use. Consequently, there are no final results in the usual sense. However, analyses conducted during the past two years have contributed significantly to our understanding of the system, given insight into the nature of potential improvements in system operation rules, and provided information that can be used to improve the operation efficiency of individual projects in the system. Coordination of hydroelectric generation and purchases of thermal energy to meet Federal contractual commitments for power supply require that time and quantities of thermal energy purchases by the Federal marketing agency be based on the relative state of the reservoir systems. Supplementary thermal energy is purchased only when it is needed to augment the energy produced by the hydroelectric plants or when anticipation of deficient streamflow dictates that purchases should be made to avert potential future shortages in hydroelectric energy production.

Operation guides of the type shown in Fig. A.3.3, based on the performance of the system during the historical hydrologic record, are generated to determine the timing and quantities of thermal purchases. System energy in storage, the parameter used in Fig. A.3.3 to indicate the state of the system in making the decision to purchase energy, is believed to be a better indicator than, say, water in storage in the system. However, it is anticipated that problems may arise with respect to this parameter because of the run-of-river plants on the Arkansas River being unable to avail themselves of the benefits of storage on the White and Red Rivers. Future studies may require modification of this parameter to reflect the consequences of the inequitable storage distributions. Applying weighting factors to the computed energy in storage in each basin before developing a system composite value for energy in storage would be one way we could modify the parameter.

A major problem in the study thus far has been an inability to quantify operational objectives for the existing system. Because comprehensive studies have not been performed in the past, the existing set of objectives may be incomplete or inharmonious. Another related problem is the lack of information on criteria and constraints associated with complementary and competitive interactions among water uses. Again, the absence of this information can be attributed primarily to the fact that past studies were less comprehensive and did not require this type of information. Current studies have been very valuable from the standpoint of indicating specifically which areas need additional study and consideration.

Another major problem associated with the study of the A-W-R system is the tremendous volume of output generated. The output from a single simulation run for a complex system, such as the A-W-R system, can easily amount to 1000 pages or more. The need for numerous simulation runs rapidly multiplies the output volume.

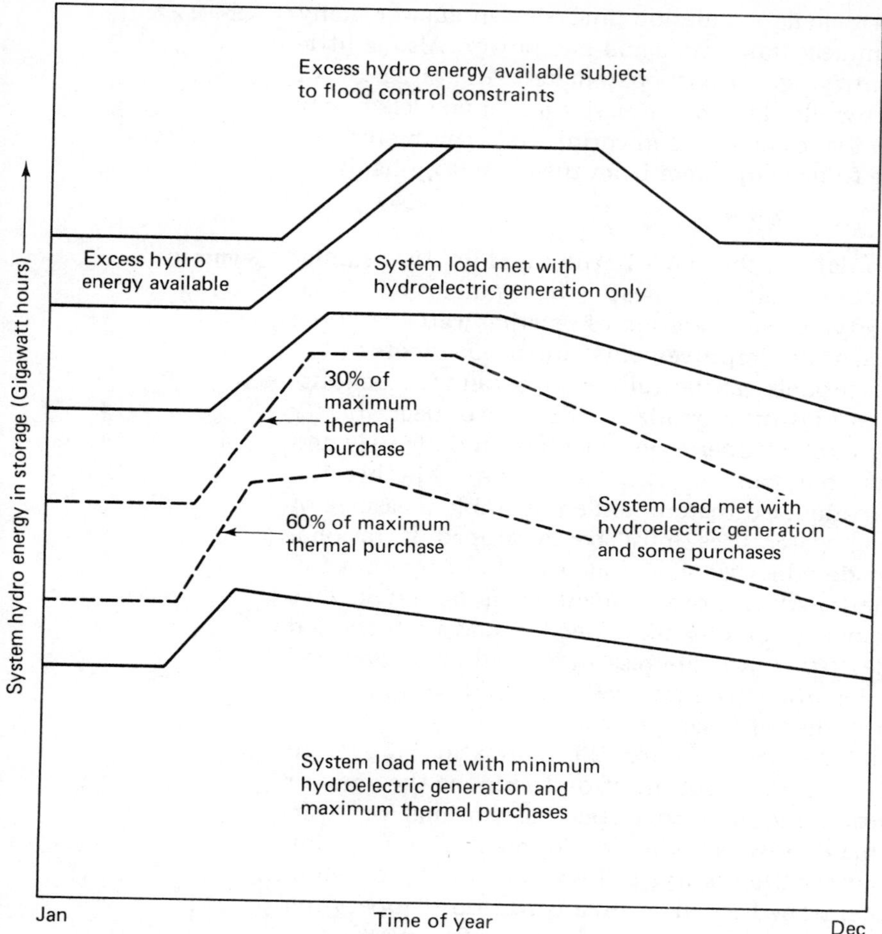

Figure A.3.3 System operating curves Arkansas-White-Red River Basins system conservation study.

It is relatively easy to produce in a week or two far more output than can be intelligently analyzed by many people in a year or so. The roots of the problem and the key to its solution are in the presimulation planning. Carelessly thought-out, limited-objectives studies with poorly documented criteria almost always result in studies of only limited utility. In the early stages of a simulated study, the proper answer to any question always seems to be "perform another simulation." However, as the unanalyzed or partially analyzed studies pile up, it becomes evident that this is not only not the proper answer—it is often a very poor answer. There is no substitute for a well-planned, properly executed, carefully documented simulation study. Resisting the temptation to perform analyses as rapidly as possible leads us to discover that a little planning for a single simulation analysis can answer many questions and save immeasurable time and manpower. Also, a little forethought in identifying outputs parameters of value or of interest is well worthwhile. It is much easier and much less costly to have the computer calculate and print out parameters of interest than to have to develop them from the output by hand.

Summary

Efforts expended to date in the A-W-R study and in other similar water resource system studies indicate that simulation is an effective tool for studying the operation of existing water resource systems. However, certain improvements are needed before the power of simulation models can be fully utilized and the benefits of the simulation study fully realized. First, we need better methods to identify explicit operation objectives and identify and quantify parameters that can be used to measure whether the objective is being satisfied. Consistent and comparable measures of value for competing water uses must be developed if rational choices are to be made when water deficiences do not permit full service to all demands. Also, more thought must be devoted to determining in advance what output is needed and its form and format, so that selectivity can suppress unwanted and unneeded results. This must be done to reduce the output volume to a manageable level in studies of large systems.

Finally, it would be desirable for the simulation model to possess some capability for self-optimization to reduce the amount of human intervention required to obtain an optimal or near-optimal operation plan. However, the development of self-optimization capability must, of necessity, follow some of the developments previously mentioned, such as quantification of operation objectives and comparable measures of value for alternative water uses.

Acknowledgment

The digital simulation model described herein was developed by Mr. Leo R. Beard, Director of The Hydrologic Engineering Center. The A-W-R system operation study has been conducted jointly by The Hydrologic Engineering Center and the Southwestern Division, Little Rock District and Tulsa District Offices of the U. S. Army Corps of Engineers. The views expressed here are those of the author and do not necessarily reflect the policies and practices of the Corps of Engineers.

Case Study A-IV

SYSTEMS SIMULATION FOR SOLID WASTE PLANNING*

Robert M. Clark and James I. Gillean†

This case study was chosen to show the reader the interplay of a management information system with a simulation effort. It also demonstrates very nicely how most studies consist of a series of independent but related experiments. The reader should particularly note the sequence of small experiments and how they build upon each other to increase the overall understanding of a very complex system.

Introduction

Adequate solid waste management planning is one of the major problems facing most medium-to-large urban communities in the United States. Increasing waste generation, difficulties in labor-management relations, decreasing land resources, increasing effluent standards, rising costs, and uncertain technology are only a few of the problems facing today's solid waste managers.

Many studies have explored the application of deterministic and simulation modeling to urban solid waste management problems [4, 7]. Several papers have applied these techniques to routing and facility location, with the use of either assumed data or data collected on a one-time basis [2, 5, 6]. These approaches have been useful for demonstrating the potential of "systems" or operations research techniques for assisting the solid waste manager in making important operational decisions. None of the studies, however, has considered the problem of obtaining continuous data and utilizing them, together with mathematical

*Condensed and reprinted from Report No. EPA-670/5-73-12, dated July, 1973, by permission of the U. S. Environmental Protection Agency, National Research Center, Office of Research and Development, Cincinnati, Ohio.

† Robert M. Clark is Sanitary Engineer, Office of Program Coordination, National Environmental Research Center, U. S. Environmental Protection Agency, Cincinnati, Ohio 45268. James I. Gillean is President, ACT Systems, Inc., 807 W. Morse Boulevard, Suite 200, Winter Park, Fla. 32789.

models, for making "on-line" decisions. This paper reports on the results of a project in Cleveland, Ohio, in which reliable, uniform, and continuous data collection combines with a dynamic simulation model to form a system for making short- and long-term management decisions. A system composed of five basic components has been established:

1 a mechanism for the collection of continuous, uniform, and reliable data on the solid waste management operation;

2 an analysis of the variables that have a significant effect on the solid waste system, such as present and expected changes in population trends and distribution, and changes in transportation systems;

3 a simulation or resource allocation model using the continuous data as input and incorporating variables that significantly affect the system;

4 a mechanism for exercising the model and utilizing its results for making immediate or long-range decisions;

5 the capability to compare the models' predictions with continuous data.

The situation in Cleveland

In most large American cities, the population is reluctant to vote for additional taxes, and yet either demands more services or refuses to relinquish the services they already have. Faced with demands for higher wages as well as increases in the purchase price of equipment, facilities, and other non-labor-related items, many cities have an eroding tax base. As middle- and upper-income families move to the suburbs, taking with them needed tax revenue, lower-income families, who produce less tax revenue but who require just as many services, take their place. The condemnation of property for highways and other nontaxable uses also reduces potential income for the city.

A similar situation developed in Cleveland. This, coupled with the defeat of a much-needed tax levy, created a financial impact felt in all city departments; nowhere was the impact more acute than in the department responsible for the collection and disposal of solid waste. The City of Cleveland's Division of Waste Collection and Disposal was moving waste from the point of generation, transporting it to the disposal point, and disposing of it with no effort required by the general citizenry.

Shortly before this financial crisis, however, the city initiated a cooperative program with the U. S. Environmental Protection

Agency (EPA), in which data from Cleveland was used as a source for a national solid waste data network. Working with the Division of Waste Collection and Disposal, EPA began collecting data on a regular basis in October, 1970 [3]. Two routes were selected for continuous evaluation, and data were obtained from the collection vehicle operator on each route in the form of daily reports.

The Commissioner of the Waste Collection and Disposal Division, faced with a number of difficult decisions regarding possible reductions in service levels, was able to use several months' worth of data available from the two routes being monitored within the city. Their six-man crews giving back-yard, once-per-week service could be compared with other systems using the two routes being evaluated within the EPA pilot network.

After careful review of the preliminary monitoring data for these sample routes, back-yard service was eliminated and the collection crew was reduced by two men, leaving one driver and three laborers. Several months later, the crew was reduced to two laborers. On these two routes, the cost per ton for waste collected for an average day dropped from close to $30.00 to approximately $13.00, with an estimated annual savings of over $4 million per year.

These national network data proved so valuable that a project was initiated to collect similar data from all the city's routes. Using an EPA local and regional planning grant, the city developed an on-line solid waste management information system [1].

Data system development

To develop the information system properly, the organization and management of the Division of Waste Collection and Disposal was completely analyzed. The division, whose purpose is to provide service to slightly less than 250,000 family units, is composed of personnel, vehicles, and facilities. The commissioner in charge of the division supervises six station superintendents with subordinate foremen, as Fig. A.4.1 shows. The number of foremen in each station varies. Station 100 has three foremen, whereas station 400 has six foremen. Each foreman is responsible for collecting waste from the routes assigned to him and may supervise a different number of crews.

The collection crew is usually composed of a driver and two waste collectors, who, upon arrival at the route, work until the vehicle is filled to capacity or until the day's effort is completed. If the vehicle is filled before the route is completed, the driver takes the load to the disposal point and returns to the route. Upon completion of the route, the driver unloads at the city's incinerator if he has a partial load, or at a private landfill if he has a full load and has time remaining in the working day to make the

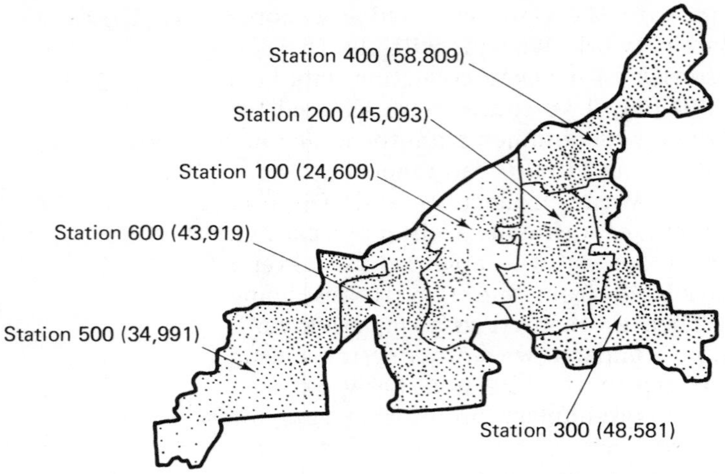

Figure A.4.1 Distribution of families within each of the six waste collection subdivisions in Cleveland, Ohio, October 1970. Each dot (•) represents 100 households.

round trip. On some occasions, he is instructed to return to the motor pool with the partial load if time becomes a critical factor.

As part of the organization study, numerous interviews were conducted with the staff of the division. Cost data were collected on use and maintenance of vehicles and on facilities and equipment for maintenance. This information provided important background data for development of the management information system. As part of the system development, the outputs needed to satisfy management requirements were determined. After the output requirements were specified, input data to produce the required outputs were indentified. A data processing system was developed, and the management information system was implemented. Information from each route foreman and station is computed on a daily average for a week. In addition to the daily averages, the foreman or station may sum the values related to the routes where deemed meaningful.

When the specified outputs were precisely established, the inputs needed to produce them were determined. Forms were developed to log the necessary data that would become the major inputs to the system. Figure A.4.2 illustrates the forms used to collect information relating to vehicle identification and maintenance, crew size, number of homes served, discharge point, and other critical data. The data system was developed so that data could be key punched directly from the daily collection forms.

DAILY COLLECTION ROUTE INFORMATION

Route________ Date________ Crew________

Vehicle: No.________ Size________ Fuel (gal)________ Oil (qt)________

No. homes served________

	Time	Miles	Weight	Discharge point*
Leave motor pool				
Start collection				
Leave route for discharge point				
At discharge point				
Arrive back on route				
Leave route for discharge point				
At discharge point				
Arrive back on route				
Leave route for discharge				
At discharge point				
Return to motor pool				

			Maintenance problem (circle number)
Lunch—start (leave route) —finish (arrive route)			1 Brakes, wheels, tires 2 Cooling or exhaust system
Maintenance—start —finish			3 Electrical system 4 Fuel system 5 Packer 6 Power or steering system 7 Other

*Enter number
1—Incinerator
2—Landfill
3—Transfer

Remarks:

Figure A.4.2 Sample daily data collection form completed by collection vehicle operator.

Simulation model development

An attempt was made to define all those nonsolid waste factors that have an impact on the solid waste management function. For example, population trends and densities and dwelling unit densities have a significant impact. Transportation networks, including changes in street mileages and the location of major arteries and expressways, must be considered. On-going or planned urban renewal projects have a significant impact on population (location, numbers, and densities) and, therefore, are significant for solid waste planning. State laws and city, county ordinances have a potential impact on solid waste management.

The demographic, transportation, and neighborhood development data were compiled along with data from the management information system to provide the following input information for the simulation model:

1 generation data, composed of population, dwelling unit, density, weight, and cost information;

2 collection data, including distances, volume, pickup time, route identification, vehicle type, crew size and costs;

3 transport data, related to distance, time, and speeds;

4 disposal data, such as distance traveled, offload time, disposal site, and related costs.

The model consists of five master programs, each of which corresponds to the basic operation in the solid waste collection and disposal activity: input data logic, truck generation logic, collection logic, transportation logic, and disposal logic. Figure A.4.3 is a flow diagram of the way the model simulates the system. The input data consist of the basic statistics that describe the system: e.g., solid waste generated, truck size, crew size, number of homes served, collection time, and transport miles are all input data to the model and are, therefore, fixed for each simulation run. The values for each of these categories may change, depending on the type of collection system being modeled.

The model then indicates the number of trucks needed each day to collect the solid waste; this number is based on the number of routes being considered, which, in turn, is based on the assumed amount of solid waste generated per day. An 8-hr working day was used as the base line for comparing various types of collection systems. When the complete cycle of collection and disposal was considered, however, a week's effort was used for evaluating the various alternatives.

The collection logic assumes a set of work rules and contains the supporting data to be used for a specified collection system. For example, when a two-man collection system, which consists of a truck, a driver, and a helper, is being considered, the assumptions are that the driver will assist the helper for approximately 60% of the time spent collecting on the route; that, for this assistance, the driver receives a wage adjustment; that, the collection rate in the model varies from 1.1 to 2.5 lb/sec/man; and that the walking rate

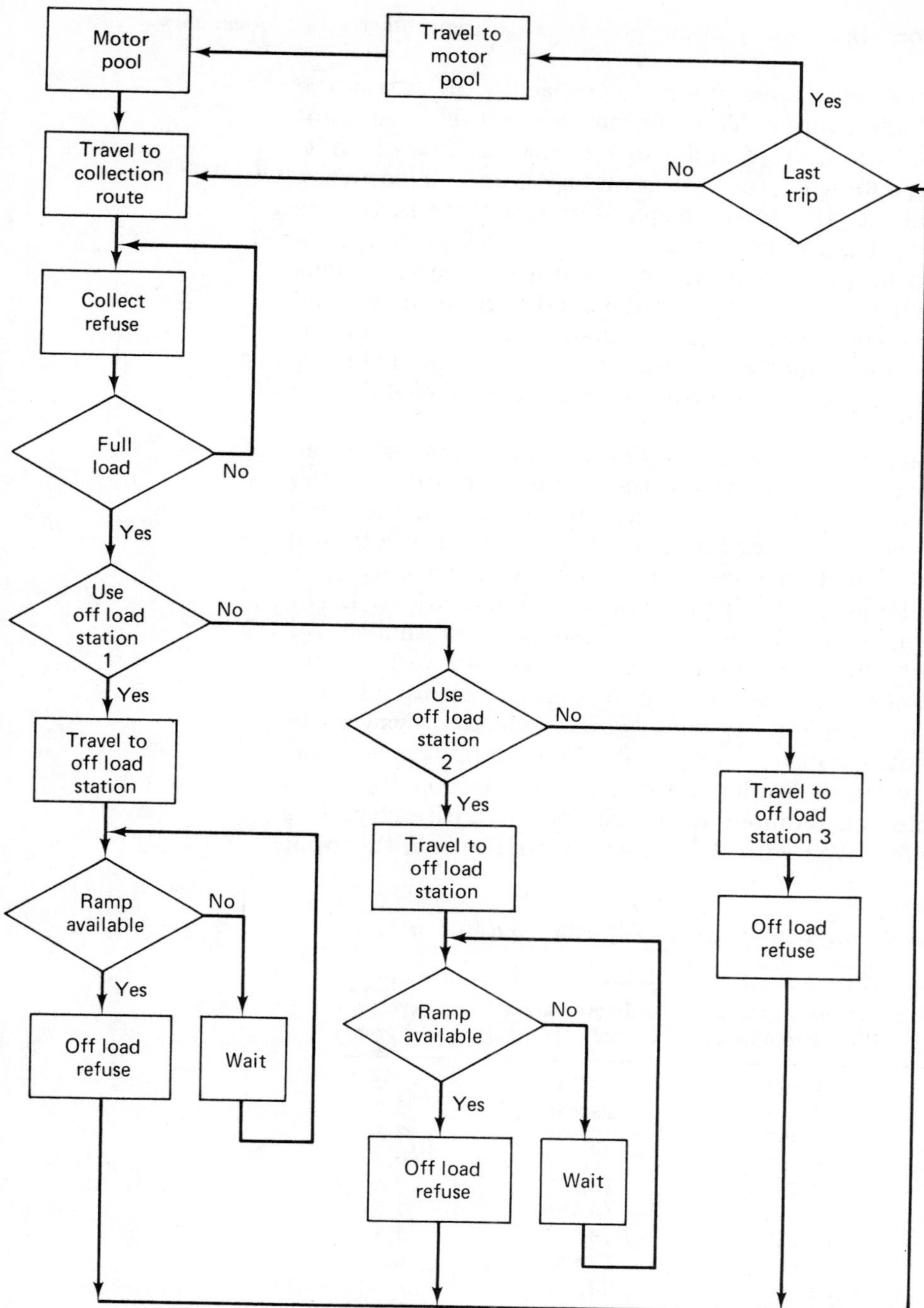

Figure A.4.3 Flow diagram of model logic.

for the collectors between pickup points was assumed to be 2 ft/sec.

When a truck has been filled to capacity, it enters the transportation logic routine. This routine assigns a disposal point and an associated transportation distance to the truck based on its position on the route when the collection task is completed. When a truck has made its trip to the disposal or unloading point, the vehicle enters the disposal logic routine. At the offload point, the truck is assigned to an unloading lane where it may enter a queue or unload immediately, depending on the conditions at the offload point. After the vehicles are empty, their return to either the motor pool or the collection route depends on their having completed their assigned service routes, or on the time of day, or on both.

The routines were used to model the performance of an individual truck and crew as well as the performance of the entire system. The first step in the modeling process was to assume a three-man crew and a rear-loading packer collecting over a typical route using the collection logic. Various dwelling densities were assumed, and a long- and a short-haul option were considered. The long-haul option was rejected as a reasonable alternative for Cleveland. Next, the collection logic was used to test various collection systems until several good ones were found. The disposal logic was used to test the good collection systems to determine various transfer station configurations. Once these configurations were determined, the truck generation logic was used to estimate the required fleet size for the entire city. This procedure was followed with the numerous variations in the input data.

Table 1 *Simulation output compared with collected data for station 100.*

Route No.	Simulated Data (lb collected/day)	Collected Data (lb collected/day)	Percent Difference
111	12,984	12,606	3.0
112	14,252	13,968	2.0
113	7,680	7,404	3.7
121	9,660	9,583	0.8
122	13,483	13,547	0.5
123	20,045	19,750	1.5
124	19,222	18,956	1.4
131	17,955	17,968	0.1
132	22,800	23,548	3.2
133	22,108	22,314	0.9
134	21,509	20,836	3.2
135*			

*Route information not usable.

A General Purpose Simulation System (GPSS) language, version V, was used to program the model. Input data and work rules reflected current operations in Cleveland. For purposes of validating the model, the current system was simulated with the use of input data describing the present operation. Results from the model were compared with data collected from the management information system (Table 1). Output data from the model includes (1) volume and/or weight of solid waste handled; (2) miles traveled; (3) collection times; (4) travel time; (5) weight handled by each disposal point; (6) queueing conditions at offload ramps; and (7) number of homes collected.

System evaluation

After the model had been conceptualized, programmed, and validated, it was used to evaluate various alternatives that might be considered for current and future use in the City of Cleveland. Part of this effort was devoted to modifications to the present system that might make it more efficient. Various collection configurations were evaluated during the course of this study. Figure A.4.4 shows a schematic diagram of some of the alternatives that were considered. For each collection configuration, the effects of dwelling unit density, pickup location, generation rate, vehicle size, and crew size were evaluated. For example, the effectiveness of a scooter system with a two-man crew, serving dense multifamily dwellings, was compared with a mother truck with a satellite vehicle (M/ST), using a two-man crew, serving a low-density single-family home neighborhood with normal generation rates. Each of the alternatives shown in the schematic diagram was considered under long-haul (40 mi) and short-haul (8 mi)

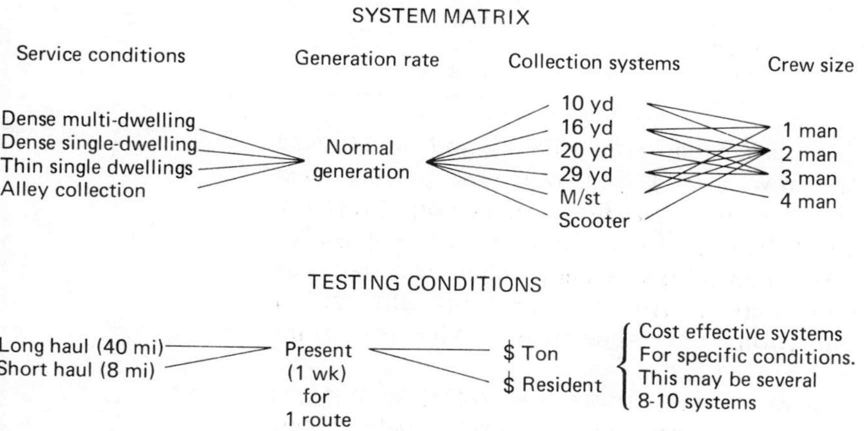

Figure A.4.4 Schematic of alternative collection configurations considered.

conditions when serving one route with once-per-week collection. Immediately, this analysis showed that the short-haul situation was far superior for all of the alternatives considered. The ratio of the total weight collected with a short haul is approximately 2:1. With these results, the long-haul alternative was eliminated from future consideration.

The criteria used to rate each of the systems were (1) ratio of pickup time to transport time; (2) dollar per ton; (3) dollar per residence; and, (4) "optimum termination point." The first three items are self-explanatory, and the final item is defined as the physical point at which the collection effort terminated in an 8-hr working day. Highest ratings were given to the vehicle-crew combinations that could complete the work load in an 8-hr day. The best system was assigned a rank of one, the next best system two, and so on. Table 2 summarizes the rankings achieved by the various systems considered.

Table 2 *Comparison of collection patterns, rating various vehicle and crew sizes.*

Ranking of System	Vehicle and Crew Size (configurations)	Negative Points Accrued
1	10—1	50
2	29—2	66
3	16—1	81
4	29—3	83
5	20—2	88
6	10—2	101
7	M/ST-1	115
8	20—3	120
9	29—4	123
10	16—2	130
11	M/ST—2	141
12	16—3	150

The 29-yard truck with a 2-man crew (29—2), and the 10-yard truck with a 1-man crew (10-1) had the best performance capability. This choice was made under average conditions for waste generation. To test the capacity of these two alternatives for collecting under extreme conditions, a peak generation day was assumed. After routes were restructured and brought within optimum length, simulation runs were made under maximum generation conditions.

Another test was made to see if these two collection systems together could accomplish the service responsibility better than either system alone. The 29-yard truck was assigned the task of collecting dense multifamily dwellings, and the 10-yard truck was given sparsely settled areas. This combined system proved to be efficient but no more so than either system alone.

Based on several subjective considerations, the 29—2 configuration was declared optimum. The 29-yard truck with a 2-man crew allows for more flexibility than a 10—1 system. Union rules and safety requirements are more consistent with a 29—2 system than with a 10—1 system.

Offloading site evaluation

After the most efficient collecting system had been identified, an evaluation was made to determine the best offloading sites. The topography of the Cleveland area dictated the possibility of dividing the city into two general route groups: the area currently in stations 500, 600, and the western part of station 100; and the area currently in stations 200, 300, 400, and the eastern part of station 100 (Fig. A.4.1). The western district developed favorably into a 36-route configuration, offloading at the Ridge Road landfill, and the eastern district into a 53-route, offloading at a landfill at 34th and Broadway (due east of the West 3rd facility and halfway into the district; see Fig. A.4.5). At this time, it was determined that a motor pool would be located at both landfills to eliminate

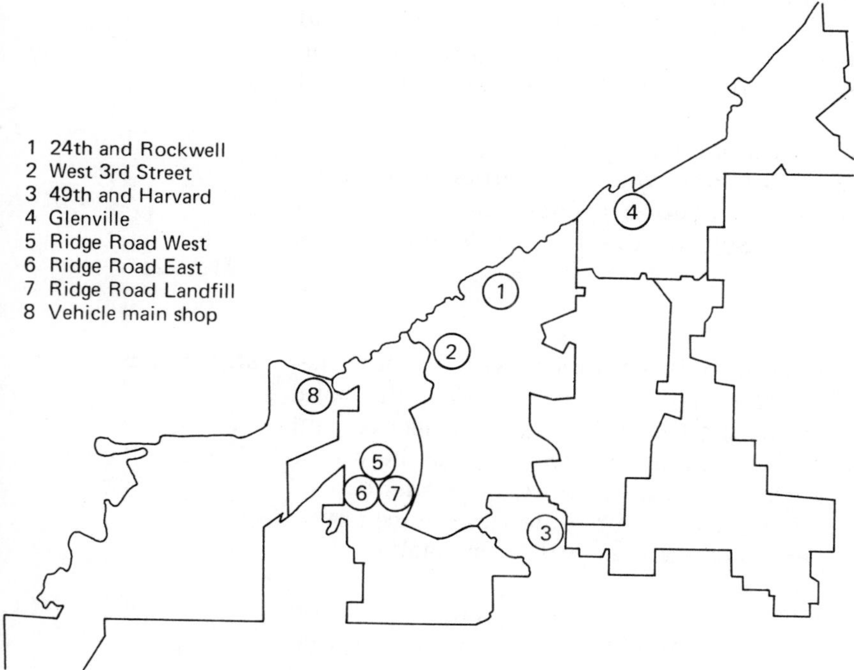

Figure A.4.5 Locating city-owned property used by the Cleveland division of waste collection and disposal.

travel time after the last offload. The constraints on the simulation runs were (1) the truck must make at least one but no more than two offload trips; (2) collections must be made at a peak generation period; (3) the number of homes must be maximum; and (4) the task must be accomplished with a 40-hr work week, with 1.5 to 2.0 hr remaining on the fifth day for general truck maintenance and a 0.5 hr period remaining on each of the other 4 days to wash the trucks. Simulation runs were made with the use of the 29—2 truck-crew configuration and with pickup times varied from 36 to 52 sec. The number of homes to be serviced was set at 83,016 in the western district and 129,470 in the eastern district. The results of these runs were within the framework of the conditions stated previously.

Another variation was examined in which the offload destination was changed from 34th and Broadway to the Garden Valley site, 2 mi northeast of the 49th and Harvard facility (Fig. A.4.5). This change naturally constituted a variation in total transport miles; however, the model indicated that the variation was not significant. Again, the results were positive, falling well within the required conditions previously stated. Careful consideration indicated the best offloading configuration would be to use the Rockside landfill as a terminal and disposal point for the proposed westside division and the Garden Valley site as a terminal and transfer station for the eastside division. The waste would be transported in vans from Garden Valley to the Rockside landfill for final disposal.

To check the results of the model, the 92-route configuration, with the 29—2 configuration and the most efficient disposal alternatives, was simulated. Simulation results indicate that this total configuration could easily handle the city's solid waste problem.

Study results

The results of the simulation study combined with the management information system were spectacular. In 1970, the total complement of employees of the Cleveland Waste Collection and Disposal Division reached a maximum of 1825. With the use of data obtained in EPA's pilot data network, the number of personnel actually collecting waste on the route was reduced 50%. Service in collection operations was changed from back-yard to curb-side.

After the management information system and simulation model were developed, the total number of routes was reduced from 224 to 138. The total personnel complement in 1973 numbered approximately 600 employees. The annual budget declined from $14.3 million in 1970 to approximately $9.0 million in 1972.

The simulation model has been used for both short-term and long-term analysis. In the short term, it was used to restructure existing routes to achieve the budget reduction. In the long term, it was used to develop the following solid waste management plan:

1 The city will be broken into two divisions for solid waste collection. These two divisions are identified as the Westside District and the Eastside District.

2 The ultimate minimum route configuration will be 92 (36 in Westside District and 56 in Eastside District).

3 Motor-pool terminals will be located at the offload sites to maintain a high collection-to-transport time ratio.

4 Large-capacity, side-loading, dual-drive, compacting-type vehicles will be purchased for use by one driver and one loader.

5 A vehicle maintenance program will be implemented whereby daily servicing and cleaning will be performed by the operating crew and more intensive maintenance will be performed by a welder-mechanic.

As the result of this program, Cleveland's budget, based on 1972 dollars, is expected to stabilize at approximately $72 million.

This study illustrates the power of this approach in effecting and documenting changes in the solid waste system. Hopefully, other communities will use Cleveland's experience as an example for their own solid waste planning activities.

REFERENCES

1. CLARK, ROBERT M., *Management Information For Solid Waste Collection*, Environmental Protection Series EPA-R2-72-012, National Environmental Research Center, Office of Research and Monitoring, U. S. Environmental Protection Agency, Cincinnati, Ohio 45268, 1972.

2. CLARK, ROBERT M., and BILLY P. HELMS, "Decentralized Solid Waste Collection Facilities," *Journal of the Sanitary Engineering Division*, American Society of Civil Engineers, Vol. 96, No. SA5, Proc. Paper 7594, Oct., 1970, pp. 1035-43.

3. CLARK, ROBERT M., JOHN M. SWEETEN, and DANIEL G. GREATHOUSE, "Basic Data For Solid Waste: A Pilot Study," *Journal of the Sanitary Engineering Division*, American Society of Civil Engineers, Vol. 98, No. SA6, Proc. Paper 9424, Dec., 1972, pp. 897-907.

4. MARKS, DAVID H., and J. C. LIEBMAN, *Mathematical Analysis of Solid Waste Collection*, Public Health Service Publication No. 2104, 1970.

5. MARKS, DAVID H., CHARLES S. ReVELLE, and JON C. LIEBMAN, "Mathematical Models of Location: A Review," *Journal of the Urban Planning and Development Division*, American Society of Civil Engineers, Vol. 95, No. UP1, Mar., 1970, pp. 81-93.

6. QUON, JIMMIE E., MASARU TANAKA, and STEPHEN J. WERSAN, "Simulation Models of Refuse Collection Policies," *Journal of the Sanitary Engineering Division*, American Society of Civil Engineers, Vol. 95, No. SA3, Proc. Paper 6626, Jun., 1969, pp. 575-92.

7. TRUITT, M. M., J. C. LIEBMAN, and C. W. KRUSE, "Simulation Model of Urban Refuse Collection," *Journal of the Sanitary Engineering Division*, American Society of Civil Engineers, Vol. 95, No. SA2, Apr., 1969.

Case Study A-V

SOLVING MACHINE INTERFERENCE BY SIMULATION*

D. R. Freeman, S. V. Hoover and J. Satia †

This case study was selected to present a complex queueing type situation. The reader should particularly note the discussion of other possible techniques and their use to establish reasonable starting conditions and an approximately optimal system configuration to cut down on computer run time. This case also demonstrates how a well-designed model can explore multiple "What if"-type questions.

Among special queuing situations, machine interference is one of considerable practical importance. In the simplest form of the problem, one operator has charge of a number of similar machines. From time to time a machine stops and does not resume production until attended by the operator. If two or more machines are stopped at the same time, only one can be attended at once, and thus, production is lost while machines are awaiting attention. This phenomenon is called *interference.*

Production is lost both while machines are being repaired and while the machines are awaiting attention. The rate of production per machine and per operator, depends on the frequency of repair, repair times and the number of machines per operator. One is, therefore, interested in finding an optimum number of operators for a given number of machines.

Many practical situations can be analyzed as interference problems. Thread cutting machines, wire drawing machines, special purpose machines, airport baggage crews and staffing nurses in hospital wards are but a few examples. There are

* Reprinted from *Industrial Engineering*, Vol. 5, No. 7, July 1973, by permission of the American Institute of Industrial Engineers, Inc., 25 Technology Park/Atlanta, Norcross, Georgia, 30071.

† Of Northeastern University, Boston, Mass.

numerous instances in textile industries that fall in this category. In each of these instances the labor cost has to be balanced against the cost of interference.

Here we present a general machine interference problem. The traditional analysis techniques are discussed and a general purpose computer simulation which estimates interference times and work loads is described. This simulator does not require the restrictive assumptions usually found in traditional analysis. The simulation model has been used for analyzing several practical problems and one such case is presented in detail.

Interference situation

A general version of the problem could be stated as: "Given a set of machines, the information about the tasks that need to be performed, and the skill levels of operators who can perform these tasks, what is the optimum number of operators of each skill level?" The parameters of the problem are:

Task characteristics: Traditional analysis usually assumes that all tasks force the machines to go down. In a real application there may be tasks that can only be done when machines are running (i.e., an inspection for running speed) and tasks that can be performed while a machine is either running or down (i.e., lunch, breaks, entering data in operator logs, etc.).

We need to specify the recurrence and performance times for each task. The recurrence time (the time between two successive arrivals) may depend upon the machine running time or absolute clock time. For example, a task requiring loading of raw material recurs as a function of actual machine run time whereas operator bookkeeping recurs as a function of real time. It is usually assumed that these times have exponential probability distributions. But this assumption must be verified, using actual data and, if necessary, more suitable probability distributions such as Normal, Erlang, Uniform or constant should be used. Some tasks can or must be performed by multiple-men crew. In this case, the performance time needs to be specified for all possible crew assignments.In addition, some tasks have slack times associated with them. That is, a task will become available (i.e., it can be performed) but will not cause interference unless it is not begun within the slack time.

Machine characteristics: All machines may not have identical task requirements. This condition may arise due to either different types of machines or similar machines with varying degrees of wear. It also arises when service types of businesses, such as hospital wards, are considered.

Operator characteristics: A family of machines may be serviced by operators with different levels of skill and job classifications.

Perhaps some tasks could be performed by several types of operators. In any event, each task must be analyzed in terms of desired skill requirements. At times, some operators may be assigned to only a subset of the family of machines. This would be the case, for instance, when a mechanic may service several families of machines whereas other operators are restricted to one family. Some operators may not be assigned to the machines continuously, but only during certain portions of a shift. For example, a crew chief may be allowed to fill in for operators who are absent on lunch periods.

Scheduling rules: When several tasks are waiting to be done, selection priorities must be exercised. These selection priorities may depend upon the estimated performance times or slack times, etc. The availability of a critical task may require interruption of some other ancillary tasks. Therefore, the interruptability characteristics of each task (whether it can interrupt other tasks and/or can be interrupted) must be specified. A priority system introduces two difficulties: identification of a good priority system, and the adherence to this priority.

Traditional techniques

The problem of determining the proper number of men to assign to a group of machines has been solved by a variety of techniques. Some of these are:

1 Use of interference tables [1]. Interference tables developed initially for textile operations, involving operators manning several machines requiring periodic attendance, are available. This technique suffers from the necessity of oversimplification of the problem. These tables assume: identical machines, one class of operators, operators available to work on all machines, single operator performing a given task, no interruptability of low priority tasks for higher priority, etc. These simplifications make use of the tables impossible for many real world applications.

2 Queueing analysis. Cox and Smith [2] present a technique of solving the problem by applying essentially finite customer, multiserver queueing models with interarrival times of machines assumed to be exponential, and repair or task performance assumed to be an exponentially distributed random variable.

These two techniques suffer from disadvantages similar to those described. Also, assumptions of exponentially distributed times may be poor. Further, first-come-first-served disciplines may not be a close match with reality.

3 Expected work load analyses. This technique takes the expected value of performance times of operators applied to the number of machines and attempts to estimate optimum manning. It suffers from all the disadvantages of the first two techniques and fails to recognize the impact of high variability in arrival or performance times in causing interference.

All three techniques fall far short of giving an accurate picture of manning results in real applications. However, they are useful for first approximations and have been used as a pre-analysis technique to give reasonable starting conditions to the simulation discussed in this paper.

Simulation program

Input and output of the simulator are shown in Fig. A.5.1. The input consists of task data and a man-machine configuration. The output summarizes running, repair, and interference times for each machine, work load and utilization of operators, and the task statistics observed during the simulation.

Input task data consist of:

Task number: A unique number for each task.

Task name: Name identifying the task.

Operator class: The operators may belong to different classes due to difference in skill level or responsibility. Each task may be performed by one or several different classes.

Task type: Specifies one of three machine conditions necessary for performance: 1, when a machine is down (such as repair tasks); 2, when it is running (such as adjustment tasks); or 3, either case (such as material handling).

Task priority: Each task may have one of the three preemptive priorities: can interrupt but cannot be interrupted, cannot

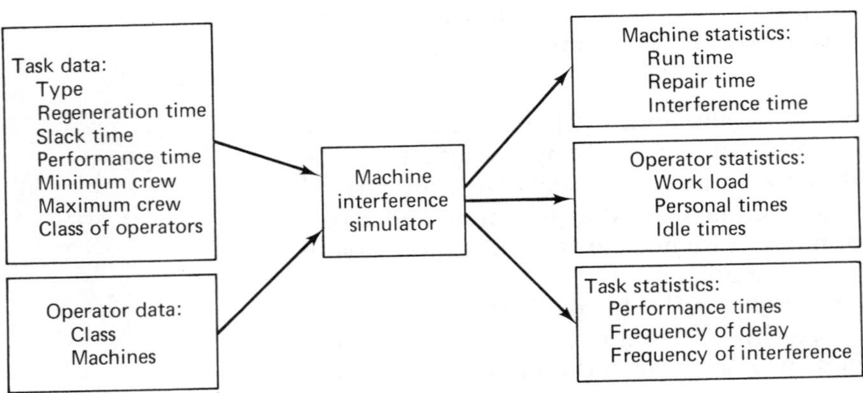

Figure A.5.1 Input and output of the simulation program.

interrupt and cannot be interrupted, and cannot interrupt but can be interrupted. In addition, a separate category is provided for tasks that are not essential and hence need not be performed if suitable manpower is not available. These tasks will never cause interference, but do increase work loads.

Regeneration time interval: Specifies the probability distribution and the associated parameters of the task interarrival time.

Slack time interval: Specifies the probability distribution and the associated parameters of the slack time. In this context, the slack time interval is defined as the time a task could wait to be started before interference occurs.

Performance time interval: The probability distribution and its associated parameters are specified for the time required to perform the task.

Regeneration method: May depend upon machine running time or upon the absolute clock time. Further, the regeneration interval may begin when the task becomes available, is begun, or is completed. Thus, six different cases may arise.

For each operator-related task further data on first occurrence times and subsequent regeneration time interval are needed. Figure A.5.2 shows the flow chart of the simulation program.

Several output modes are available which provide different levels of detail. We present the summary output in Table 1. For the period simulated, the machine summary shows the average values and range of run, repair, and interference times. The operator summary shows the work loads, personal, and the idle times experienced by each operator during the simulation. The frequency of occurrence, delay, and interference of each task are given in the task summary. In addition, average range of performance times are calculated for each task. This information can be used to recheck the input.

Several possible rules could be used to prioritize the available tasks to the operators. In this respect the problem is similar to designing a dispatching rule for a job shop. The simulator uses the slack times to determine the relative priorities. The task with the minimum slack time is considered first for possible assignment of the operators. For tasks with equal slack times, the one with the smallest mean performance time is given higher priority. More complex rules can be incorporated and output results analyzed for possible improved performance. However, the problems in implementing these rules increase as the complexity of the rules increases.

The simulator has been used in both batch and time shared modes. It was found easier to input the data and perform the sensitivity analysis in the time shared mode. Of course, the cost of using the time shared mode is higher. These usual trade-offs have to be considered in selecting the appropriate mode.

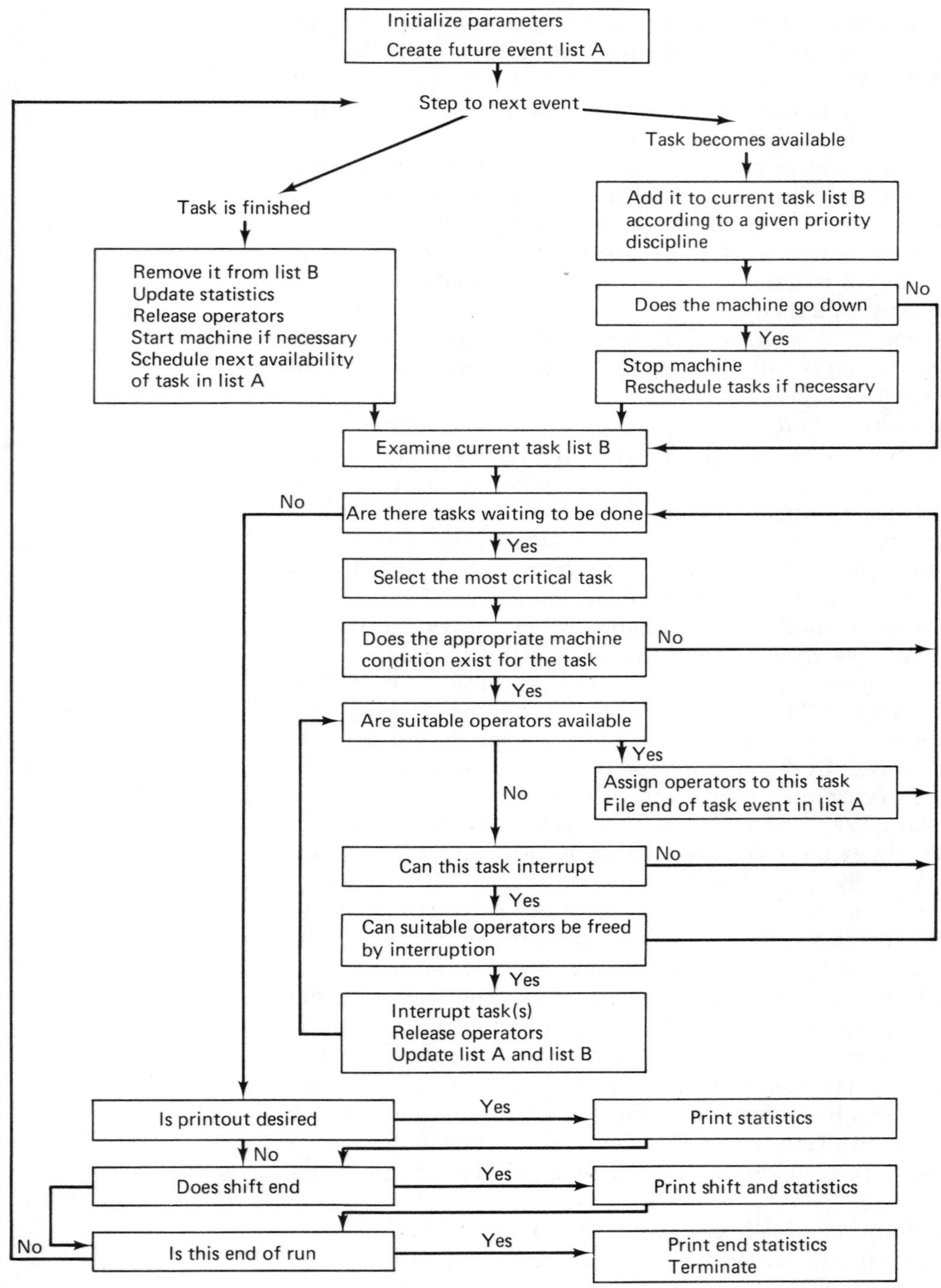

Figure A.5.2 Flow chart of simulation program.

Use in several applications has proved the versatility of the simulator. Conditions such as partial operator availability or shutting down machines during personal breaks can be taken into consideration by suitable modification of the task data. The simulator has been used for as many as 25 machine related tasks for a family of six machines. The storage and computational time requirements are modest (25,000 integer words, 5 to 10 minutes per run for the above problem on a CDC3300).

The simulator is also applicable to the problems that are not directly perceived as machine interference situations but can be modeled as finite source, finite server systems.

Table 1 *A summary output of the simulation run.*

OUTPUT SUMMARY
MACHINE STATISTICS

AVE RUN TIME	377.689	PERC RUN TIME	78.685
RANGE MIN	278.944	MAX 408.563	
AVE DOWN TIME	102.311	PERC DOWN TIME	21.315
AVE REP TIME	83.835	PERC REP TIME	17.466
RANGE MIN	45.655	MAX 139.464	
AVE TNT TIME	18.454	PERC TNT TIME	3.845
RANGE MIN	5.180	MAX 53.656	

Operator Statistics

Oper No	*Operator Class*	*Workload*	*Percent Workload*	*Perstime*	*Percent Perstime*	*Idletime*	*Percent Idletime*
1	1	287.95	56.46	90.00	17.65	132.03	25.89
2	1	255.71	50.14	90.00	17.65	164.27	32.21
3	2	263.38	51.84	90.00	17.65	156.60	30.71
4	2	208.59	40.90	90.00	17.65	211.39	41.45
5	3	173.57	34.03	90.00	17.65	246.41	48.32
6	3	90.67	17.78	90.00	17.65	329.31	64.57

Summary by Task

Task No	*Task Type*	*Freq Pershift*	*Ave Perf Time*	*Percent Tasktime*	*Max Perf Interval*	*Perctimes Delayed*	*Maxdelay Interval*	*Freq of Intrfrnc*
1	2	15.45	2.30	18.90	3.90	32.61	6.86	4
2	1	9.79	3.99	20.74	4.97	17.45	4.33	0
3	3	21.25	2.00	22.60	2.46	44.90	18.17	1
4	3	34.42	.79	14.37	1.49	30.99	17.52	3
5	2	4.62	9.52	23.38	50.82	18.92	39.74	0

Comparison of techniques

The cost of the analysis increases as it becomes more sophisticated. Rather severe assumptions and semi-empirical approximations are required to apply queueing theory analysis. Therefore, we have always found it easier to implement the decisions determined by using this simulation model. We can also print out the detailed sequence of events for specified time periods. This information can be communicated easily to management, which increases the confidence in the validity of the model's behavior and final results.

In several instances, queueing theory analysis cannot be applied. One instance is the situation where ancillary tasks are the major tasks and several operators with the same job classification are needed. But, wherever possible, queueing theory analysis should be utilized prior to a computer simulation examination; it helps in identifying an approximately optimal man-machine configuration.

Example

A new product is about to be put into production. The machines which make the product require periodic servicing by machine attendants. Some of the service tasks are scheduled while the others occur randomly. There are six machines. Three types of attendants are assigned to the machines: operator, helper, and mechanic. Due to the skill levels and union contracts, all service tasks cannot be performed by all three types of attendants.

Data collection

The data regarding interarrival time, slack time, performance time, crew size, and other task characteristics need to be collected. Usually, two sources of data are available: pilot production runs, and engineering estimates derived from technical considerations. The data can be collected in two ways: for each task separately, and/or use sample histories of production runs as shown in Figure A.5.3.

If a sample history is used, the run time between successive occurrences of each task needs to be extracted. The usual statistical tests can be used to determine the probability distribution and their associated parameters. Table 2 shows the considerations used to determine the information about this situation. Table 3 tabulates this information for the example problem.

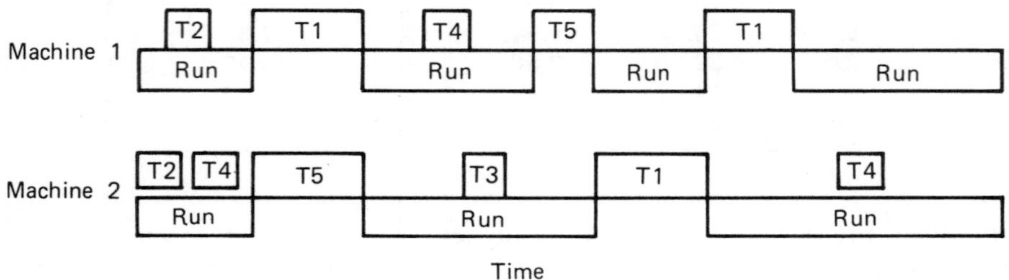

Figure A.5.3 Sample history. Note: Tk is the performance time for task k.

Table 2 *Possible sources of data items.*

Data Item	Source
Number of machines in family	Geographic considerations and managerial decisions
Task Information	
Type	Technical considerations
Interrupt characteristics	Technical considerations
Interarrival time	Sample histories
Performance time	Sample histories
Slack time	Technical and operational considerations
Method of regeneration	Technical considerations
Crew size and efficiency	Experimentation
Skill level of operators	Analysis of skill requirements union considerations company policy
Personal Tasks	Union considerations

Traditional analysis

The total work load, using expected interarrival and performance times, suggests that 3 to 6 men will be required for a family of 6 machines. When queueing theory analysis was performed, the optimal staffing level consisted of 3 operators/helpers and 2 mechanics for a family of 6 machines. The downtime was estimated to be 51 percent. Hence, we shall use the simulation model to examine the staffing levels similar to this solution.

Simulation analysis

The simulation analysis was performed for a family of 3 and 6 machines. The results are tabulated in Table 4. It can be seen that the minimum cost solution is 2 operators, 2 helpers, and 2 mechanics. The work loads for this staffing level are presented earlier in the sample output, Table 1.

Post simulation analysis

Several issues need to be resolved before the results of the analysis can be implemented.

What is the optimal number of machines in a family? The total cost will always be minimized by pooling all six machines. But there are other considerations involved in the choice of optimal group size. The tangible factor is the extra walking distances involved as the group size increases. In addition, as the number of machines increases, the sense of responsibility and involvement is diffused. It can be seen that a reduction in the group size from 6 to 3 results in an increase of $17.50 in the cost per hour. This

Table 3 *Task data for example problem.*

Task Number	Type of Attendant	Machine Condition	Can Interrupt	Cannot be Interrupted	Regeneration Time Probability Distribution		Slack Time Probability *Distribution*		Performance Time Probability Distribution		Crew[2] Size	Efficiency Factor	Method: Scheduling Based on Machine Run Time Where Task Is:
					Type[1]	*Mean (parameters)*	*Type*[1]	*Mean (parameters)*	*Type*[1]	*Mean (parameters)*			
1	Operator	Down	Yes	No	E	25.0	D	0.0	N	2.5 Std Dev 0.5	1 2	1.6	Finished
2	Operator or helper	Up	Yes	No	E	40.0	D	30.0	T	4.0 Range (3.0, 5.0)	1 1	1.0	Available
3	Helper	Up or down	No	No	D	18.0	D	20.0	T	2.0 Range (1.5, 2.5)	1 1	1.0	Available
4	Operator or helper	Up or down	No	Yes	U	8.0 Range (3.0, 13.0)	U	3.0 Range (2.0, 4.0)	U	1.0 Range (0.5, 1.5)	1 1	1.0	Started
5	Mechanic	Down	Yes	No	E	80.0	D	0.0	E	10.0	1 1	1.0	Finished

Note 1: D=deterministics, N=normal, U=uniform, E=Erlang, and T=triangular.

Note 2: The man-minutes required to perform a task, when maximum crew size is used, is given by performance time for minimum crew multiplied by an efficiency factor. For variable crew size the required man-minutes are determined by linear interpolation between the man-minutes for maximum and minimum crew size.

Table 4 *Simulation results.*

Staffing Level	Downtime	Labor Cost per Hour	Downtime Cost per Hour	Total Cost per Hour
Group Size: 6 Machines				
20p 2H 1M	0.280	26.0	50.40	76.40
10p 2H 1M	0.345	21.0	62.10	83.10
30p 2H 1M	0.280	31.0	50.40	81.40
20p 1H 1M	0.320	22.0	57.60	79.60
20p 3H 1M	0.280	30.0	50.40	80.40
30p 1H 1M	0.325	27.0	58.50	85.50
10p 3H 1M	0.300	25.0	54.00	79.00
10p 2H 2M	0.290	29.0	52.20	81.20
20p 2H 2M	0.213	34.0	38.34	72.34
20p 1H 2M	0.286	30.0	51.48	81.48
10p 3H 2M	0.280	33.0	50.40	83.40
30p 1H 2M	0.261	35.0	46.98	81.98
30p 3H 1M	0.255	35.0	45.90	80.90
30p 2H 2M	0.205	39.0	36.90	75.90
20p 3H 2M	0.203	38.0	36.54	74.54
30p 3H 2M	0.195	43.0	35.10	78.10
Group Size: 3 Machines				
10p 1H 1M	0.336	17.0	30.24	47.24
10p 2H 1M	0.266	21.0	23.94	44.94
20p 1H 1M	0.277	23.0	24.93	47.93
20p 2H 1M	0.236	26.0	21.24	47.24
20p 2H 1M	0.208	34.0	18.72	52.72

The simulation model was run until the estimated downtime stabilized to within 0.5%.

Cost of Labor per Hour ($)

Operator	5
Helper	4
Mechanic	8

Cost of downtime per machine per hour = $30.

increase needs to be balanced against other tangible and intangible factors.

What will the effect be if we relax the restriction on the job description of mechanics to help operators? In general, of course, any relaxation in the restrictions can only result in an improved performance. The simulation model was run again, allowing idle mechanics to help operators. The downtime is reduced from 21.3 to 19.7% and the total cost is reduced by $23 per shift. In other words, the removal of this restriction improves performance more than the addition of another operator. The effect of other restrictions can be analyzed in a similar fashion.

Should we educate our people in terms of giving priorities to some tasks? Any priority scheme will, perhaps, use slack time as a criterion. Two different disciplines, first-come-first-served, and shortest mean processing time, were used for tasks with the same slack time. The differences in the run times were not significant in this case. It may be difficult to enforce any particular discipline and, therefore, the simulation analysis may only point out the variations to be expected in the interference time.

Task number 4 is the most frequently occurring task. What will be the effect if the slack time for this task changes? Any reduction in the slack time may result in an increase in the interference, and vice versa. The simulation model was run using several values of the slack. Figure A.5.4 is a graph of average interference time versus average slack time. It can be seen that an increase in average slack time beyond 3 minutes has no effect on interference time.

We usually experience about 5% absenteeism. What will be the effect on interference? The effect of a reduced level of staffing when an operator, helper, or mechanic is absent can be determined using Table 4. A 7.5% increase in interference time will be experienced if an operator is absent.

What is the cost of this type of analysis? Two types of costs are involved: developmental cost of the general purpose simulator,

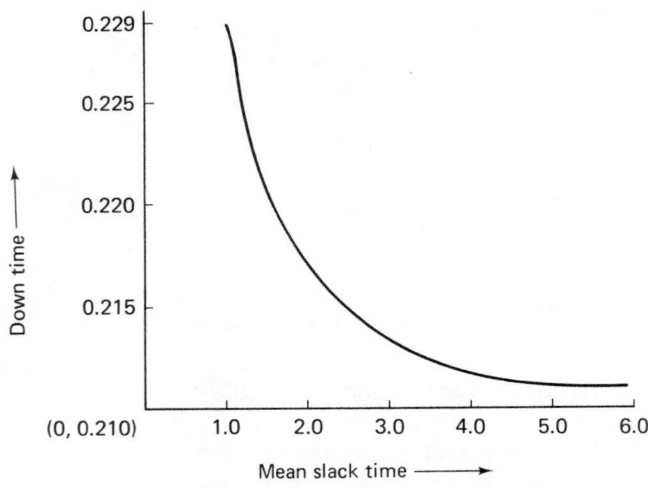

Figure A.5.4 **Graph of mean slack time of task 4 vs. downtime.**

and the cost of the analysis using the simulator. The latter amounts to approximately \$50 per run. Since about 20 runs were performed, the analysis cost is approximately \$1,000. The developmental cost of such a simulator will be in the range of from \$6,000 to \$8,000. Assigning 50% of the developmental cost to the project, the total costs amount to \$4,000. The annual personnel costs are in the range of \$120,000 to \$160,000. Therefore, the analysis costs are 2 to 3% of the direct cost of the decision involved.

REFERENCES

1. MAYNARD, H. B., *Industrial Engineering Handbook*, McGraw-Hill Book Company, New York, 1972.
2. COX, D. R., and W. L. SMITH, *Queues*, John Wiley & Sons, Inc., New York, New York, 1961.

Case Study A-VI

THE OVERHEAD CRANE*

J. Douglas DeMaire†

This case study was chosen to illustrate how one group went about developing confidence in the results of their simulation model. The involvement of various levels of management throughout the study and the process of validation used helped gain acceptance of the final results.

As a result of market growth and predicted future growth, the manufacturing capacity of a copper and copper alloy casting plant at Olin Brass was projected to become insufficient to meet market demand.

The casting operation employs a semicontinuous direct-chill casting process. The function of the casting operation is to provide large cast bars of non-ferrous alloys for further processing by rolling mills and fabricators. The bars are cast from various combinations of virgin and scrap raw materials.

The casting operation consists of a number of subsystems which include melting furnaces, holding furnaces, casting pits, and various supporting material handling equipment. Perhaps the most critical piece of material handling equipment is the overhead crane which has the ability to move the entire length of the plant.

When the plans to add casting equipment were developed, the question arose as to whether the crane, which had been completely adequate till then, would be able to handle the subsequent increase in tasks.

Workload on the crane is critical to the casting plant operation. Certain types of crane delays occur as a result of crane task interference (more than one task requiring service simultaneously). These delays often cause an extension of the casting cycle and

* Reprinted with permission from *Industrial Engineering*, Mar., 1973.

† Of the Olin Corporation, East Alton, Ill.

subsequently a loss in total casting capacity. With an estimated 20–25% increase in crane workload projected as a result of the additional casting equipment being planned, a significant question began to crystallize. This question became a statement of the problem this simulation study was called upon to solve. "*How much casting capacity would be lost, without modification to the overhead crane operating system, as a result of the new casting equipment being brought on line?*"

Case Study A-VI

Simulation necessary

Several subquestions required answers to help answer this major question and to provide other decision-making information to manufacturing management:

1 What magnitude of crane task interference would occur?

2 What new workload would result for the crane?

3 Would interference be great enough to justify extensive research to alleviate same?

4 Was it possible that no real problem existed?

After some analysis by manufacturing management, it became evident that the crane operating system was extremely complex and almost impossible to analyze reliably by conventional methods. Assistance in analyzing the situation was requested by manufacturing management. It was specifically requested that a simulation model of the crane operating system be constructed.

The fact that manufacturing had specifically requested the simulation study proved to be extremely valuable throughout the study. It helped elicit good cooperation with the manufacturing operating personnel, from the design phases to the conclusion of the project. The involvement of manufacturing personnel at all phases of the study served to inform them about the technique of simulation and generated a trust in the method and its results. The results of the study were viewed as reliable inputs to the decision-making process by manufacturing.

The parameters

Manufacturing involvement in this phase of the project was critical. In essence, they described the system and their description was abstracted into a simulation model. It is important to note that a model was not constructed and submitted for manufacturing approval: the model was developed with manufacturing as an integral part of the development.

A series of meetings were held with operating and management personnel ranging from first line supervision through director of

Table 1 *Each crane task is detailed by dividing it by breakpoints into sub-tasks. Crane operation may be preempted at any breakpoint in a major task if a higher priority task is waiting.*

Task	Description	Breakpoint Times* (minutes)	Priority
D.C. casting bar removal	Cast bars must be removed from casting units and moved to other equipment.	11 (No Breakpoints)	10 (Top)
Coil change	As a result of failure, heating coils must be replaced.	10—5—20—40—20—5—10	9
Replace charge weigh line	During charge weigh line failure, crane must move materials to casting floor ordinarily done by charge weigh line.	6	8
Replace gantry crane	During gantry failure, crane must handle gantry crane's jobs on the casting floor.	6	8
D.C. mold carriage	Replacing the mold carriage on D.C. Casting Units.	6—6—6—6	7
D.C. mold liner change	Replacing the mold liners on D.C. Casting Units.	5—5	7
D.C. unit change	Completely changing 5 melt furnaces and hold furnace associated with D.C. Casting	145—10—10—10—10—10—10—10—10—60—10—10—10—10—120—6—6—6—6—10—10—10—10—120—20—15—30	6
Unplanned melt changes	Changing only 1 of the 5 melt furnaces on a casting unit as the result of some failure.	10—5—10—40—10—15—20	6

5-melt change	Replacing the 5 melt furnaces.	10—5—10—40—10 (5 Times) Then 13—10 (5 Times)	6
Random tasks	Small jobs such as moving materials to casting floor.	5	5
Ascast bar removal	Removing bars from another type of casting unit.	5	5
Book mold casting mold changes	Replacement of molds on "book mold" casting unit.	20—25	4
Ascast mold change	Replacement of molds on "ascast" casting unit.	6—6	4
Ascast furnace change	Replacement of melting furnace on ascast unit.	15—15—10—10—10	4
Changing wertli	Molten metal must be carried to this production unit by the crane.	10	4
Wertli failures	Given certain types of failures on this production unit, overhead crane is required to help recover the unit.	10—20—10	4

* Underlined times are times crane is required. Times not underlined must elapse before the next step is done; however, the crane is free to do another job during the non-underlined times.

manufacturing. The initial meeting was somewhat unstructured. Generalities about the project were explained, and "rough" descriptions of system parameters were developed. At subsequent meetings, parameters were refined and expanded until all parametric information was developed.

Development of parameters consisted of accumulating information such as descriptions, detailed times for executing tasks, and descriptive information relative to the distribution of occurrences of the various tasks.

Each crane task (16 discrete types) was detailed by subdividing it by breakpoints into subtasks (see Table 1). For example, one task might be broken down into 5 subtasks, each subtask with a specified time for completion. Tasks were detailed in this manner in order that task priorities might be handled in the most realistic manner possible.

After several meetings with manufacturing, a sufficiently complete set of crane activities and their descriptions was compiled. At this point, a study of several months' past production and maintenance history was done, using the information previously developed as a basic structure for gathering the data. The study was used to develop occurrence frequencies for the tasks, based on history.

The results of the historical study were combined with manufacturing's best "operating feel" to yield the best estimate of the system parameters involved in the simulation. At this time all crane tasks were identified and defined, frequency distributions for occurrences for each type task were complete, the nature of the occurrence distribution was known, and priorities had been assigned to the tasks.

Making the model

Formulation of the model combined the parametric information with the necessary logic to represent properly the interactions of the various components of the model.

A major part of the program logic required was knowledge about logical sequences of occurrence—that is, what events could logically occur at the same time and what events must occur at mutually exclusive times. Through cooperation with operating personnel, a matrix was developed that fixed which events could not happen simultaneously (see Table 2).

The other major part of logic development used in the model dealt with generating tasks by the proper distributions. It became clear that the crane tasks that had been defined could be classified into three types with respect to the nature of the occurrence distributions. These classifications were:

1 Totally random occurrences.

Table 2 *Information obtained from operating personnel produced this matrix which indicates (by X's) row and column events that cannot occur simultaneously.*

	1	2	3	4	5	6	7	8	9	10	11	12	13	14	15	16
1. D.C. bar removal	X	X			X	X	X		X							
2. Coil change	X	X					X	X	X							
3. Change/weigh failure			X													
4. Gantry Failure				X												
5. D.C. mold carriage	X				X		X									
6. D.C. mold liner	X					X	X	X								
7. D.C. unit change	X	X			X	X	X	X								
8. Unplanned melt changes		X				X	X	X	X							
9. 5-Melt change	X	X						X	X							
10. Random task										X						
11. Ascast bar removal											X		X	X		
12. Book mold mold changes												X				
13. Ascast mold changes											X		X			
14. Ascast furnace changes											X			X		
15. Charge wertli															X	X
16. Wertli failure															X	X

2 Totally random occurrences within some known but varying cycle time.

3 Fixed time occurrences.

The simulation program was developed, or structured, in three main sections that correspond to the three occurrence types.

The totally random events are exactly what the name implies. There is a known fixed number of occurrences of this type event, but the time of an occurrence is generated in a truly random way. For example, furnace failures are considered to be totally random. Their occurrences are equally likely at any logical time throughout the simulation.

Removing cast bars from the casting equipment is an example of an event that is random within a known cycle time. Occurrence time for this task is assigned by a random distribution with a cycle time assigned by a known distribution.

A unit change on a casting unit is an example of a fixed time occurrence. The time for this task to occur was assigned before the simulation and its occurrence was forced to concur with this assignment.

It is worthwhile to note that the fixed occurrence tasks are in reality not fixed. They are referred to as fixed here, because they were dealt with in the actual simulation program as though they were fixed. These "fixed" occurrence times were preassigned by pseudo-random generation. They were handled as "fixed" in the model to assure logical execution sequences.

As mentioned briefly before, each of the sixteen crane tasks were subdivided by breakpoints into subtasks. This was done to facilitate the most realistic application of job practices. Since many of the crane tasks are relatively long in duration and consist of several distinct steps, it was decided to allow the crane to be pre-empted at any breakpoint in a major task if a higher priority task was waiting. The pre-empted task or tasks would be completed as priorities and crane workload allowed. In this manner the model operated the crane very much as it was physically operated. Queue statistics on all task and task breakpoints were tabulated for analysis.

An advantage capitalized on in this study was that initially, the simulation model simulated an existing setup. Given the vast knowledge the manufacturing personnnel had gained by experience, it was possible to calibrate or benchmark the accuracy of the model before adding the unknown element, the additional equipment. The manufacturing people gained additional confidence in the model, since they could see that it was simulating the existing operation with a high degree of accuracy.

Programming considerations

The programming problem most basic to the success of the project was, of course, representing the crane operating system in the most realistic manner possible. To achieve this goal, certain types of activities were fixed in time prior to the simulation run. This preassignment was done using a combination of randomness and logical pattern knowledge for event occurrences.

In the simulation program, these preassigned tasks were handled as follows: A function was simulated each day which contained code for the proper preassigned tasks to take place on that day. These functions, in conjunction with the various generate statements, kept transactions coming at the crane in a logical pattern throughout the simulation.

Both tasks associated with preassigned events and random events contained breakpoints as explained in the article. These breakpoints allowed the crane to be pre-empted within a job, if a higher priority task came due. The task time distributions associated with the various tasks were handled by creating a function for each task which included advance times for service and free time for the crane within a job execution. For example, a job could consist of the following:

5 min	crane time
6 min	crane time
10 min	non-crane time
7 min	crane time

During the 10-minute non-crane time, the crane was considered to be free for executing other tasks.

Various tables were set up to collect data in addition to the standard data output. For example, a table was generated for assigned cycle times and for realized cycle times. Comparing these two distributions was helpful in demonstrating the effect of workload on casting cycle time realization.

Tables were also kept showing interference patterns. These tables indicated which tasks were interfering with each other. This information was collected to help in designing new procedures for the actual crane operating system to minimize interference in the future design.

The simulation model was written using RCA's Flow Simulator language. Flow Simulator is a language very similar to GPSS and was selected basically because the computer available was an RCA Spectre 70, on which the simulation language available was Flow Simulator.

The smallest unit of time considered in the simulation was one minute: the total simulated time period was one month. The simulation took approximately 25 minutes to simulate one month and required 220K of core.

Benchmarking the model

Several simulation runs were made of the "before" situation. The results were tabulated for management review. Due to the structure of the model, it was possible to determine many significant operating parameters that could be measured against reality or known past performance. For example, it was possible to report production for the month, not only in total, but in some detail with respect to product mix. The fact that production generated by the model was totally believable, when compared to actual history, played a major role in convincing management that the model was valid. It was also possible to demonstrate that all prescribed tasks had been completed by the crane throughout the month. Again, as shown in Table 3, task occurrences closely paralleled actual experience. The model was designed so that the additional equipment could be easily incorporated. The approach from the beginning was to model the present situation, gain the confidence of management by demonstrating the model's accuracy, and insert new equipment and measure its effect. This approach was taken to maximize the probability of acceptance of the study's results. The method proved to be a good approach.

Simulating the proposal

The equipment that was projected to be added to the casting operation was incorporated into the existing and accepted model of the present casting operation.

Table 3 *This comparison of historical and simulated crane task frequencies indicates that the simulation model closely parallels actual experience.*

	Percent error in simulated frequency
1. D.C. Bar Removal	Not Applicable
2. Coil Changes	0
3. Charge/Weigh Failures	30
4. Gantry Failures	30
5. D.C. Mold Carriage Change	0
6. D.C. Mold Liner Change	0
7. D.C. Unit Change	0
8. Unplanned Furnace Changes	40
9. 5-Melt Changes	0
10. Random Tasks	4
11. Ascast Bar Removal	20
12. Book Mold Mold Changes	0
13. Ascast Mold Change	0
14. Ascast Furnace Changes	0
15. Charging Wertili	20
16. Wertli Failures	10

Simulation runs were made with the revised model. By comparing the "before" and "after" simulation run results, it was possible to achieve the project objectives.

The comparisons of before-and-after run results allowed predictions to be made on the critical characteristics of the crane operating system. It enabled the estimation of increased casting delay due to crane task interference increases.

Valid conclusions

Conclusions were drawn by comparing the simulation of the present casting operation, an accepted valid picture of reality, to the simulation results of the proposed casting operation with the additional equipment.

The major conclusion drawn showed that due to increased crane task interference, a 2.8% loss in capacity could be expected. This percentage was, in turn, converted into total pounds of production loss to be expected. This fact then became the major contribution of the simulation to the decision process. The

Table 4 *The simulation model shows the percent changes in crane functions caused by the proposed addition to the casting plant. These changes are caused by increased crane task interference.*

	Proposed situation (percent change)
Average number of casting cycles per day per machine	−2.8
Crane workload	+15
Zero wait time bar removals	−17
Average wait time for all bar removals	+44
Average wait time for bar removals that had to wait	No change
Percent of bar removals when more than one production unit was ready to have bars pulled at the same time	+7.9
Percent of occurrence of bar removals that exceeded allowable delay	+4

magnitude of the problem had been quantified and it was then up to management to determine what action was required.

The important contribution this study made was that management was in a position to make a much more informed decision than it would have been without the simulation. Critical to this fact was that management accepted the results of the simulation as valid.

From the detailed structure of the model, it is possible to conclude that the differences measured between the two situations are attributable strictly to the increase in crane task interference. Table 4 shows some critical relative statistics between the two situations simulated.

Management involvement

It is possible to greatly enhance the probability of management acceptance of a simulation if management involvement and participation in the simulation are forced to a maximum. This involvement should spread over all levels of management involved and should be sustained throughout the study.

The final acceptance and utilization of a simulation study are less related to the technical excellence of the study than they are to the confidence it is possible to gain from management if the project is executed properly. This statement is not intended to say that technical excellence is not a requirement for ultimate success, but to stress that management participation cannot be overlooked or underestimated if simulation's full potential is ever to be realized.

appendix B

RANDOM VARIATE GENERATION

B.1 Introduction

If we are dealing with a stochastic model, the simulation must be capable of generating random variates either from empirical tables or from theoretical probability distributions with desired parameters. As we discussed in Section 2.9, we accomplish this by the use of random numbers and Monte-Carlo sampling. Obviously, if our simulation model is computerized, we must have a means of (1) obtaining uniformly distributed random numbers and (2) using these random numbers to generate random variates with the desired characteristics. Most computers will already have subroutines for accomplishing this as a part of their program library. If so, the easiest approach for the analyst is to use the routines that are already part of the computer system he is using (if they are suitable). Occasionally, if an organization is just getting started in the use of simulation, the analyst will find that either the subroutines have not been made a part of the library or they have bugs which have not been worked out. For that reason, Appendix B presents FORTRAN computer programs that the analyst may find useful.

Before the advent of computers, random numbers were generated by physical devices, such as roulette wheels, special die, or mixing numbers up in some device and drawing them out by hand one at a time. Such devices are adequate if only a few numbers are needed. As greater use was made of random numbers, however, researchers turned to electronic devices for quicker generation. Undoubtedly, the most widely known use of such a device was at RAND Corporation, where an electronic pulse generator driven by a noise source was used. One million random numbers generated by this device were published in a book [21] and are also available on magnetic tape from RAND.

The use of such physical devices has two difficulties. First, it is difficult to provide and maintain a physical device such that a computer can call up a random number from it whenever it needs it. Second, the numbers generated by such devices are not reproducible; therefore, the use of correlated simulation runs (discussed in Section 5.16) is not possible. Of course, we can get around both these difficulties by first generating the numbers and then storing them in an array in the computer's memory or on tape to be called up when needed. Unfortunately, such a procedure is slow and uses up valuable memory storage; therefore, it is not practical for problems requiring a large number of random numbers.

Several schemes have been devised for generating pseudorandom numbers through the use of recursive mathematical relations. We say they are pseudorandom because even though the numbers generated might pass all our statistical tests for randomness and distribution, they are, in fact, completely deterministic. By this we mean that if we start each run of the generator with the same inputs (constants and seeds), we get the same sequence of numbers as an output. This, of course, allows us to use the correlated sampling techniques discussed in Section 5.16, but it also means we must be careful not to get the same sequence when we don't want it.

Although a great deal of the theoretical work has gone into developing random number and random variate generators, we do not intend to cover the topic here owing to lack of space. An excellent survey covering the period until 1962 is presented in a paper by Hull and Dobell [5], which contains about 150 citations in the bibliography. A very comprehensive treatment of the subject is Jansson's [8] monograph written in 1966, which contains about 300 references. Naylor *et al.* [17] contains two chapters devoted to the subject as well as FORTRAN computer subroutines.

B.2 Midsquare method

One of the first arithmetic procedures used to generate sequences of uniformly distributed, pseudorandom numbers was the midsquare method. In the technique, proposed by Von Neumann and Metropolis in 1946, each new number of a sequence is produced by taking the middle "m" digits of a number obtained by squaring an "m" digit number. The midsquare method follows these steps:

Section B.2
Midsquare method

1 Choose a four-digit number.

2 Square that number and add zeros to the left if necessary to make eight digits.

3 Select the middle four digits as the random number to be used.

4 Square the four-digit number selected in step 3 (again adding zeros to the left if necessary to make eight digits).

5 Repeat steps 3 and 4 until desired number of random numbers is obtained.

To illustrate the method, suppose we select $X_0 = 2152$ as the first number. The following sequence would then be generated:

$$
\begin{array}{llll}
X_0 = 2152 & (X_0)^2 = 04631104 \\
X_1 = 6311 & (X_1)^2 = 39828721 \\
X_2 = 8287 & (X_2)^2 = 68674369 \\
X_3 = 6743 & (X_3)^2 = 45468049 \\
X_4 = 4680 & (X_4)^2 = 21902400 \\
X_5 = 9024 & \text{etc.}
\end{array}
$$

Unfortunately, this method is difficult to analyze, relatively slow, and statistically unsatisfactory. For example, the relationship between the starting number and the length of the sequence generated before it begins to repeat itself (referred to as the period) is difficult to analyze ahead of time. Quite often, the resulting sequence is too short. Even worse, the samples may not have any randomness. Consider, for example, what happens if we select $X_0 = 4500$:

$$
\begin{array}{ll}
X_0 = 4500 & (X_0)^2 = 20250000 \\
X_1 = 2500 & (X_1)^2 = 06250000 \\
X_2 = 2500 & (X_2)^2 = 06250000 \\
X_3 = 2500 & \text{etc.}
\end{array}
$$

In view of these shortcomings, we no longer recommend the midsquare method, included here for historical interest only. Its simplicity of description and ease of understanding led to its earlier popularity. Presently, almost all computer library codes for generating sequences of uniformly distributed random numbers use some variation of the congruential methods developed by Lehmer [10].

B.3 Multiplicative congruent method

The congruent method is an arithmetic procedure for the generation of a finite sequence of uniformly distributed numbers. Several recursive relationships can be used and numerous congruential methods have been developed. Each uses the fundamental congruent relationship,* and the three popular ones are identified as the multiplicative, mixed, and additive methods. We first discuss the multiplicative method.

The basic relationship for the multiplicative congruent method is

$$X_{i+1} = aX_i \pmod{m}$$

where a and m are nonnegative numbers. This expression tells us to take the last random number X_i, multiply it by the constant a, and take the result modulo m (i.e., divide aX_i by m and treat the remainder as X_{i+1}). Therefore, to generate a sequence of numbers X_i, we need a starting number or seed (X_0), a multiplier (a), and a modulus (m). For any pseudorandom number generator, only a finite number of distinct integers can be generated, after which the sequence repeats itself. The period or length of the sequence (P) depends upon the particular computer and the modulus chosen, whereas the statistical properties of the generated sequence depend upon the choice of the starting value (seed) and the multiplier. The choice for a, X_0, and m is thus influenced by the desire for a maximum period and a minimum degree of correlation between the numbers generated.

The proper choice of the modulus m depends upon the computer number system being used. The most natural choice for m is one that equals the capacity of the computer word. For a binary machine, m would equal 2^b, where b is the number of bits in the computer word. In a decimal machine, m would equal 10^d, where d is the number of digits in the computer word. The maximum period (obtainable when a and X_0 are properly chosen)

* Two integer numbers A and B are said to be *congruent modulo m* (m being an integer) if and only if there is an integer k such that $A - B = km$—in other words, if $(A - B)$ is divisible by m and if A and B leave identical remainders when divided by the absolute value of m. This defining relationship is expressed as $A \equiv B \pmod{m}$, which is read, "A is congruent to B modulo m."

is, then,

$$P = 2^{b-2} = \frac{m}{4} \text{ for binary systems with } b > 2$$

$$P = (5)\,10^{d-2} = \frac{m}{20} \text{ for decimal systems with } d > 2$$

This period or sequence length is realizable only if X_0 and a are chosen in a certain way. Again, the choice depends upon the type of computer being used. For the binary case, a is selected as

$$a = 8\,T \pm 3$$

where T may be any positive integer and X_0 is selected as any positive but *odd* integer. In the decimal case, a is selected based upon

$$a = 200\,T \pm Q$$

where again T may be any positive integer and Q is one of the following values: ±(3, 11, 13, 19, 21, 27, 29, 37, 53, 59, 61, 67, 69, 77, 83, or 91). The seed X_0 for the decimal case may be any *odd* but positive integer that is not divisible by 2 or 5.

To satisfy the requirement of an uncorrelated sequence, we must further restrict the choice of the multiplier, a. Several authors recommend that to minimize correlation, the multiplier should be chosen as approximately equal to the square root of the modulus, m [5, 7, 8]. Unfortunately, this advice appears now to be poor. Coveyou and Mac Pherson [2], as well as Van Gelder [27], have shown that if $a \approx m^{1/2}$, we may indeed produce very small correlation between adjacent pairs, but unfortunately get strong correlations between triplets. Based upon these and other results, the multiplier, a, apparently should probably *not* be some fraction of m. Rather, it should be chosen such as to be five or more digits and not contain long strings of zeros or ones.

By following the preceding recommendations, the analyst should be able to obtain a more than sufficient number of random digits for most simulations. Note, however, that even though choosing m to be 2^b or 10^d makes the calculations fast and simple, it does not allow the longest possible periods. If the preceding procedure does not provide a sufficiently long sequence or period, one can sacrifice speed by using the procedure suggested in a paper by Hutchinson [6]. By choosing m as the largest prime number less than 2^b and, a as a primitive root of m, the maximum period can be extended from $m/4$ to $(m-1)$.

We can perhaps summarize the process of computing random numbers between 0 and 1 by the following:

1 Choose any number less than nine digits and designate it as X_0, the starting value. This seed should be chosen randomly, perhaps by using a table of random digits, such as Table C-I.

2 Multiply this by a number designated, a, of at least five digits.

3 Multiply the product from step 2 by a fraction or decimal number that is equal to $1/m$. We use multiplication because it is the faster operation for a computer.

4 Choose the decimal portion of the answer from step 3 as a random number $0 \leqslant X \leqslant 1$.

5 Drop the decimal point from the number obtained in step 4 and use it for the X is be multiplied by a in step 2.

6 Repeat steps 2 through 5 until the desired number of random numbers is obtained.

Such a procedure generates random numbers between 0 and 1. This facilitates their use in other random variate generators, as we shall see in later sections. To change from the decimal fraction to an integer requires only multiplication by an appropriate power of 10.

To generate a uniformly distributed random number using the multiplicative congruential method requires very few coding steps. Therefore, no great penalty is paid by repeating the coding wherever a random number is needed in the program. However, most computer programmers prefer to use a subroutine that can be used as a common source of all random numbers.

The general form of such a subroutine called RANDUM written in FORTRAN is shown in Fig. B.1. Note that the analyst needs to substitute actual values for the constants a, m, and m^{-1} at the time the subroutine is keypunched.

This subroutine is executed by issuing the following statement whenever a random number is needed:

```
CALL RANDUM (IX,IY,RN)
```

The value of the initial seed (an odd integer of less than 9 decimal digits) will have to have been read in as IX as a part of the run input. We could derive any number of independent random number sequences from this subroutine by defining different seeds, IX1, IX2, etc., and providing a different seed value for each. After each execution of the subroutine, the output IY must be substituted for the input by the statement IX = IY, which prepares the subroutine for the next execution. If this is done, then either IX or IY can be taken as the output of random integer numbers, if needed.

```
      SUBROUTINE RANDUM (IX,IY,RN)
    1 IY = IX*a
    2 IF(IY)3,4,4
    3 IY = IY + m
    4 RN = IY
    5 RN = RN*m^-1
    6 IX = IY
    7 RETURN
      END
```

Figure B.1 Generator for uniformly distributed random numbers. *Note:* Actual values for the constants a, m, and m^{-1}, must be supplied by analyst.

The rationale of the subroutine is straightforward. Statement 1 computes aX_i using fixed-point integer arithmetic. This product will consist of $2b$ bits from which the high-order bits are discarded and the low-order bits will remain as X_{i+1}. The integer multiplication instruction in FORTRAN automatically discards the high-order bits.

Statements 2 and 3 of the subroutine guard against a negative value of IY resulting from discarding the high-order bits through statement 1. The value added at statement 3 is the largest integer that can be entered in the machine. This will usually have to be entered as $2^{b-1} + 1$. Statements 4 and 5 produce the floating-point output. At the conclusion of statement 5, we have a floating-point random number between 0 and 1 called RN.

EXAMPLE B.3.1

The subroutine for a computer with a 32-bit word size and using $a = 5^{13} =$ 1,220,703,125 would be keypunched as follows:

```
      SUBROUTINE RANDUM (IX,IY,RN)
    1 IY = IX* 1220703125
    2 IF(IY) 3,4,4
    3 IY = IY + 2147483647 + 1
    4 RN = IY
```

```
5 RN = RN* 0.4656613E - 9
6 IX = IY
7 RETURN
  END
```


B.4 The Lehmer generator

A slightly slower routine using division, double precision accuracy, a prime value for the modulus, and a equal to a positive primitive root of m (of at least 4 digits) is presented in Mize and Cox [15, p. 87]. This subroutine is based upon a suggestion by D. H. Lehmer and was extensively tested by Hutchinson [6].

EXAMPLE B.4.1

For a computer with 36 bits, the largest prime number less than 2^{36} has been determined to be $2^{35} - 31$, which is used for m. The primitive roots of m are 5^5 and 5^{13}. In the following subroutine taken from reference [15], the constant $a = 5^5 = 3125$ (also of the form $8T \pm 3$, where $T = 391$), and $m = 2^{35} - 31 = 34{,}359{,}738{,}337$.

```
SUBROUTINE RANDUM (IX,RN)

IX = IX * 3125
RN = FLOAT (IX)/34359738337.D0
RETURN
END
```

Notice that the DO is required to specify double precision accuracy and that IX is automatically changed after each execution. The floating-point pseudorandom number, RN (between 0 and 1), is available to the called program for whatever purpose we need it. The period or length of the sequence is equal to $m - 1$ or $2^{35} - 32$.

B.5 Other congruent methods

Literally dozens of pseudorandom number generators based upon congruent relationships have been proposed and tested. Thomson [25] proposed the use of what is usually called *mixed congruential* generators. These are generators that use the relationship

$$X_{i+1} = aX_i + C(\text{mod}\, m)$$

Green, Smith, and Klem [4] discuss another method, known as the *additive congruential method*. This technique is based upon the recursion formula,

$$X_{i+1} = (X_i + X_{i-1})(\text{mod}\, m)$$

When $X_0 = 0$ and $X_1 = 1$ in the additive method, we get the special case called the Fibonacci sequence. Knuth [9] discusses a more complex additive method.

MacLaren and Marsaglia [11], as well as Westlake [28] and others, have proposed the combination of two congruential generators with one shuffling the sequence from the other. MacLaren and Marsaglia also used a stored table of numbers from the RAND Corporation table of random digits [21] in combination with a mixed congruential generator in a similar manner. As each number is used from the table, it is replaced by a scrambled version of itself provided by the mixed congruential generator.

The goal of all of these proposals has been to develop generators that were either faster and/or had better characteristics for the resulting sequences. Various generators proposed by Coveyou and MacPherson [2], Van Gelder [27], Jansson [8], Gorenstein [3], Maclaren and Marsaglia [11], Hull and Dobell [5], Green *et al.* [4], Smith [23], and others have been extensively tested. The results show that although each of the proposed methods has advantages under certain conditions, each also is very sensitive to the selection of seeds and constants. In my opinion, very little is to be gained under most circumstances from using anything other than one of the multiplicative congruent algorithms. On most computers they are as fast or faster, and are not nearly as likely to give problems (if the seed and constants are chosen as recommended in Section B.4).

B.6 Testing for randomness

Since we know that we are generating a sequence of pseudorandom rather than truly random numbers whenever we use a deterministic computer algorithm, we ask, "How close to random are they in their behavior?" A bewildering array of statistical tests has been proposed and used. In fact, if one really performed all the tests that have been proposed as necessary, one could perform an unlimited number of tests and never get around to using the generator. Among the commonly used tests are the following:

1 *Frequency Tests.* Uses either the chi-square or Kolmogorov-Smirnoff test to compare the distribution of the set of numbers generated against a uniform distribution.

2 *Serial Test [13].* Tallies the frequency of occurrence of all possible combinations of 2, 3, 4, etc., digits and then runs a chi-square test against expectations.

3 *Gap Test [22]*. Counts the number of digits that appear between repetitions of a particular digit and then uses a chi-square test against expectations.

4 *Runs Test [22]*. Tests the number of runs above and below some constant (usually the mean) or runs up and down. The test involves counting the actual number of occurrences of runs of different lengths and comparing these counts to expectations by chi-square.

5 *Spectral Test [2]*. Measures the independence of adjacents sets of n numbers based on Fourier analysis. Knuth [9] characterizes this test as the most discriminating known.

6 *Poker Test [22]*. Analogous to testing poker hands, this test counts combinations of five or more digits for all digits different, one pair, two pairs, three of a kind, full house, etc., and tests against expected occurrences.

7 *Autocorrelation Tests [22]*. Tests the correlation between Xn and $Xn + k$, where k is the lag in the generation order ($k = 1, 2, 3 \ldots$).

8 *d^2 or Distance Test [8]*. Successive pairs of random numbers are regarded as coordinates for points in the unit square, and the square of the distance between the two points is tested against theoretical probabilities given by a set of equations.

9 *Order Statistic Tests [11]*. Tests the maximum or minimum value of n consecutive numbers or the range of n consecutive values.

The preceding list in no way covers all the tests that have been proposed. It is obviously neither feasible nor necessary for the analyst to run all these tests every time he runs a simulation. In fact, no set of tests can guarantee that the sequence generated is totally suitable anyway. Given any finite set of tests, there will always be a sequence of numbers that will pass all of the tests but is totally unacceptable for some particular application. It is always possible that it will have patterns that remain undetected despite extensive testing. It is therefore probably better to pick a generator that passes the more straightforward tests (such as the first four) and any tests that may be relevant to the particular problem at hand. Unless a new generator is being used, very little testing is required.

B.7 The uniform distribution

The uniform distribution (Fig. B.2) is a continuous probability density function, which is constant over the interval from a to b and zero otherwise.

$$\text{Density:} \quad f(x) = \frac{1}{b - a}$$

$$\text{Mean:} \quad \mu = \frac{b + a}{2}$$

$$\text{Variance:} \quad \sigma^2 = \frac{(b - a)^2}{12}$$

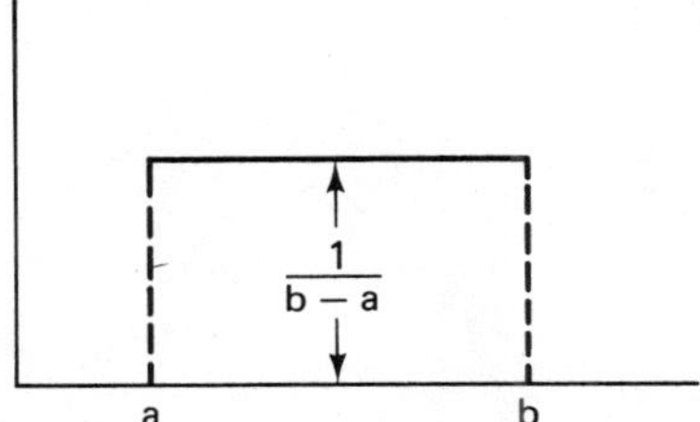

Figure B.2 Uniform distribution.

To simulate a uniform distribution over the range from a to b, we use the inverse transformation of the density function:

$$x = a + (b - a)RN, \qquad 0 \leqslant RN \leqslant 1$$

A suitable subroutine is shown in Fig. B.3. The subroutine is executed by issuing the following statement whenever a uniform variate is needed:

```
CALL UNIFRM(A,B,X)
```

The values for A and B (the range of interest) will have to be read in as a part of the run input.

```
SUBROUTINE UNIFRM(A,B,X)
CALL RANDUM (IX,IY,RN)
X = A + (B-A)*RN
RETURN
END
```

Figure B.3 Generator for uniform variates on interval from A to B.

B.8 The Poisson distribution

The Poisson distribution can describe a number of real world processes. If we take a series of independent Bernoulli trials (go-no go, success-failure, etc.), each having a small probability of an event occurring, then, as n increases, the probability of x occurrences is given by the Poisson distribution (Fig. B.4). Thus, many phenomena—such as the number of fires, airplane crashes, tornadoes, ship sinkings, etc.—per time period are described by the Poisson distribution:

$$\text{Density:} \quad f(x) = \frac{\lambda^x e^{-\lambda}}{x!}$$

$$\text{Mean:} \quad \mu = \lambda$$

$$\text{Variance:} \quad \sigma^2 = \lambda$$

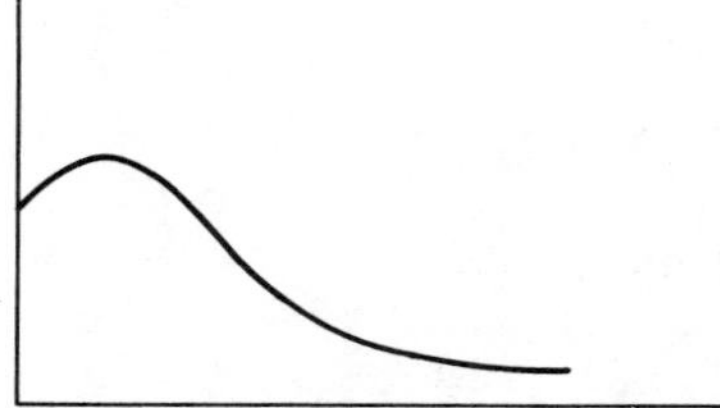

Figure B.4 Poisson distribution.

The Poisson distribution is a discrete distribution (the variable can take on only integer values including zero) with both mean and variance equal to λ. Lambda (λ) can have any positive value and need not be an integer. To generate Poisson variates, we take advantage of the well-known relationship between the exponential and Poisson distributions. That is, if the time between arrivals during a certain time period is exponentially distributed, then the number of arrivals per period is Poisson distributed. Thus, one method of generating Poisson variates involves generating exponentially distributed time arrivals with expected value equal to 1 and summing them until their total exceeds λ. A faster method is presented by Tocher [26] based upon generating random uniform variates (r_i) on the interval from zero to one until the following relationship holds:

$$\prod_{i=0}^{x} r_i \geqslant e^{-\lambda} > \prod_{i=0}^{x+1} r_i$$

A FORTRAN subroutine utilizing this method is shown in Fig. B.5. The program is executed by issuing the command

```
CALL POISSN (LAMBDA,X)
```

The value for the mean (LAMBDA) will have to be read in as a part of the run input.

```
      SUBROUTINE POISSN (LAMBDA,X)
1     X = 0.0
2     A = EXP(-LAMBDA)
3     S = 1.0
4     CALL RANDUM(IX,IY,RN)
5     S = S*RN
6     IF(S-A)9,7,7
7     X = X + 1.0
8     GO TO 4
9     RETURN
      END
```

Figure B.5 Generator for Poisson variates with a mean of LAMBDA.

B.9 The exponential distribution

If the probability that an event will occur during a small time interval (Δt) is very small, and if the occurrence of this event is independent of the occurrence of other events, then the time interval *between* the occurrence of events is exponentially distributed (Fig. B.6). Thus, for example, if in a queueing situation the arrival of customers is Poisson distributed with mean = λ, then the time between arrivals is exponentially distributed with mean equal θ, where $\theta = 1/\lambda$. Many other phenomena are also exponentially distributed, such as length of time of telephone conversations, life of many electronic components, arrival of orders at a firm, arrival of aircraft at an airport, etc. The ability to generate random variates that are exponentially distributed is therefore very important.

To simulate an exponential distribtuion, we shall call the mean THETA (θ) and use the inverse transformation of the density function:

$$X = -\theta \ln RN$$

One possible subroutine is shown in Fig. B.7. The subroutine is executed by issuing the following command whenever an exponentially distributed variate is needed:

$$\text{Density:} \quad f(x) = \lambda e^{-\lambda x}$$

$$\text{Mean:} \quad \mu = \frac{1}{\lambda}$$

$$\text{Variance:} \quad \sigma^2 = \frac{1}{\lambda^2}$$

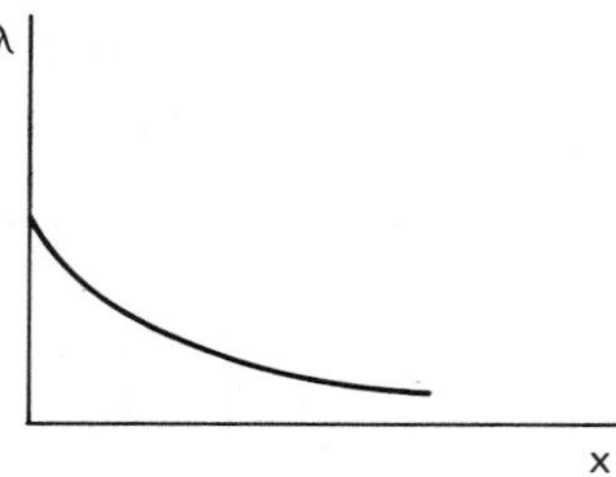

Figure B.6 Exponential distribution.

```
CALL EXPONT(THETA,X)
```

The value for the mean (THETA) will have to be read in as a part of the run input.

```
  SUBROUTINE EXPONT (THETA,X)
1 CALL RANDUM (IX,IY,RN)
2 X = -THETA*ALOG(RN)
3 RETURN
  END
```

Figure B.7 Generator for exponential variates with mean of THETA.

The computation of the natural logrithm on digital computers requires the use of some approximation technique, such as a power series expansion. Therefore, even though it is very easy to program, the inverse transformation method is not particularly fast. A faster (although harder to program) method is given by MacLaren, Marsaglia, and Bray [12].

B.10 The normal distribution

The normal or Gaussian distribution (Fig. B.8) is undoubtedly one of the most important and widely used continuous distributions. It is symmetrical about its mean value and is characterized by its mean value μ and standard deviation σ. A number of different methods of generating normally distributed pseudorandom numbers have been reported and are generally used. All these methods utilize the transformation $Z = (X - \mu)/\sigma$, so that the distribution generated is normally distributed with mean equal 0 and standard deviation 1. We then convert to the desired normal distribution by using the relationship $X = \mu + (RNN)\sigma$.

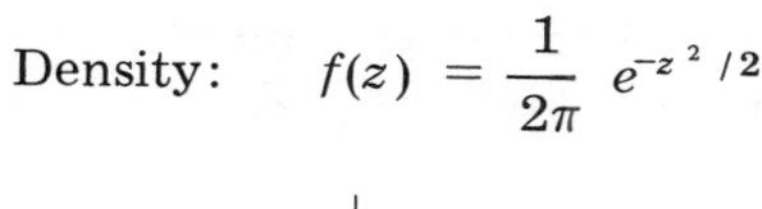

Density: $f(z) = \frac{1}{2\pi} e^{-z^2/2}$

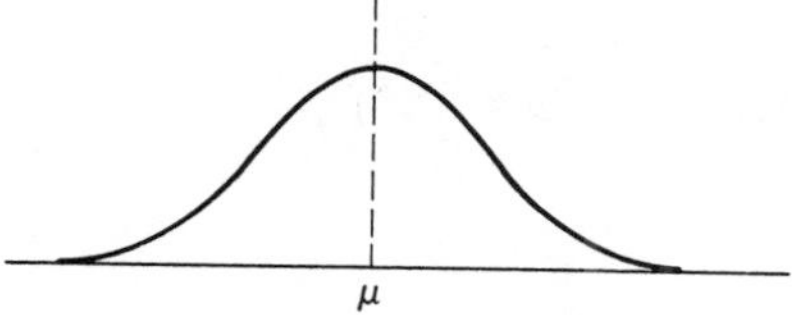

Figure B.8 Normal distribution.

An exact inverse method, proposed by Box and Muller [1], is accurate, easy to program, and reasonably fast. The method involves generating a *pair* of standardized normal numbers ($\mu = 0$, $\sigma = 1$) from two standard random numbers (r_1 and r_2 on the interval from 0 to 1), by letting

$$X_1 = -2 \ln r_1 \cos(2\pi r_2)$$

$$X_2 = -2 \ln r_1 \sin(2\pi r_2)$$

Marsaglia and Bray [14] give a variation of the Box-Muller method that is not only as accurate and easy to program but also faster. In this method, we generate the two random numbers r_1 and r_2. Then, setting $V_1 = -1 + 2r_1$ and $V_2 = -1 + 2r_2$, we compute $S = V_1^2 + V_2^2$. If $S \geqslant 1$, we start over again. If $S < 1$,

$$X_1 = V_1 \sqrt{\frac{-2 \ln S}{S}}$$

$$X_2 = V_2 \sqrt{\frac{-2 \ln S}{S}}$$

By the Marsaglia-Bray variation we avoid having to calculate sines and cosines, although we must still calculate square roots and natural logrithms. An estimated 127 pairs of random numbers (r_1 and r_2) will be needed to generate 100 pairs of random normal numbers. [13, p. 154]. The FORTRAN program based upon the

Marsaglia-Bray method is shown in Fig. B.9. We executed the subroutine by issuing the following command:

```
CALL NORMAL (EX,STD,X1,X2)
```

The value for the mean (EX) and the standard deviation (STD) will have to be read in as a part of the run input. At the end of each call we will have *two* random, normally distributed numbers, X1 and X2, from a normal population with mean = EX and standard deviation = STD.

```
SUBROUTINE NORMAL (EX,STD,X1,X2)
CALL RANDUM (IX,IY,RN)
V1 = 2*RN-1
CALL RANDUM (IX,IY,RN)
V2 = 2*RN-1
S = V1*V1+V2*V2
IF(S.GE.1)GO TO 1
RNN1 = V1*SQRT((-2.0*ALOG (S))/S)
RNN2 = V2*SQRT((-2.0*ALOG (S))/(S)
X1 = EX + RNN1*STD
X2 = EX + RNN2*STD
RETURN
END
```

Figure B.9 Generator for normal variates.

Another popular random normal variate generator is based upon the central limit theorem. If we take n samples from a distribution with variance σ^2 and mean μ, then the sum of the n samples is asymptotically normally distributed with mean $n\mu$ and variance $n\sigma^2$ (when n is large). Now, if we take our n samples from a uniform distribution over the interval zero to one, then $\mu = ½$ and $\sigma^2 = 1/12$, and, from the sum of our n samples, we would get a variable X that was normally distributed with mean $= n/2$ and variance $= n/12$. Thus, if we chose $n = 12$, the variance of X is equal to 1, and if we subtract 6 from the sum, the mean is equal to zero. Therefore, if r_i is a uniformly distributed random number on the interval from zero to one, we can calculate a randomly distributed normal variate with $\mu = 0$ and $\sigma^2 = 1$ by

$$X = \sum_{i=1}^{12} r_i - 6$$

A FORTRAN program using this technique is shown in Fig. B.10. There are several disadvantages to this particular generator.

First, it takes 12 uniformly distributed random numbers for each normal variate generated. Thus, if a large number of random variates is required, one could run out of uniform random numbers. Second, this method does very poorly in generating the tails of the normal distribution. In the testing of this technique by Jansson [8] and Muller [16], it was found that beyond about $\mu \pm 2\sigma$ this generator deviated from the desired characteristics rather widely. Teichroew [24] has extended the usefulness of this technique out to about $\pm 3\sigma$ by modifying the technique as follows. Compute

$$R = \frac{\left(\sum_{i=1}^{12} r_i - 6\right)}{4}$$

Then set

$$X = ((\langle[(C_1 R^2 + C_2)R^2 + C_3]R^2 + C_4\rangle R^2 + C_5)R$$

where

$$C_1 = 0.029899776, \quad C_2 = 0.008355968,$$
$$C_3 = 0.076542912, \quad C_4 = 0.252408784,$$
$$\text{and} \quad C_5 = 3.949846138$$

```
      SUBROUTINE NORMAL (EX,STD,X)
    1 A = 0.0
    2 DO 4 I = 1,12
    3 CALL RANDUM (IX,IY,RN)
    4 A = A + RN
    5 X = EX + (A-6.0)*STD
    6 RETURN
      END
```

Figure B.10 Central limit theorem generator for random normal variates.

B.11 The gamma distribution

The gamma distribtuion (Fig. B.11) is one of the most useful continuous distributions available to the simulation analyst. If the variables from some random phenomenon cannot assume negative values and generally follow a unimodal distribution, then the chances are excellent that a member of the gamma family can adequately simulate the phenomenon. The gamma distribution is defined by two parameters, α and β, where α is the shape parameter and β is the scale parameter. As the two parameters are varied, the gamma density can assume a wide variety of shapes, making it one of the most versatile of distributions for simulation purposes.

$$\text{Density:} \quad f(x) = \frac{\beta^{\alpha} X^{(\alpha-1)} e^{-\beta x}}{(\alpha-1)!},$$

where $\alpha > 0$, $\beta > 0$, $x \geqslant 0$

$$\text{Mean:} \quad \mu = \frac{\alpha}{\beta}$$

$$\text{Variance:} \quad \sigma^2 = \frac{\alpha}{\beta^2}$$

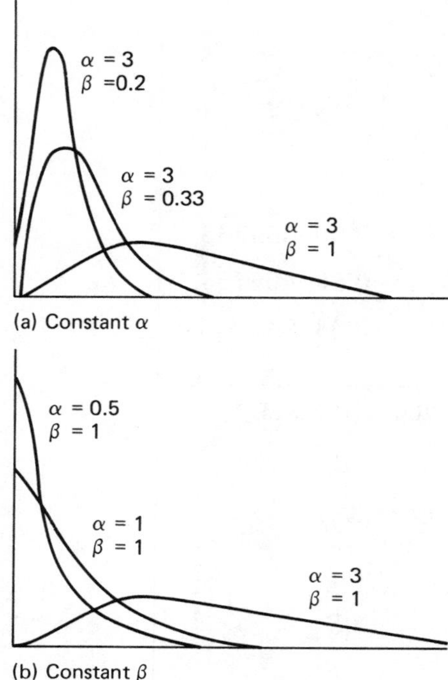

Figure B.11 Gamma distribution.

One of the problems that limited the use of the gamma distribution in the past was the lack of a good generator if α was not an integer. Don T. Phillips [18] has proposed a very versatile two-parameter gamma generator that overcomes this problem. The Phillips generator is shown in Fig. B.12. At this point, you should again remember the existence of Phillip's computer program, which will take a set of empirical data, calculate the values of α and β that give the best gamma fit, and then statistically test the goodness of fit [19]. We can also use the following equations to determine the value of the parameters:

$$\alpha = \frac{(\text{mean})^2}{\text{variance}}, \qquad \beta = \frac{\text{mean}}{\text{variance}}$$

The subroutine shown in Fig. B.12, is executed by issuing the command:

CALL GAMMA (ALPHA,BETA,START,X)

The values for ALPHA, BETA, and the initial value of START (any integer less than 1.5) must be supplied as inputs to the run.

If α is an integer, an alternative generator that is easier to program is presented by Naylor *et al.* [17]. In comparative tests run by Phillips [18], it was found that for $\alpha < 3$, Naylor's

```
      SUBROUTINE GAMMA(ALPHA,BETA,START,X)
      IF(START.GT.1.5) GO TO 60
      X3 = 1.0
      IF (ALPHA.LE.1.5) GO TO 1
      IF(ALPHA.LE.19.0) GO TO 2
      GO TO 3
    1 B = 0.24797+(1.34735740*ALPHA)-(1.00004204*ALPHA**2)+(0.53203176*ALP
     1HA**3)-(0.13671536*ALPHA**4)+(0.01320864*ALPHA**5)
      GO TO 4
    2 B = 0.64350+(0.45839602*ALPHA)-(0.02952801*ALPHA**2)+(0.00172718*ALP
     1HA**3)-(0.00005810*ALPHA**4)+(0.00000082*ALPHA**5)
      GO TO 4
    3 B = 1.33408+(0.22499991*ALPHA)- (0.00230695*ALPHA**2)+(0.00001623*ALP
     1HA**3)-(0.00000006*ALPHA**4)
    4 Y = 1.0+(1.0/B)
      START = 5.0
   12 IF(Y-1.0)110,50,15
   15 Y = Y-1.0
      X3 = X3*Y
      GO TO 12
  110 GY = 1.0+Y*(-0.5771017+Y*(0.985854+Y*(-0.8764218+Y*(0.8328212+
     1Y*(-0.5684729+Y*(0.2548205+Y*(-0.05149930)))))))
      X3 = X3*GY/Y
   50 A = (X3/(ALPHA*BETA))**B
      B = 1.0/B
      A = 1.0/A
   60 CALL RANDUM(IX,IY,RN)
      X = (-A*ALOG(RN))**B
      RETURN
      END
```

Figure B.12 Phillips [18] gamma variate generator with parameters ALPHA and BETA. *Note:* ALPHA = α; BETA = β; START = any real integer initially less than 1.5; and, X = random gamma variate.

subroutine is slightly faster. However, as α increases, Phillips' algorithm showed increasingly better performance (almost 3½ times faster at $\alpha = 10$).

One of the things making the gamma distribution so interesting is the fact that it is closely related to a number of other useful distributions. For example, if α=1 and β is a constant, θ, the exponential density function, is generated with an expected value of Θ. If α is an integer value K, then the gamma distribution is commonly referred to as an Erlang-K distribution. Furthermore, if $\beta = 1$, then as α becomes large, the gamma distribution approaches the normal distribution. If we set $\alpha = \nu/2$ and $\beta = 2$ (where ν is the degrees of freedom), we get the chi-square distribution. We can also use the Phillips' generator to obtain beta variates, F distribution variates, and student t variates [18].

B.12 Other random variate generators

The preceding random variate generators will generally meet the needs of most simulations. Naylor *et al.* [17] provide FORTRAN generators for all the previous distributions, plus the following: geometric, binomial, negative binomial, hypergeometric, and normally distributed, autocorrelated variates. Polge et al. [20] present an algorithm for generating a time series with any desired degree of autocorrelation from any arbitrary probability distribution.

REFERENCES

1. BOX, G. E. P., and M. E. MULLER, "A Note on the Generation of Random Normal Deviates," *Annals of Mathematical Statistics*, Vol. 29, pp. 610-11, 1958.

2. COVEYOU, R. R., and R. D. MacPHERSON, "Fourier Analysis of Uniform Random Number Generators," *Journal of the Association for Computing Machinery*, Vol. 14, No. 1, Jan., 1967.

3. GORENSTEIN, S., "Testing a Random Number Generator," Communications of the ACM, Vol. 10, No. 2, Feb., 1967.

4. GREEN, B. F., J. E. K. SMITH, and L. KLEM, "Empirical Tests of an Additive Random Number Generator," *Journal of the Association for Computing Machinery*, Vol. 6, 1959, pp. 527-37.

5. HULL, T. E., and A. R. DOBELL, "Random Number Generators," *SIAM Review*, Vol. 4, No. 3, 1962.

6. HUTCHINSON, D. W., "A New Uniform Pseudorandom Number Generator," *Communications of the ACM*, Vol. 9, No. 6, Jun., 1966.

7. IBM Corp., "Random Number Generation and Testing," *Reference Manual C20-8011*, White Plains, N. Y., 1959.

8. JANSSON, B., *Random Number Generators*, Almqvist & Wiksell, Stockholm, 1966.

9. KNUTH, D. E., *The Art of Computer Programming Vol. 2: Seminumerical Algorithms*, Addison-Wesley Publishing Co., Inc., Reading, Mass., 1969.

10. LEHMER, D. H,. "Mathematical Methods in Large-Scale Computing Units," *Annals Computer Laboratory, Harvard University*, Vol. 26, 1951.

11. MacLAREN, M. D., and G. MARSAGLIA, "Uniform Random Number Generators," *Journal of the Association for Computing Machinery*, Vol. 12, No. 1, Jan., 1965.

12. MacLAREN, M. D., G. MARSAGLIA, and T. A. BRAY, "A Fast Procedure for Generating Exponential Random Variables," *Communications of the ACM*, Vol. 7, No. 5, May, 1964.

13. MAISEL, H., and G. GNUGNOLI, *Simulation of Discrete Stochastic Systems*, Scientific Research Associates, Inc., Chicago, Ill., 1972.

14. MARSAGLIA, G., and T. A. BRAY, "A Convenient Method for Generating Normal Variables," *SIAM Review*, Vol. 6, No. 3, pp. 260-64, 1964.

15. MIZE, J. H., and J. G. COX, *Essentials of Simulation*, Prentice-Hall, Inc., Englewood Cliffs, N. J., 1968.

16. MULLER, M. E., "A Comparison of Methods for Generating Normal Deviates on Digital Computers," *Journal of the Association for Computing Machinery*, Vol. 6, pp. 376-83, 1959.

17. NAYLOR, T. H., J. L. BALINTFY, D. S. BURDICK, and K. CHU, *Computer Simulation Techniques*, John Wiley & Sons, Inc., New York, 1966.

18. PHILLIPS, D. T., "Generation of Random Gamma Variates from the Two-Parameter Gamma," *AIIE Transactions*, Vol. III, No. 3, Sept., 1971.

19. PHILLIPS, D. T., *Applied Goodness of Fit Testing*, O.R. Monograph Series, No. 1, AIIE-OR-72-1, American Institute of Industrial Engineers, Atlanta, Ga., 1972.

20. POLGE, R. J., E. M. HOLLIDAY, and B. K. BHAGAVAN, "Generation of Pseudo-Random Sets with Desired Correlation and Probability Distribution," *Simulation*, Vol. 20, No. 5, May, 1973.

21. RAND Corporation, *A Million Random Digits with 100,000 Normal Deviates*, The Free Press, New York, 1955.

22. SCHMIDT, J. W., and R. E. TAYLOR, *Simulation and Analysis of Industrial Systems*, Richard D. Irwin, Inc., Homewood Ill., 1970.

23. SMITH, C. S., "Multiplicative Pseudo-Random Number Generators with Prime Modulus," *Journal of the Association for Computing Machinery*, Vol. 18, No. 4, Oct., 1971.

24. TEICHROEW, D., "Distribution Sampling with High Speed Computers," Ph.D. Thesis, University of North Carolina, Chapel Hill, N. C., 1953.

25. THOMSON, W. E., "A Modified Congruence Method of Generating Pseudo-random Numbers," *The Computer Journal*, Vol. 1, No. 2, Feb., 1958.

26. TOCHER, K. D., *The Art of Simulation*, D. Van Nostrand Co., Inc., Princeton, N. J., 1963.

27. VAN GELDER, A., "Some New Results in Pseudo-Random Number Generation," *Journal of the Association for Computing Machinery*, Vol. 14, No. 4, Oct., 1967.

28. WESTLAKE, W. J., "A Uniform Random Number Generator Based on the Combination of Two Congruential Generators," *Journal of the Association for Computing Machinery*, Vol. 14, No. 2, Apr., 1967.

appendix C

STATISTICAL TABLES

APPENDIX C-I

Random digits

3513	2883	5675	8101	7526	2251	5765	4334	4112	5664
6976	882	2501	7066	9890	5741	5434	928	6666	1641
9847	4311	7238	4632	2642	4166	1213	9783	7778	8621
1722	2741	945	998	7480	5897	8665	5313	9289	7013
3874	2020	7257	5356	6822	2726	4055	5105	6567	2650
6793	6904	283	9564	4834	2927	4052	7966	1325	6253
5587	7242	3170	3843	4524	2555	4612	4678	6555	7228
4368	1157	7628	5352	3463	2608	4477	3327	25	9661
7740	9490	7272	8223	3880	9328	962	1810	2252	7136
2544	1043	3355	742	4256	8859	4846	9342	2434	526
1247	2749	5269	6867	3781	884	1270	9660	6531	2243
4679	7881	5173	108	4089	3558	4543	5234	515	5086
1275	3774	1169	3049	7772	9190	5186	8405	3757	6896
7556	3275	1644	384	7509	1596	1993	7590	7600	7283
5297	6238	9747	2344	6337	6919	4481	4618	7370	2662
9636	3854	6403	3730	4748	4918	6778	6404	7413	6846
4353	4503	7840	6510	8497	2388	7857	5648	3172	8196
627	9998	4341	6063	7311	9291	9951	6083	6935	6861
8749	743	5716	7602	4171	6603	2080	3047	9563	9952
3644	2297	979	5202	2397	7564	3808	4767	4330	3075
9475	9175	9776	6077	8476	6164	698	8712	5988	7513
1186	9497	6306	2359	7398	3158	2361	5744	3214	7581
6565	1153	7834	6623	9231	5779	1589	7522	829	7277
6196	1679	4311	753	5720	7541	3767	4733	4491	4348
5667	4871	8221	5485	8918	4139	4576	199	7	8252
9448	2418	9474	5075	5187	5440	5955	6771	7026	1217
4063	3423	3971	3013	2337	6903	382	162	7529	3713
4518	3687	1460	5569	276	1533	6711	6466	8391	2152
7392	4977	3333	5294	1228	537	2158	8113	9247	2466
1571	7234	9261	463	9424	2381	9463	5352	6938	3461
8226	8799	7860	7972	7086	769	838	8101	1064	3469
1237	6197	6048	517	8665	7335	6021	107	6454	7756
8451	900	9334	7905	3425	9400	5373	8834	2847	7572
9810	711	5972	9432	2847	2191	7518	5384	4643	9398
4602	3030	6755	3260	8760	3224	495	3958	9288	103
7028	1236	4165	3860	5677	9317	4807	4992	6684	5176
896	8789	4673	8931	1532	8808	9060	5086	8975	8075
7677	3382	1204	6779	9839	8019	9561	5185	5121	3969
7719	594	4086	9168	8233	6885	7211	1301	2909	5744
8276	7959	3268	7978	8449	8894	7320	3868	7327	9145
8928	1254	7176	1763	5995	98	6630	8902	3733	2282
92	15	9250	5397	9883	5916	3751	9260	1797	7443
8484	3912	7116	7486	873	7862	9314	5122	6901	5307
9730	611	6098	1085	1627	0	5353	2116	4353	8086
7810	4083	4207	8492	3085	2082	4723	9596	5065	4023

APPENDIX C-II

Random normal numbers $\mu = 0, \sigma = 1.$

1.38	-1.73	0.33	0.24	-0.29	1.93	-1.40	0.51	0.71
-0.77	-0.69	1.32	-0.93	-1.14	-0.06	-0.35	1.04	0.10
-0.48	0.06	0.43	-0.24	2.63	0.65	-0.38	1.25	0.43
-0.92	0.53	0.51	0.65	-0.40	0.68	0.05	-0.45	-0.00
-0.89	-0.04	-0.01	1.64	1.46	-0.14	0.36	0.43	-0.08
-1.02	0.64	-0.13	0.22	-0.94	-0.64	-0.00	-0.31	0.45
1.61	0.42	1.19	-1.98	1.04	-0.69	-0.83	-0.78	-1.13
0.40	-0.38	-0.92	-1.11	-0.01	1.10	-1.83	-0.57	0.93
0.74	-1.36	-1.60	-0.35	-1.78	1.17	-0.73	-0.87	0.80
-1.96	-0.93	0.91	-0.72	0.08	-0.61	-0.89	1.04	-3.33
1.39	-0.61	-0.98	0.76	-1.22	0.99	-1.29	-0.12	-1.15
1.70	-0.83	-0.91	0.43	-1.11	0.04	-1.96	-0.06	0.47
-0.51	0.93	-0.67	0.92	-0.72	-0.47	0.15	0.30	0.71
-1.85	-0.52	-0.79	0.66	1.70	-1.02	1.82	-0.42	-0.38
-0.13	-0.13	-0.55	-0.07	-2.08	-0.35	-0.17	0.20	-0.51
-1.99	-1.14	-0.42	0.46	-0.61	-0.78	0.77	0.33	0.76
0.15	0.02	0.10	-0.33	-0.52	0.18	0.38	0.23	0.56
-0.38	0.32	0.11	-0.65	0.30	-0.36	0.54	0.33	-1.26
0.62	0.32	-1.42	1.01	0.10	-0.20	0.71	-1.25	-0.52
-1.09	-0.09	0.17	1.75	-0.86	0.93	-0.98	0.41	-0.48
1.08	0.64	-0.42	0.66	0.24	-0.08	-1.10	0.74	1.83
-0.70	0.59	-1.50	-0.13	0.81	0.43	0.21	-0.09	-0.98
1.76	-0.25	1.08	0.06	-0.50	-0.46	0.06	0.64	0.10
-1.45	0.10	2.04	-1.40	0.34	0.02	-1.94	-0.41	-0.99
-0.82	0.11	-1.78	0.23	-0.47	0.13	-0.78	1.24	0.56
-0.36	-0.14	-0.16	1.33	-0.28	-1.01	-0.63	-0.52	2.00
0.01	-0.23	-1.82	0.69	-0.12	0.51	-0.82	-0.44	0.85
-1.36	-0.20	0.92	-1.10	-0.23	1.90	-0.32	-0.21	1.08
-0.06	-0.73	1.03	-0.72	-0.10	-2.27	0.41	0.69	1.34
-2.39	-0.74	-0.84	-0.60	0.07	0.50	1.10	-1.25	-0.19
-1.40	-0.49	-0.35	-0.91	1.57	-1.08	-0.09	-0.27	-1.98
0.45	1.90	0.13	-0.90	-0.23	-1.44	0.24	-0.48	-0.89
0.38	-0.43	0.57	0.27	1.44	-0.79	1.10	0.60	1.34
-1.22	1.50	1.03	-1.71	0.50	-0.26	-1.96	-0.97	-0.88
2.36	-0.32	0.86	-0.02	-2.56	0.55	-0.33	-0.03	-0.35
0.82	0.89	0.53	-0.81	2.23	-0.48	-0.32	2.86	1.08
0.12	1.00	-1.01	0.22	0.81	0.62	1.48	-0.93	-2.71
0.26	-1.18	0.62	-0.20	0.17	0.01	-0.59	-0.85	0.18
-0.33	0.56	-0.28	-1.31	-1.06	-1.28	-0.57	-0.36	0.55
3.14	-0.20	1.54	-1.24	-2.39	-1.11	-0.52	-0.57	0.50
1.31	-0.90	-0.04	-0.29	0.25	-1.41	1.32	-0.26	0.47
0.73	-0.96	-1.38	-0.73	-1.17	-0.79	-1.02	1.53	-0.21
-1.52	0.97	0.69	0.07	0.08	0.34	-1.26	-1.34	-0.22
0.38	-0.76	1.17	-2.07	0.94	-0.07	1.71	2.44	1.45

APPENDIX C-III

Percentile values (χ_p^2) for the chi-square distribution, with ν degrees of freedom (shaded area = p).

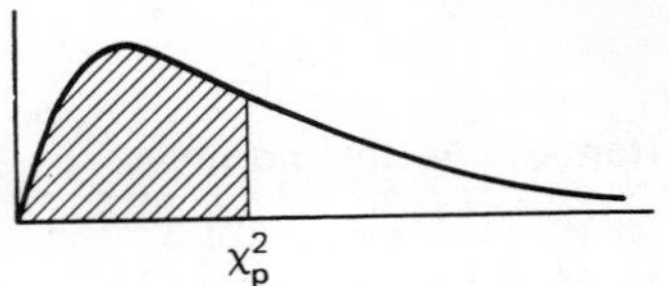

ν	$\chi^2_{0.995}$	$\chi^2_{0.99}$	$\chi^2_{0.975}$	$\chi^2_{0.95}$	$\chi^2_{0.90}$
1	7.88	6.63	5.02	3.84	2.71
2	10.60	9.21	7.38	5.99	4.61
3	12.84	11.34	9.35	7.81	6.25
4	14.96	13.28	11.14	9.49	7.78
5	16.7	15.1	12.8	11.1	9.2
6	18.5	16.8	14.4	12.6	10.6
7	20.3	18.5	16.0	14.1	12.0
8	22.0	20.1	17.5	15.5	13.4
9	23.6	21.7	19.0	16.9	14.7
10	25.2	23.2	20.5	18.3	16.0
11	26.8	24.7	21.9	19.7	17.3
12	28.3	26.2	23.3	21.0	18.5
13	29.8	27.7	24.7	22.4	19.8
14	31.3	29.1	26.1	23.7	21.1
15	32.8	30.6	27.5	25.0	22.3
16	34.3	32.0	28.8	26.3	23.5
17	35.7	33.4	30.2	27.6	24.8
18	37.2	34.8	31.5	28.9	26.0
19	38.6	36.2	32.9	30.1	27.2
20	40.0	37.6	34.2	31.4	28.4
21	41.4	38.9	35.5	32.7	29.6
22	42.8	40.3	36.8	33.9	30.8
23	44.2	41.6	38.1	35.2	32.0
24	45.6	43.0	39.4	36.4	33.2
25	49.6	44.3	40.6	37.7	34.4
26	48.3	45.6	41.9	38.9	35.6
27	49.6	47.0	43.2	40.1	36.7
28	51.0	48.3	44.5	41.3	37.9
29	52.3	49.6	45.7	42.6	39.1
30	53.7	50.9	47.0	43.8	40.3
40	66.8	63.7	59.3	55.8	51.8
50	79.5	76.2	71.4	67.5	63.2
60	92.0	88.4	83.3	79.1	74.4
70	104.2	100.4	95.0	90.5	85.5
80	116.3	112.3	106.6	101.9	96.6
90	128.3	124.1	118.1	113.1	107.6
100	140.2	135.8	129.6	124.3	118.5

APPENDIX C-IV

Area under the standard normal curve from $-\infty$ to z.

z	0.00	0.01	0.02	0.03	0.04	0.05	0.06	0.07	0.08	0.09
0.0	0.5000	0.5040	0.5080	0.5120	0.5160	0.5199	0.5239	0.5279	0.5319	0.5359
0.1	0.5398	0.5438	0.5478	0.5517	0.5557	0.5596	0.5636	0.5675	0.5714	0.5753
0.2	0.5793	0.5832	0.5871	0.5910	0.5948	0.5987	0.6026	0.6064	0.6103	0.6141
0.3	0.6179	0.6217	0.6255	0.6293	0.6331	0.6368	0.6406	0.6443	0.6480	0.6517
0.4	0.6554	0.6591	0.6628	0.6664	0.6700	0.6736	0.6772	0.6808	0.6844	0.6879
0.5	0.6915	0.6950	0.6985	0.7019	0.7054	0.7088	0.7123	0.7157	0.7190	0.7224
0.6	0.7257	0.7291	0.7324	0.7357	0.7389	0.7422	0.7454	0.7486	0.7517	0.7549
0.7	0.7580	0.7611	0.7642	0.7673	0.7704	0.7734	0.7764	0.7794	0.7823	0.7852
0.8	0.7881	0.7910	0.7939	0.7967	0.7995	0.8023	0.8051	0.8078	0.8106	0.8133
0.9	0.8159	0.8186	0.8212	0.8238	0.8264	0.8289	0.8315	0.8340	0.8365	0.8389
1.0	0.8413	0.8438	0.8461	0.8485	0.8508	0.8531	0.8554	0.8577	0.8599	0.8621
1.1	0.8643	0.8665	0.8686	0.8708	0.8729	0.8749	0.8770	0.8790	0.8810	0.8830
1.2	0.8849	0.8869	0.8888	0.8907	0.8925	0.8944	0.8962	0.8980	0.8997	0.9015
1.3	0.9032	0.9049	0.9066	0.9082	0.9099	0.9115	0.9131	0.9147	0.9162	0.9177
1.4	0.9192	0.9207	0.9222	0.9236	0.9251	0.9265	0.9279	0.9292	0.9306	0.9319
1.5	0.9332	0.9345	0.9357	0.9370	0.9382	0.9394	0.9406	0.9418	0.9429	0.9441
1.6	0.9452	0.9463	0.9474	0.9484	0.9495	0.9505	0.9515	0.9525	0.9535	0.9545
1.7	0.9554	0.9564	0.9573	0.9582	0.9591	0.9599	0.9608	0.9616	0.9625	0.9633
1.8	0.9641	0.9649	0.9656	0.9664	0.9671	0.9678	0.9686	0.9693	0.9699	0.9706
1.9	0.9713	0.9719	0.9726	0.9732	0.9738	0.9744	0.9750	0.9756	0.9761	0.9767

Area under the standard normal curve from $-\infty$ to z (*cont.*).

z	0.00	0.01	0.02	0.03	0.04	0.05	0.06	0.07	0.08	0.09
2.0	0.9772	0.9778	0.9783	0.9788	0.9793	0.9798	0.9803	0.9808	0.9812	0.9817
2.1	0.9821	0.9826	0.9830	0.9834	0.9838	0.9842	0.9846	0.9850	0.9854	0.9857
2.2	0.9861	0.9864	0.9868	0.9871	0.9875	0.9878	0.9881	0.9884	0.9887	0.9890
2.3	0.9893	0.9896	0.9898	0.9901	0.9904	0.9906	0.9909	0.9911	0.9913	0.9916
2.4	0.9918	0.9920	0.9922	0.9925	0.9927	0.9929	0.9931	0.9932	0.9934	0.9936
2.5	0.9938	0.9940	0.9941	0.9943	0.9945	0.9946	0.9948	0.9949	0.9951	0.9952
2.6	0.9953	0.9955	0.9956	0.9957	0.9959	0.9960	0.9961	0.9962	0.9963	0.9964
2.7	0.9965	0.9966	0.9967	0.9968	0.9969	0.9970	0.9971	0.9972	0.9973	0.9974
2.8	0.9974	0.9975	0.9976	0.9977	0.9977	0.9978	0.9979	0.9979	0.9980	0.9981
2.9	0.9981	0.9982	0.9982	0.9983	0.9984	0.9984	0.9985	0.9985	0.9986	0.9986
3.0	0.9987	0.9987	0.9987	0.9988	0.9988	0.9989	0.9989	0.9989	0.9990	0.9990
3.1	0.9990	0.9991	0.9991	0.9991	0.9992	0.9992	0.9992	0.9992	0.9993	0.9993
3.2	0.9993	0.9993	0.9994	0.9994	0.9994	0.9994	0.9994	0.9995	0.9995	0.9995
3.3	0.9995	0.9995	0.9995	0.9996	0.9996	0.9996	0.9996	0.9996	0.9996	0.9997
3.4	0.9997	0.9997	0.9997	0.9997	0.9997	0.9997	0.9997	0.9997	0.9997	0.9998

APPENDIX C-V

Values of students t distribution with ν degrees of freedom (shaded area $= p = 1 - \alpha$).

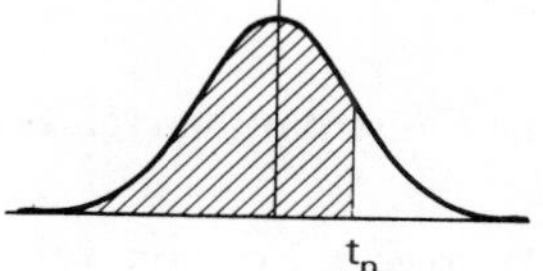

ν	$t_{0.995}$	$t_{0.99}$	$t_{0.975}$	$t_{0.95}$	$t_{0.90}$
1	63.66	31.82	12.71	6.31	3.08
2	9.92	6.96	4.30	2.92	1.89
3	5.84	4.54	3.18	2.35	1.64
4	4.60	3.75	2.78	2.13	1.53
5	4.03	3.36	2.57	2.02	1.48
6	3.71	3.14	2.45	1.94	1.44
7	3.50	3.00	2.36	1.90	1.42
8	3.36	2.90	2.31	1.86	1.40
9	3.25	2.82	2.26	1.83	1.38
10	3.17	2.76	2.23	1.81	1.37
11	3.11	2.72	2.20	1.80	1.36
12	3.06	2.68	2.18	1.78	1.36
13	3.01	2.65	2.16	1.77	1.35
14	2.98	2.62	2.14	1.76	1.34
15	2.95	2.60	2.13	1.75	1.34
16	2.92	2.58	2.12	1.75	1.34
17	2.90	2.57	2.11	1.74	1.33
18	2.88	2.55	2.10	1.73	1.33
19	2.86	2.54	2.09	1.73	1.33
20	2.84	2.53	2.09	1.72	1.32
21	2.83	2.52	2.08	1.72	1.32
22	2.82	2.51	2.07	1.72	1.32
23	2.81	2.50	2.07	1.71	1.32
24	2.80	2.49	2.06	1.71	1.32
25	2.79	2.48	2.06	1.71	1.32
26	2.78	2.48	2.06	1.71	1.32
27	2.77	2.47	2.05	1.70	1.31
28	2.76	2.47	2.05	1.70	1.31
29	2.76	2.46	2.04	1.70	1.31
30	2.75	2.46	2.04	1.70	1.31
40	2.70	2.42	2.02	1.68	1.30
60	2.66	2.39	2.00	1.67	1.30
120	2.62	2.36	1.98	1.66	1.29
∞	2.58	2.33	1.96	1.645	1.28

APPENDIX C-VIa

The F distribution (values of $F_{0.01}$).*

Degrees of Freedom for Numerator

Degrees of Freedom for Denominator	1	2	3	4	5	6	7	8	9
1	4,052	5,000	5,403	5,625	5,764	5,859	5,928	5,982	6,023
2	98.5	99.0	99.2	99.2	99.3	99.4	99.4	99.4	99.4
3	34.1	30.8	29.5	28.7	28.2	27.9	27.7	27.5	27.3
4	21.2	18.0	16.7	16.0	15.5	15.2	15.0	14.8	14.7
5	16.3	13.3	12.1	11.4	11.0	10.7	10.5	10.3	10.2
6	13.7	10.9	9.78	9.15	8.75	8.47	8.26	8.10	7.98
7	12.2	9.55	8.45	7.85	7.46	7.19	6.99	6.84	6.72
8	11.3	8.65	7.59	7.01	6.63	6.37	6.18	6.03	5.91
9	10.6	8.02	6.99	6.42	6.06	5.80	5.61	5.47	5.35
10	10.0	7.56	6.55	5.99	5.64	5.39	5.20	5.06	4.94
11	9.65	7.21	6.22	5.67	5.32	5.07	4.89	4.74	4.63
12	9.33	6.93	5.95	5.41	5.06	4.82	4.64	4.50	4.39
13	9.07	6.70	5.74	5.21	4.86	4.62	4.44	4.30	4.19
14	8.86	6.51	5.56	5.04	4.70	4.46	4.28	4.14	4.03
15	8.68	6.36	5.42	4.89	4.56	4.32	4.14	4.00	3.89
16	8.53	6.23	5.29	4.77	4.44	4.20	4.03	3.89	3.78
17	8.40	6.11	5.19	4.67	4.34	4.10	3.93	3.79	3.68
18	8.29	6.01	5.09	4.58	4.25	4.01	3.84	3.71	3.60
19	8.19	5.93	5.01	4.50	4.17	3.94	3.77	3.63	3.52
20	8.10	5.85	4.94	4.43	4.10	3.87	3.70	3.56	3.46
21	8.02	5.78	4.87	4.37	4.04	3.81	3.64	3.51	3.40
22	7.95	5.72	4.82	4.31	3.99	3.76	3.59	3.45	3.35
23	7.88	5.66	4.76	4.26	3.94	3.71	3.54	3.41	3.30
24	7.82	5.61	4.72	4.22	3.90	3.67	3.50	3.36	3.26
25	7.77	5.57	4.68	4.18	3.86	3.63	3.46	3.32	3.22
30	7.56	5.39	4.51	4.02	3.70	3.47	3.30	3.17	3.07
40	7.31	5.18	4.31	3.83	3.51	3.29	3.12	2.99	2.89
60	7.08	4.98	4.13	3.65	3.34	3.12	2.95	2.82	2.72
120	6.85	4.79	3.95	3.48	3.17	2.96	2.79	2.66	2.56
∞	6.63	4.61	3.78	3.32	3.02	2.80	2.64	2.51	2.41

*This table is reproduced from M. Merrington and C. M. Thompson, "Tables of Percentage Points of the Inverted Beta (F) Distribution," *Biometrika*, Vol. 33, 1943, by permission of the *Biometrika* trustees.

The F distribution (values of $F_{0.01}$) (*cont.*).

10	12	15	20	24	30	40	60	120	∞
6,056	6,106	6,157	6,209	6,235	6,261	6,287	6,313	6,339	6,366
99.4	99.4	99.4	99.4	99.5	99.5	99.5	99.5	99.5	99.5
27.2	27.1	26.9	26.7	26.6	26.5	26.4	26.3	26.2	26.1
14.5	14.4	14.2	14.0	13.9	13.9	13.7	13.7	13.6	13.5
10.1	9.89	9.72	9.55	9.47	9.38	9.29	9.20	9.11	9.02
7.87	7.72	7.56	7.40	7.31	7.23	7.14	7.06	6.97	6.88
6.62	6.47	6.31	6.16	6.07	5.99	5.91	5.82	5.74	5.65
5.81	5.67	5.52	5.36	5.28	5.20	5.12	5.03	4.95	4.86
5.26	5.11	4.96	4.81	4.73	4.65	4.57	4.48	4.40	4.31
4.85	4.71	4.56	4.41	4.33	4.25	4.17	4.08	4.00	3.91
4.54	4.40	4.25	4.10	4.02	3.94	3.86	3.78	3.69	3.60
4.30	4.16	4.01	3.86	3.78	3.70	3.62	3.54	3.45	3.36
4.10	3.96	3.82	3.66	3.59	3.51	3.43	3.34	3.25	3.17
3.94	3.80	3.66	3.51	3.43	3.35	3.27	3.18	3.09	3.00
3.80	3.67	3.52	3.37	3.29	3.21	3.13	3.05	2.96	2.87
3.69	3.55	3.41	3.26	3.18	3.10	3.02	2.93	2.84	2.75
3.59	3.46	3.31	3.16	3.08	3.00	2.92	2.83	2.75	2.65
3.51	3.37	3.23	3.08	3.00	2.92	2.84	2.75	2.66	2.57
3.43	3.30	3.15	3.00	2.92	2.84	2.76	2.67	2.58	2.49
3.37	3.23	3.09	2.94	2.86	2.78	2.69	2.61	2.52	2.42
3.31	3.17	3.03	2.88	2.80	2.72	2.64	2.55	2.46	2.36
3.26	3.12	2.98	2.83	2.75	2.67	2.58	2.50	2.40	2.31
3.21	3.07	2.93	2.78	2.70	2.62	2.54	2.45	2.35	2.26
3.17	3.03	2.89	2.74	2.66	2.58	2.49	2.40	2.31	2.21
3.13	2.99	2.85	2.70	2.62	2.53	2.45	2.36	2.27	2.17
2.98	2.84	2.70	2.55	2.47	2.39	2.30	2.21	2.11	2.01
2.80	2.66	2.52	2.37	2.29	2.20	2.11	2.02	1.92	1.80
2.63	2.50	2.35	2.20	2.12	2.03	1.94	1.84	1.73	1.60
2.47	2.34	2.19	2.03	1.95	1.86	1.76	1.66	1.53	1.38
2.32	2.18	2.04	1.88	1.79	1.70	1.59	1.47	1.32	1.00

APPENDIX C-VIb

The F distribution (values of $F_{0.05}$).*

Degrees of Freedom for Numerator

Degrees of Freedom for Denominator	1	2	3	4	5	6	7	8	9
1	161	200	216	225	230	234	237	239	241
2	18.5	19.0	19.2	19.2	19.3	19.3	19.4	19.4	19.4
3	10.1	9.55	9.28	9.12	9.01	8.94	8.89	8.85	8.31
4	7.71	6.94	6.59	6.39	6.26	6.16	6.09	6.04	6.00
5	6.61	5.79	5.41	5.19	5.05	4.95	4.88	4.82	4.77
6	5.99	5.14	4.76	4.53	4.39	4.28	4.21	4.15	4.10
7	5.59	4.74	4.35	4.12	3.97	3.87	3.79	3.73	3.68
8	5.32	4.46	4.07	3.84	3.69	3.58	3.50	3.44	3.39
9	5.12	4.26	3.86	3.63	3.48	3.37	3.29	3.23	3.18
10	4.96	4.10	3.71	3.48	3.33	3.22	3.14	3.07	3.02
11	4.84	3.98	3.59	3.36	3.20	3.09	3.01	2.95	2.90
12	4.75	3.89	3.49	3.26	3.11	3.00	2.91	2.85	2.80
13	4.67	3.81	3.41	3.18	3.03	2.92	2.83	2.77	2.71
14	4.60	3.74	3.34	3.11	2.96	2.85	2.76	2.70	2.65
15	4.54	3.68	3.29	3.06	2.90	2.79	2.71	2.64	2.59
16	4.49	3.36	3.24	3.01	2.85	2.74	2.66	2.59	2.54
17	4.45	3.59	3.20	2.96	2.81	2.70	2.61	2.55	2.49
18	4.41	3.55	3.16	2.93	2.77	2.66	2.58	2.51	2.46
19	4.38	3.52	3.13	2.90	2.74	2.63	2.54	2.48	2.42
20	4.35	3.49	3.10	2.87	2.71	2.60	2.51	2.45	2.39
21	4.32	3.47	3.07	2.84	2.68	2.57	2.49	2.42	2.37
22	4.30	3.44	3.05	2.82	2.66	2.55	2.46	2.40	2.34
23	4.28	3.42	3.03	2.80	2.64	2.53	2.44	2.37	2.32
24	4.26	3.40	3.01	2.78	2.62	2.51	2.42	2.36	2.30
25	4.24	3.39	2.99	2.76	2.60	2.49	2.40	2.34	2.28
30	4.17	3.32	2.92	2.69	2.53	2.42	2.33	2.27	2.21
40	4.08	3.23	2.84	2.61	2.45	2.34	2.25	2.18	2.12
60	4.00	3.15	2.76	2.53	2.37	2.25	2.17	2.10	2.04
120	3.92	3.07	2.68	2.45	2.29	2.18	2.09	2.02	1.96
∞	3.84	3.00	2.60	2.37	2.21	2.10	2.01	1.94	1.88

*This table is reproduced from M. Merrington and C. M. Thompson, "Tables of Percentage Points of the Inverted Beta (F) Distribution," *Biometrika*, Vol. 33, 1943, by permission of the *Biometrika* trustees.

The F distribution (values of $F_{0.05}$) (*cont.*).

10	12	15	20	24	30	40	60	120	∞
242	244	246	248	249	250	251	252	253	254
19.4	19.4	19.4	19.4	19.5	19.5	19.5	19.5	19.5	19.5
8.79	8.74	8.70	8.66	8.64	8.62	8.59	8.57	8.55	8.53
5.96	5.01	5.86	5.80	5.77	5.75	5.72	5.69	5.66	5.63
4.74	4.68	4.62	4.56	4.53	4.50	4.46	4.43	4.40	4.37
4.06	4.00	3.94	3.87	3.84	3.81	3.77	3.74	3.70	3.67
3.64	3.57	3.51	3.44	3.41	3.38	3.34	3.30	3.27	3.23
3.35	3.28	3.22	3.15	3.12	3.08	3.04	3.01	2.97	2.93
3.14	3.07	3.01	2.94	2.90	2.86	2.83	2.79	2.75	2.71
2.98	2.91	2.85	2.77	2.74	2.70	2.66	2.62	2.58	2.54
2.85	2.79	2.72	2.65	2.61	2.57	2.53	2.49	2.45	2.40
2.75	2.69	2.62	2.54	2.51	2.47	2.43	2.38	2.34	2.30
2.67	2.60	2.53	2.46	2.42	2.38	2.34	2.30	2.25	2.21
2.60	2.53	2.46	2.39	2.35	2.31	2.27	2.22	2.18	2.13
2.54	2.48	2.40	2.33	2.29	2.25	2.20	2.16	2.11	2.07
2.49	2.42	2.35	2.28	2.24	2.19	2.15	2.11	2.06	2.01
2.45	2.38	2.31	2.23	2.19	2.15	2.10	2.06	2.01	1.96
2.41	2.34	2.27	2.19	2.15	2.11	2.06	2.02	1.97	1.92
2.38	2.31	2.23	2.16	2.11	2.07	2.03	1.98	1.93	1.88
2.35	2.28	2.20	2.12	2.08	2.04	1.99	1.95	1.90	1.84
2.32	2.25	2.18	2.10	2.05	2.01	1.96	1.92	1.87	1.81
2.30	2.23	2.15	2.07	2.03	1.98	1.94	1.89	1.84	1.78
2.27	2.20	2.13	2.05	2.01	1.96	1.91	1.86	1.81	1.76
2.25	2.18	2.11	2.03	1.98	1.94	1.89	1.84	1.79	1.73
2.24	2.16	2.09	2.01	1.96	1.92	1.87	1.82	1.77	1.71
2.16	2.09	2.01	1.93	1.89	1.84	1.79	1.74	1.68	1.62
2.08	2.00	1.92	1.84	1.79	1.74	1.69	1.64	1.58	1.51
1.99	1.92	1.84	1.75	1.70	1.65	1.59	1.53	1.47	1.39
1.91	1.83	1.75	1.66	1.61	1.55	1.50	1.43	1.35	1.25
1.83	1.75	1.67	1.57	1.52	1.46	1.39	1.32	1.22	1.00

APPENDIX C-VII

Kolmogorov-Smirnov critical values.

Degrees of Freedom (N)	One Sample Test*			Two Sample Test†	
	$D_{0.10}$	$D_{0.05}$	$D_{0.01}$	$D_{0.05}$	$D_{0.01}$
1	0.950	0.975	0.995		
2	0.776	0.842	0.929		
3	0.642	0.708	0.828		
4	0.564	0.624	0.733	1.000	1.000
5	0.510	0.565	0.669	1.000	1.000
6	0.470	0.521	0.618	0.833	1.000
7	0.438	0.486	0.577	0.857	0.857
8	0.411	0.457	0.543	0.750	0.875
9	0.388	0.432	0.514	0.667	0.778
10	0.368	0.410	0.490	0.700	0.800
11	0.352	0.391	0.468	0.636	0.727
12	0.338	0.375	0.450	0.583	0.667
13	0.325	0.361	0.433	0.538	0.692
14	0.314	0.349	0.418	0.571	0.643
15	0.304	0.338	0.404	0.533	0.600
16	0.295	0.328	0.392	0.500	0.625
17	0.286	0.318	0.381	0.471	0.588
18	0.278	0.309	0.371	0.500	0.556
19	0.272	0.301	0.363	0.474	0.526
20	0.264	0.294	0.356	0.450	0.550
25	0.24	0.27	0.32	0.40	0.48
30	0.22	0.24	0.29	0.37	0.43
35	0.21	0.23	0.27	0.34	0.39
Over 35	$\frac{1.22}{\sqrt{N}}$	$\frac{1.36}{\sqrt{N}}$	$\frac{1.63}{\sqrt{N}}$	$1.36\sqrt{\frac{n_1 + n_2}{n_1 n_2}}$	$1.63\sqrt{\frac{n_1 + n_2}{n_1 n_2}}$

*Used for testing goodness of fit of a sample to a theoretical distribution where N = sample size.

†Used to determine if two samples are from the same distribution. For small samples (up to 35), $N = n_1 = n_2$.

INDEX

D

E

F

G

H

I

J

K

L

M

N

O

P

Q

R

S

T

U

V

W

Y

Z